ACS SYMPOSIUM SERIES 427

Protein Purification

From Molecular Mechanisms to Large-Scale Processes

Michael R. Ladisch, EDITOR
Purdue University

Richard C. Willson, EDITOR
University of Houston

Chih-duen C. Painton, EDITOR
Mallinckrodt Medical, Inc.

Stuart E. Builder, EDITOR
Genentech, Inc.

Developed from a symposium sponsored
by the Division of Biochemical Technology
at the 198th National Meeting
of the American Chemical Society,
Miami Beach, Florida, September 10-15, 1989

.. Chemical Society, Washington, DC 1990

Library of Congress Cataloging-in-Publication Data

Protein purification: from molecular mechanisms to large-scale processes
Michael R. Ladisch, editor . . . [et al.].

p. cm.—(ACS symposium series, ISSN 0097–6156; 427)

"Developed from a symposium sponsored by the Division of Biochemical Technology at the 198th National Meeting of the American Chemical Society, Miami Beach, Florida, September 10–15, 1989."

Includes bibliographical references.

ISBN 0–8412–1790–4

1. Proteins—Purification—Congresses. 2. Proteins—Biotechnology—Congresses. I. Ladisch, Michael R., 1950– . II. American Chemical Society. Division of Biochemical Technology. III. Series

TP248.65.P76P765 1990
660′.63—dc20 90–35551
CIP

The paper used in this publication meets the minimum requirements of American National Standard for Information Sciences—Permanence of Paper for Printed Library Materials, ANSI Z39.48–1984.

♾

PRINTED IN THE UNITED STATES OF AMERICA

ACS Symposium Series

M. Joan Comstock, *Series Editor*

Foreword

The ACS SYMPOSIUM SERIES was founded in 1974 to provide a medium for publishing symposia quickly in book form. The format of the Series parallels that of the continuing ADVANCES IN CHEMISTRY SERIES except that, in order to save time, the papers are not typeset but are reproduced as they are submitted by the authors in camera-ready form. Papers are reviewed under the supervision of the Editors with the assistance of the Series Advisory Board and are selected to maintain the integrity of the symposia; however, verbatim reproductions of previously published papers are not accepted. Both reviews and reports of research are acceptable, because symposia may embrace both types of presentation.

Contents

Preface

PROTEIN PURIFICATION: *From Molecular Mechanisms to Large-Scale Processes* had its genesis during informal discussions in Houston, Texas, in April 1989, based upon Richard Willson's suggestion of this project. The thought was that a relevant cross-disciplinary treatment of large-scale protein purification could be possible given the rapid progression of several recombinant protein products from laboratory to large-scale, and the willingness of industry to present some of the fundamental aspects of parameters that have an impact on practical considerations of protein purification. This was reflected, in part, by the papers which had already been submitted and organized as part of the Miami Beach program (Jim Swartz, Genentech, Program Chairman) of the Division of Microbial and Biochemical Technology, now known as the Division of Biochemical Technology. Specifically, two sessions on separations: *Large-Scale Protein Purification* (M. R. Ladisch and C.-D. Painton, chairpersons) and *New Advances in Protein Purification* (R. C. Willson and S. E. Builder, chairpersons) included approximately 50% of the papers from industrial contributors. Attendance at these sessions reached as many as 250 people.

Upon embarking on this project, Dr. Chih-duen Painton (of Mallinckrodt) and Dr. Stuart Builder (of Genentech) were quickly enlisted to assist in the development of this book. The cooperation of all the contributors was tremendous, the response of the reviewers impressive, and the assistance of the ACS Books Department outstanding. The result is that we can bring this volume to you in a timely fashion.

MICHAEL R. LADISCH
Purdue University
West Lafayette, IN 47907

February 12, 1990

Chapter 1

Large-Scale Protein Purification

Introduction

Richard C. Willson[1] and Michael R. Ladisch[2]

[1]Department of Chemical Engineering, University of Houston, Houston, TX 77204
[2]Laboratory of Renewable Resources Engineering and Department of Agricultural Engineering, Purdue University, West Lafayette, IN 47907

Large scale protein purification is the final production step, prior to product packaging, in the manufacture of therapeutic proteins, specialty enzymes, and diagnostic products. The art and science of protein purification evolves from laboratory scale techniques which are often adapted and scaled up to satisfy the need for larger amounts of extremely pure test quantities of the product for analysis, characterization, testing of efficacy, clinical or field trials, and, finally, full scale commercialization. Development of appropriate strategies for protein recovery and purification differs from development of separation techniques for more traditional chemical or agricultural processing technologies by the broadness of cross-disciplinary interactions required to achieve scale-up. The uncompromising standards for product quality, as well as rigorous quality control of manufacturing practices embodied in current good manufacturing practices (cGMP's), provide further challenges to the scale-up of protein purification. Analysis of electrokinetic, chromatographic, adsorptive, and membrane separation techniques suggests that if yield recovery is paramount, documented purity is critical, and both must ultimately be attained within certain cost constraints. Examples of purification of insulin and proinsulin, IgM, recombinant interferon-alpha, interleukins, histidine containing peptides, lutenizing hormone releasing hormone, and bovine growth hormone illustrate conceptual approaches used in successful industrial processes.

Bio-separation processes have a significant impact on the economics of producing proteins for animal and human health care products (*1*). The recovery sequence for isolating product from a fermentation broth has changed little over the last ten years and consists essentially of (*2*):

1. removal of insolubles;
2. primary product isolation;
3. purification; and
4. final product isolation.

0097–6156/90/0427–0001$06.00/0

The advent of manufacture of recombinant proteins and peptides for pharmaceutical, diagnostic, and agricultural applications has, however, changed the way in which separation techniques are chosen, purification strategies are developed, and economics applied to purification scale-up (see Chapters 2, 3, and 14). Purification costs are particularly important determinants in production costs of diagnostic reagents, enzymes and animal care products. Even the most promising human pharmaceuticals are ultimately subject to cost constraints. Within this context, purity and product activity are still the primary goal.

Removal of trace contaminants that are difficult to detect is becoming a key issue as new recombinant and cell culture production technologies are phased in (Chapter 2). Examples of such contaminants include pyrogens, viruses, and transforming DNA, inaccurately translated or glycosylated forms of the protein, degradation and oxidation products, aggregates and conformational isomers which are similar to the desired product. Process validation is therefore quite complex and requires many different types of analytical procedures as shown in Table I.

The detection of trace contaminants also presents many challenges (see, for example, Chapters 2, 11, 14, and 15). It is particularly important that any fractionation-based analytical method used in product characterization employ a separation mechanism different from those already used in the purification process. Otherwise, a contaminant which has co-purified with the product through the preparative process could escape detection. This concept represents an orthogonal protein separation strategy, also used in large scale processes where several purification steps based on different principles would be used (*1*). For example, a purification sequence might include ion exchange, hydrophobic interaction chromatography (HIC), and affinity chromatography. Technical issues for each of these steps are the effect of overload on protein retention (Chapter 7) attaining high throughput at reasonable pressure drop (Chapter 8), prediction of protein retention in HIC as a function of salt type (chaotrope vs. kosmotrope) and salt concentration (Chapter 6), and selection of appropriate affinity chromatography techniques for attaining high final protein purity (Chapters 10 and 11).

Novel affinity methods can reduce the number of separation steps by enabling highly selective separation from a relatively impure starting materials. Recent developments in this context include chelating peptide-immobilized metal affinity chromatography for fusion proteins (Chapter 12) and immobilized metal chelates attached to water soluble polymers for use in two phase extraction (Chapter 10).

Purification Strategies

The primary factors determining a preferred separation method depend on parameters of size, ionic charge, solubility and density as illustrated in Figure 1 (from reference *5*). This applies to both small molecule and protein separations. Recovery and separation of proteins covers this entire range: i.e., microbial cells (ca. 1 to 5 microns), inclusion bodies (ca. 0.1 to 0.5 microns), protein aggregates (ca. 10 nm to 200 nm), as well as proteins and peptides themselves (less than ca. 20 nm). This book emphasizes protein properties and purification and consequently focusses on chromatography and partitioning in liquid systems. Analogies between traditional chemical separation principles and those applicable to proteins are apparent (Chapter 3). However, the structure and function of proteins results in product molecules which differ from variants by as little as one amino acid out of 200. Separations also need to accomplish removal of other macromolecules (such as DNA and RNA) which could compromise product efficacy at trace levels. This requires a long list of special analytical techniques, many of which are based on use of recombinant technology, to validate product purity and process operation (Table I) (*3,4*).

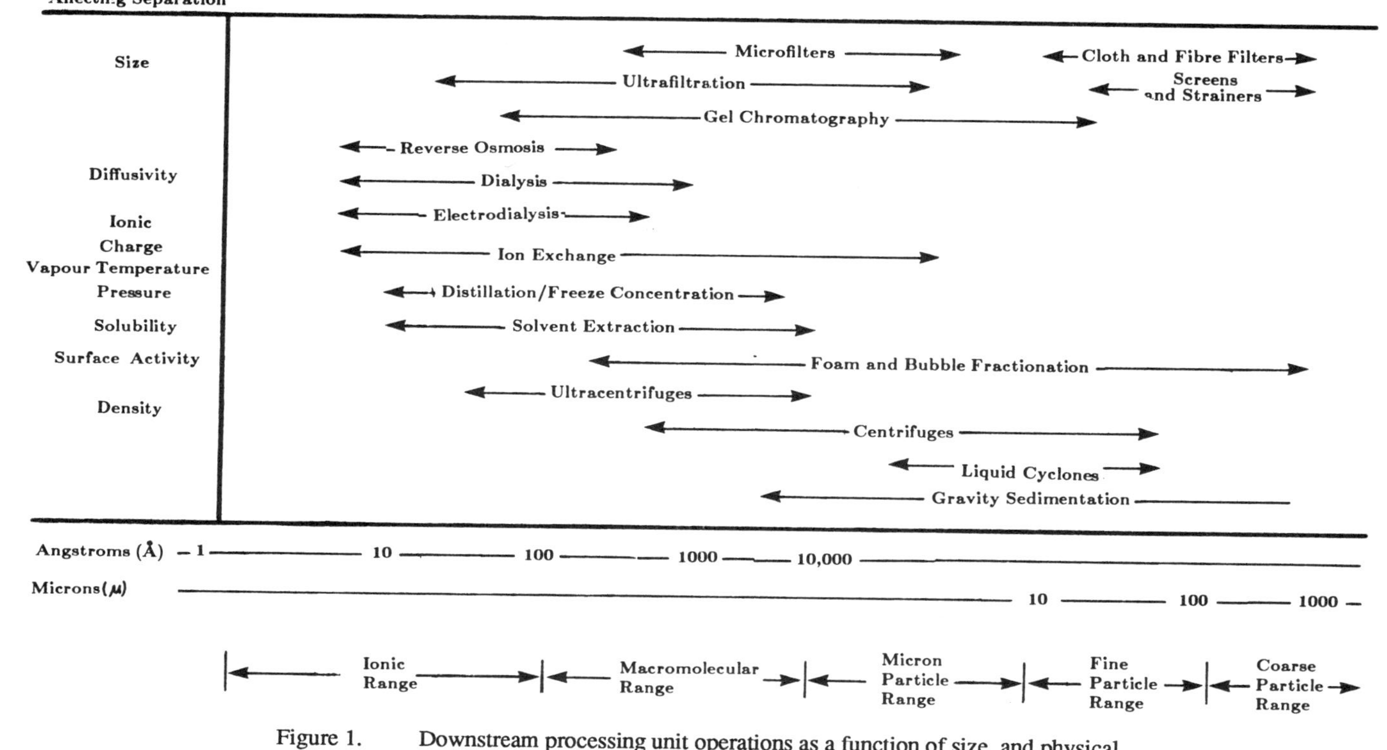

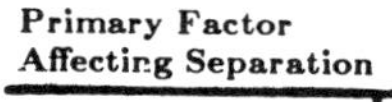

Figure 1. Downstream processing unit operations as a function of size, and physical properties (reprinted with permission, from reference 5, copyright of The Nature Press, MacMillan Publishers, Ltd., all rights reserved).

Table I. Example: Assessment of Product Purity and Processing Conditions

Methods	Objective	Procedure
ANALYTICAL		
Electrophoresis	Detect contaminating proteins	Carry out electrophoresis (examples for bovine growth hormone in Chapter 2, human serum albumin and hemoglobin in Chapter 10, recombinant IFN and IL-1, IL-2 in Chapter 11, signal peptide in Chapter 14; principles discussed in Chapters 15 and 16). Use Coomassie blue. For more sensitive detection follow with destaining, and silver staining to detect protein bands. Also combine silver staining and immunoblotting.
Liquid chromatography	Detect contaminating proteins	Select appropriate stationary phase (recall orthogonal separation principle), inject samples look for more than one peak (examples for insulin, insulin A, insulin B in Chapter 7; IgM in Chapter 14).
ELISA	Detect host proteins	**E**nzyme **L**inked **I**mmuno**S**orbent **A**ssay. Competitive binding assay (see ref. *3* for monoclonal antibodies; Chapter 14 for IgM (human) monoclonal antibodies).
Immunoassays	Detect contaminants (i.e., antibodies leached from affinity columns); small epitope size makes antibodies poor for variants ("see" small area).	Based on specific antigen-antibody reactions (discussed in Chapters 2, 11, and 14).
Amino acid sequence, composition	Determine terminal sequence of a newly produced protein for comparison against a standard	Carry out controlled hydrolysis of protein. Analyze for amino acids or peptides using appropriate liquid chromatography techniques. (See Chapter 12 for discussion of lutenizing hormone releasing hormone (LHRG) and His-Trp proinsulin.)

Tryptic mapping	Map protein based on analysis and identification of peptide fragments	Protease digestion. Reverse phase chromatography of resulting peptides (See Chapter 13 for discussion of proteases used for site-specific cleavage applied to fusion proteins).
Hybridization	Detect transforming DNA sequences	Binding of complementary nucleic acid sequences to find specific sequences of DNA or RNA (see reference 4 for background on techniques).
NMR, MS, Light Scattering, Analytical Ultracentrifugation Size Exclusion Chromatography	Characterization of biophysical properties, and state of aggregation	These cover a wide range of analytical techniques discussed in biochemistry and biology textbooks (see, for example, ref 4).
Rabbit pyrogen test	Detect presence of pyrogens	Standard Assay.
Antibody Production Test	Detect viral contamination	Check for production of anti-virus antibodies by pathogen free mice immunized with product samples.
PROCESS		
Blank Runs	Detect contaminants due to host	Host organism, not producing the recombinant products, put through purification process at full production scale, followed by isolation of contaminants (if any) not eliminated by the purification process.
Tracer Studies	Demonstrate elimination of certain contaminants	Add radiolabeled contaminants, viruses to crude product to demonstrate their elimination after the product is processed through the purification train.
Repeated Processing of product	Detect if changes occur during purification due to product/stationary phase interactions and other purification operations	Run product through the purification sequence many times.
Protein and activity material balance of eluting peaks from chromatography system	Verify that protein recovery is complete	Detect proteins, eluting from column, based on activity, protein content using methods based on different detection principles (see Chapter 7 for example with β-gal/BSA separation).

Process validation is another key consideration in developing a purification strategy. Process validation refers to establishing documented evidence which assures that a specific process will consistently produce a product meeting its predetermined specifications and is based on FDA guidelines (D. Julien, Triad Industries, at Purdue University Workshop on Chromatographic Separations and Scale-Up, October 3, 1989).

The industrial perspective appears to be to keep the separation strategy as simple as possible (Chapter 2). Genetic engineering can simplify purification by increasing product titer and providing a molecular structure which is in a proactive form or is otherwise modified *in vivo* to enhance purification efficiency (Chapter 11). Process development should thus include a cross-disciplinary approach which considers the engineering of the organism's traits to fit scale-up constraints in fermentation or cell culture, as well as in purification. Similarly, purification conditions should be developed, if possible, to help overcome limitations in the organism's protein production and transport mechanisms with particular emphasis on traits which are difficult to alter by genetic manipulation. Some aspects of purification development are so product specific that the necessary skills are best assembled in an industrial setting.

Individual purification steps need to be addressed in a generic context so that a fundamental, mechanistic knowledge base for each type of separation technique will ultimately be developed. This type of research is also cross-disciplinary, by definition, given the large number of factors other than the absence of contaminating molecules which impact the definition of purity. Examples are: protein refolding from inclusion bodies; protein secretion and expression in novel host organisms; operational aspects of immunoaffinity chromatography and preparative chromatography of complex mixtures; post-translational modifications and immunogenicity; and process integration and validation. In addition to biotechnology production companies, equipment suppliers and instrumentation companies also benefit from such a knowledge base. These companies have a critical function in developing new separations apparatus, chromatographic adsorbents, analytical instrumentation, and process monitoring and control equipment.

New Approaches Through Cross-disciplinary Collaboration

Product quality requirements for part-per-million impurity levels has led to new emphasis on high-resolution methods capable of removing subtly-altered forms of the desired product. At the same time, the prospect of gram- and even kilogram-scale production is driving the application of refined forms of classical large-scale methods to new problems. The need for practical solutions is helping to focus efforts of investigators from many disciplines on complex problems of protein purification. Polymer chemists and biochemists address the longstanding need for materials possessing hydrophilic, protein-compatible surfaces, which are sufficiently rugged to be useful as adsorbents and filtration media. Advances in the chemistry of separations materials, which frequently derive from parallel interests in biomedical device technology, have continually allowed the development of new separation methods. Fundamental studies on understanding mechanisms of protein interactions with their environment have fostered development of new separation schemes and/or improved operating conditions. Two examples given in this book are on the effect of amino acid sequence on peptide and protein partitioning in two phase aqueous systems (Chapter 4) and protein-polyelectrolyte complexation (Chapter 5).

The first tool used in development of a modern protein purification method is frequently not a centrifuge or filtration apparatus, but a DNA synthesizer. As illustrated by Chapters 12 and 13 of this volume, molecular geneticists and microbial physiologists can help to define the nature of the purification problem. The potential influence of genetic and culture manipulations has rapidly expanded to include not only host-related factors such as expression level and cellular location, but also characteristics of the protein molecule which influence its purification behavior (see Table I for illustrations).

Increasing attention to trace contaminants has driven the introduction of increasingly sophisticated instruments and bioassay techniques for the monitoring of protein purification processes. Analytical chemists also identify intractable contaminants for potential elimination from the host genome. Chemists, together with physical biochemists and chemical engineers, are promoting advances in both the understanding and application of electrophoretic and electrokinetic separation techniques (see Chapters 15 and 16). Biochemists, biologists, and biochemical engineers are the groups most directly involved the development of protein purification methods. As they jointly develop new methods, each group applies its unique collection of skills and experience. Engineers contribute the quantitative simulation and optimization of processes, and understanding of economics, and familiarity with larger-scale operations and automated process control (Chapters 2, 3, and 6-10). Biochemists know the sometimes unforgiving properties of proteins and biological materials, possess the accumulated experience of decades of research-scale purifications (Chapter 11 and 14), while biologists understand the utility of biological approaches to what appear at first to be engineering problems (Chapters 12 and 13).

Overview of This Volume

The book is divided into several distinct sections. Chapters 2 and 3 give overviews of separation strategies. The subsequent chapters present research results on phase equilibrium behavior in aqueous two phase systems (Chapters 4 and 5). New engineering approaches to analysis of mass transfer and chromatography (Chapters 6-9); affinity based separations (Chapters 10-13); a case study on IgM human monoclonal antibodies (Chapter 14) and electrically driven separations (Chapters 15-16).

Strategies for Large Scale Protein Purification. Chapter 2 by S. V. Ho describes the impact of the composition of the process stream as it passes from initial composition to final purity on the efficiency with which different classes of contaminants can be removed by various methods. This results in the division of the overall process of purification into several general stages. This division serves to limit the complexity of the design process, as each possible unit operation is normally useful in only a limited number of stages.

Based on his experience in large-scale protein purification, the author highlights some practical considerations which are illustrated with case studies. The case studies illustrate the surprising effectiveness of scaleable, classical methods such as precipitation and extraction when cleverly applied and carefully optimized. This is a theme which is further illustrated in later chapter, which has important implications for the development of truly large-scale processes.

The Challenge of Separations in Biotechnology. In Chapter 3 by E. N. Lightfoot, the initial concentration step is presented as dominating processing costs, with many methods of initial concentration become progressively less economic at lower product concentration. The operating costs of these processes are proportional to the increasing volume of inerts processed. Adsorptive separation processes can escape this unfavorable trend.

The author addresses the balance between the essential mass transfer functions of adsorptive separation equipment, and the closely-related momentum transfer processes which govern drag and pressure drop. Differences between mass and momentum transfer can be used to optimize the former without unnecessarily magnifying the latter. The design of adsorptive protein separation processes is also discussed.

The Effect of Amino Acid Sequence on Peptide and Protein Partitioning in Aqueous Two-Phase Systems. Chapter 4 by A. D. Diamond, K. Hu, and J. T. Hsu presents a structural approach for the prediction of partition coefficients of peptides and proteins in aqueous two-phase systems. By analyzing the partition behavior of many pairs of dipeptides of the same composition but opposite sequences, the authors regress a set of parameters characteristic of the influence of each residue type of partition behavior. These parameters can be used in a group-contribution equation (with corrections for the effects of N- and C-termini) to predict the partition coefficient of a peptide from its sequence alone.

The method performs well for di- and tripeptides similar to those from which the parameters were regressed. It also predicts qualitatively the dependence of partition coefficient on structure for larger molecules. While the method in its current state cannot accommodate the effects of secondary and tertiary structure, it represents an initial approach to prediction and correlation of partition data on a structural basis. Further development of such predictive methods will be essential if the design of protein separations is to be put on the rational, predictive basis characteristic of more established processes.

Protein Separation via Polyelectrolyte Complexation. Chapter 5 by M. A. Strege, P. A. Dubin, J. S. West, and C. D. Flinta analyzes the complexation and coacervation of proteins by the polycation poly(dimethylallylammonium chloride. The authors demonstrate the existence of stable, soluble intrapolymer complexes, and estimate the number of protein molecules bound per polymer chain as a function of pH and free protein concentration. They also point out that the selectivity of precipitation by polyelectrolyte coacervation can be surprisingly high. Finally, they demonstrate that the process appears to be sensitive to the nonuniform distribution of charges on a protein's surface, rather than simply to its net charge. Polyelectrolyte coacervation, therefore, may allow the initial steps of a process to achieve a much greater selectivity than has traditionally been expected.

Mechanisms of Protein Retention in Hydrophobic Interaction Chromatography. Chapter 6 by B. F. Roettger, J. A. Myers, F. E. Regnier, and M. R. Ladisch discusses hydrophobic interaction chromatography (HIC) which separates proteins and other biological molecules based on surface hydrophobicity. Adsorption and desorption is influenced by the type of salt employed in the mobile phase, as well as its concentration. HIC differs from reversed phase chromatography in that proteins separated at HIC conditions elute in their active conformation due to mild elution conditions and use of salts which stabilize the proteins. Elution occurs in a decreasing gradient. In comparison, proteins in reversed phase chromatography are adsorbed to a more strongly hydrophobic stationary phase and increasing gradients of organic solvents are required for elution. Conformational changes of the proteins may occur, and can account for different elution orders.

This chapter described experimental results which give preferential interactions of ammonium salts with HIC supports as determined by densimetric techniques. Preferential interactions of the ammonium salts of $SO_4^=$, $C_2H_3O_2^-$, Cl^- and I^- with the supports and proteins were found to explain adsorption behavior. A predictive equation which relates the capacity factor for a polymeric sorbent to the lyotropic number (i.e., reflects salt type) and salt molality is reported. The result suggests that protein retention can be estimated as a function of salt type and concentration.

Separation and Sorption Characteristics of an Anion Exchange Stationary Phase for β-Galactosidase, Bovine Serum Albumin, and Insulin. Chapter 7 by M. R. Ladisch, K. L. Kohlmann, and R. L. Hendrickson discuss anion exchange chromatography. Anion exchange media are widely used in the chromatography of proteins, with many examples given throughout this book. The theory for anion exchange chromatography at low concentrations is well established. This chapter addresses the impact of adsorption of one protein on the retention of another. This occurs at mass overload conditions, which results from operating in the nonlinear region of competitive adsorption isotherms of at least one of the components involved.

The Craig distribution model, which has previously been reported in the literature, was found to be a useful starting point in this work. Batch equilibrium studies, carried out with BSA and β-galactosidase show the polymeric, derivatized adsorbent used by the authors to have a relatively high loading capacity (200 mg protein/g dry weight), and that adsorption of β-galactosidase could affect subsequent adsorption of the BSA. The Craig distribution concept would thus suggest that altered peak retention could result for BSA, if the concentration of the β-gal were high enough.

Dynamic Studies on Radial-Flow Affinity Chromatography for Trypsin Purification. Chapter 8 by W. C. Lee, G. J. Tsai, and G. T. Tsao discuss a new engineering approach to analysis of mass transfer in radial flow chromatography. The development and application of mathematical modeling for simulation of radial-flow affinity chromatography is demonstrated. When combined with experiment, the model allows the estimation of parameters governing the process, and identification of rate-determining steps. This analysis will be useful in scale-up, and in the development of other separations using this technology.

Affinity chromatography may be particularly amenable to improvement by changes in the geometry of solid/liquid contacting devices. This is because the strong, specific protein/adsorbent interactions involved can often achieve a high degree of purification in the equivalent of a single theoretical plate. Even very short liquid paths through the adsorbent bed, therefore, may allow effective separations. The viability of this notion is further illustrated by the recent commercialization of membrane-based affinity separations.

Impact of Continuous Affinity-Recycle Extraction (CARE) in Downstream Processing. Chapter 9 by N. F. Gordon and C. F. Cooney describes further development and simulation of the Continuous Affinity-Recycle Extraction (CARE) process recently developed in their laboratory (7). Based on solid/liquid contacting in well mixed vessels, this method allows adsorptive purification to be used at an earlier process stage than possible with conventional chromatography, because of its tolerance for particulates and viscous cell debris. Distribution of the adsorbent among several vessels allows adsorption efficiency to approach that of a column of equivalent size. In the present work, the authors describe the extension of CARE to separations based on ion-exchange adsorption, and directly compare the method with column chromatography for the purification of β-galactosidase from crude *E. coli* lysates. Numerical simulation of the CARE process is used to evaluate the trade-offs among performance measures such as degree of product concentration and purification, yield, and throughput.

Novel Metal Affinity Protein Separations. Chapter 10 by S. S. Suh, M. E. Van Dam, G. E. Menschell, S. Plunket, and F. H. Arnold discusses two methods on metal-affinity separation recently introduced by the authors. These are metal-affinity aqueous two-phase extraction and metal-affinity precipitation. Both methods can be implemented using the metal chelator iminodiacetic acid (IDA) covalently attached to polyethylene glycol (PEG), but they depend on different mechanisms to achieve separation.

Metal-affinity partitioning in aqueous two-phase systems involves the use of PEG molecules singly derivatized with IDA. In a PEG-based aqueous two-phase system, this molecule partitions strongly into the PEG-rich phase. In the presence of metal ions such as Cu(II), selective interactions of IDA-bound copper atoms with proteins containing exposed surface histidine or cysteine residues enhance the partitioning of these proteins into the PEG-rich phase. Metal-affinity partitioning is based on similar interactions, but uses bis-chelates chelating two copper atoms at the ends of the PEG chain. Interaction of the chain ends with two different protein molecules produces a crosslink between them leading to precipitation by mechanisms functionally similar to the immunoprecipitation of multivalent antigens with bivalent antibodies.

Recovery of Recombinant Proteins by Immunoaffinity Chromatography. Chapter 11 by P. Bailon and S. K. Roy covers practical applications of affinity chromatography. Immunoaffinity chromatography is described as a predecessor to affinity chromatography, with the first well characterized immuno-adsorbent prepared by chemically bonding the antigen ovalbumin to a solid matrix for use in isolating antibodies to ovalbumin (*9*). This chapter presents an overview, as well as results from experimentation, on the preparation, use, and stability of immunoadsorbents. There appears to be considerable scientific background which is required to obtain a working immunoaffinity column. First the monoclonal antibodies, which bind the target protein, must be selected. However, it is noted that often antibodies which show high affinity in solid phase immunoassays exhibit little or no affinity when immobilized on a stationary phase and vice versa. The practical approach of binding the antibody on a small scale, followed by directly testing the immobilized antibody is suggested.

The chapter presents a most useful survey and discussion of procedures for preparing the immunoaffinity supports and gives reference to published procedures and commercially available materials which can be used for this purpose. Residual immunoreactivity, effect of coupling pH, activated group and antibody coupling density, detection and prevention of antibody leaching, stabilization, and even solubilization and renaturation of recombinant proteins are covered in a concise, yet complete manner. The practical matter of the FDA's stringent regulations on testing monoclonal antibodies for polynucleotides, retroviruses, and ecotropic murine leukemia virus is also mentioned. These descriptions, backed up with demonstrated separations of recombinant IFN-alpha, IFN-gamma, IL-1, and IL-2, give insight into an industrial perspective of immunoaffinity chromatography.

Chelating Peptide-Immobilized Metal Ion Affinity Chromatography. Chapter 12 by M. C. Smith, J. A. Cook, T. C. Furman, P. D. Gesellchen, D. P. Smith, and H. Hsiung is on the use of genetic manipulation to alter the retention properties of proteins in immobilized metal affinity chromatography (IMAC), by the N-terminal addition of chelating peptides (CP) with high metal affinity is describe is described. Development of the method, which they term chelating peptide immobilized metal affinity chromatography (CP-IMAC) first required the identification of small peptides with the necessary high metal affinity. This was done by screening approximately fifty candidate peptides for retention behavior on IMAC columns, resulting in the selection of one di- and two tripeptides for further study. As *E. coli* expression often results in addition of an N-terminal methionine residue which could inhibit CP interaction with IMAC supports, the IMAC retention behavior of N-methionyl analogs of the candidate peptides was also studied. Although CP affinity for Co(II) was abolished in the methionyl analogs, they retained nearly full affinity for Ni(II) and (in one case), Cu(II). These results establish the applicability of CP-IMAC to proteins expressed in *E. coli.*

CP-IMAC has been applied to purification of recombinant human insulin-like growth factor-II (IGF-II). A synthetic DNA sequence was used to extend the N-terminus of the protein to include a CP sequence, connected to IGF-II via a specific proteolytic cleavage site. During purification from a crude *E. coli* lysate, this chimeric protein bound strongly to a Cu(II) IMAC column, along with only a small minority of the contaminating host proteins. After elution by pH change, native IGF-II was liberated from the chimeric protein by enzymatic cleavage to remove the chelating peptide. A second cycle of Cu(II) IMAC efficiently removed the contaminants which had been retained along with the CP-protein in the first step. The large change in retention behavior induced in the desired protein by CP removal illustrates a major advantage of the use of removable affinity handles.

Site-Specific Proteolysis of Fusion Proteins. Chapter 13 by P. Carter presents a generic issue in the use of gene fusions to assist protein purification-removal of the affinity handle to recover the native sequence. As illustrated by Smith et al. (Chapter 12), affinity handles added by gene fusion can greatly facilitate purification. Even if not exploited as part of the purification strategy (to discriminate against contaminants of constant retention behavior), removal of affinity handles is normally required since foreign sequences may result in immunogenicity.

This chapter reviews the state of the art in selective removal of affinity handles by chemical and enzymatic means. The difficulties which can result from inaccessibility of the cleavage site are described. These range from adventitious cleavage by host proteases to misfolding. Highly-specific proteolytic enzymes which have been employed for selective protein cleavage, noting commercial sources and practical aspects of their use, are surveyed. An example is given by the serine protease subtilisin BPS from *Bacillus amyloliquifaciens*. The substrate specificity of this enzyme is too broad to be useful for selective removal of affinity handles. However, a mutant enzyme in which the histidine in the catalytic triad was replaced by alanine (H64A) is highly specific for histidine-containing substrates, apparently because the substrate histidine can partially substitute for the missing catalytic group (*8*). This behavior has been called "substrate-assisted catalysis." The H64A subtilisin mutants are available for research purposes upon request to the authors.

Purification Alternatives for IgM (Human) Monoclonal Antibodies. Chapter 14 by G. B. Dove, G. Mitra, G. Roldan, M. A. Shearer, and M. S. Cho gives a case study on purification of monoclonal IgM's from tissue culture of human B lymphocyte cell lines. The process described gave a purification sequence in which final protein purity was greater than 99%, DNA clearance was greater than 1,000,000 and virus clearance was 100,000 times. Contaminants which must be removed from the IgM's include the residual media components (albumin, transferrin, insulin and other serum proteins) as well as nucleic acids, viruses, and cellular products. DNA removal was achieved by passing the DNA containing product stream over an immobilized enzyme column in which DNA hydrolyzing enzymes decrease the size of the DNA from a molecular weight of 1,000,000 to 100,000 to 10,000. This procedure alone increased DNA clearance from 10x (without DNAse digestion) to 10,000x when the treated stream was fractionated over a size exclusion chromatography column. This is but one example of the separation techniques which are discussed in a purification sequence of filtration, precipitation, and size exclusion, anion, cation, hydroxylapitite, and immunoaffinity chromatography. This chapter provides fascinating insights into purification development for a large protein.

Analytical, Preparative and Large-Scale Zone Electrophoresis. Chapter 15 by C. F. Ivory, W. A. Gobie, and T. P. Adhi is a comprehensive and readable summary of electrophoretic techniques which integrates key theoretical considerations with clear diagrams and descriptions of basic analytical, preparative, and large-scale electrophoretic systems which separate proteins on the basis of differences in molecular weights, mobilities and/or isoelectric points. Numerous illustrations of these separation mechanisms are given.

According to the authors, a convincing demonstration that zone electrophoresis provides high resolution on a large scale will open the way to full scale bioprocessing applications. Samples in the 1 μg to 100 mg range might be handled by electrochromatography while zone electrophoresis may offer significant advantages over other electrokinetic methods at loadings of greater than 1 gm. Capillary zone electrophoresis is shown to be able to attain efficiencies in excess of 500,000 plates/meter but, unfortunately is limited to microgram size protein samples. This chapter gives the reader a sense of the rapid progress being made in practical large scale applications of electrophoretic separations. This is illustrated by recycle isoelectric focusing and recycle continuous flow electrophoresis, which are techniques that have the potential to process proteins in the 100 g/hr range. The authors make a convincing case that this area of technology has an exciting future. In the meantime, this chapter and the succeeding one bring the reader clear descriptions of the state of the art.

Applied Electric Fields for Downstream Processing. Chapter 16 by S. Rudge and P. Todd gives clear descriptions which illustrate principles of how electric fields may be applied to drive or enhance rate processes in downstream processing. These include consideration of thermodynamics at charged interfaces; the mathematics and physical chemistry of surface charge and the double layer; and the electrokinetics in transport processes. The authors present relevant scaling rules and use these rules to delineate physical processes which can occur in a closed system to cause backmixing. Their analysis shows heating is the single most important limitation to electrokinetic scale-up. Approaches to overcome heating and mixing effects are discussed.

The scaling of mass transfer in electrophoretic systems compared to chromatographic systems is a particularly interesting part of this chapter. The authors explain why the transport rate in electrophoresis is 1,000 to 10,000 times greater than ordinary liquid chromatography while attaining the same relative equilibrium associated with chromatography. This observation is drawn from comparison of electrophoretic and chromatographic Peclet numbers which reflect the ratios of transport velocity to the rate of diffusive mass transfer. This type of analysis is also incorporated in subsequent discussions of processing applications including demixing of emulsions, cell separations, density gradient column electrophoresis, continuous flow electrophoresis, and analytical applications of electrokinetics for process monitoring.

Conclusions

Large scale protein purification protocols are moving from the developmental laboratory to the pilot plant and to commercial production. While purity, regardless of cost, may be the goal during the early phases of the product discovery and development process, production economics are a necessary consideration as scale-up is pursued. For chromatographic separations, a preliminary cost estimate must consider stationary and mobile phase costs as well as the impact of throughput and support stability on these costs (*10*). Since purity at the commercial scale must usually be the same, if not better, than that initially obtained at the laboratory scale, the economic element becomes a key constraint in choosing large scale purification strategies and optimizing their operational conditions, while maintaining uncompromising standards of product purity. The chapters in this volume present insights, examples, and engineering approaches from industry, and fundamental models and engineering analysis from university researchers with both discussing many novel approaches and exciting new ideas for obtaining high purity products with large scale separations.

Acknowledgments

One of the authors (ML) acknowledges the support of NSF Grant BCS-8912150 which supported parts of the material in this work. The authors thank A. Velayudhan and G. J. Tsai for their helpful suggestions and comments during the preparation of this manuscript.

Literature Cited

1. Knight, P. *Bio/Technology* 1989, *8*, 777.
2. Belter, P. A.; Cussler, E. L., Hu, W-S. *Bioseparations*; J. Wiley & Sons: New York, NY, 1988.
3. MacMillan, J. D.; Velez, D.; Miller, L. In *Large Scale Cell Culture Technology*; Lydersen, B. J., Ed.; Hanser: Munich, 1987.
4. Alberts, B.; Bray, D.; Lewis, J.; Raff, M.; Robert, K.; Watson, J. D. *Molecular Biology of the Cell, 2nd edition*; Garland Publishing: New York, NY, 1989.
5. Atkinson, B.; Mavituna, F. *Biochemical Engineering and Biotechnology Handbook*; MacMillan Publishers, Inc.: Surrey, England, 1983.
6. Kroeff, E. P.; Owens, R. A.; Campbell, E. L.; Johnson, R. D.; Marks, H. I. *J. Chromatogr*, 1989, *461*, 45-61.
7. Pungor, E.; Afeyan, W. G.; Gordon, N. F.; Cooney, C. L. *Bio/Technology* 1987, 5(6), 604-608.
8. Carter, P.; Wells, J. A. *Science* 1987, *237*, 394.
9. Campbell, D. H.; Lusher, E.; Lerman, L. S. *Proc. Nat'l. Acad. Sci. USA* 1951, 37, 575-8.
10. Ladisch, M. R. In *Advanced Biochemical Engineering*; Bungay, H. R. and Belfort, G., Eds.; J. Wiley and Sons: New York, NY, 1987; pp. 219-327.

RECEIVED March 1, 1990

Chapter 2

Strategies for Large-Scale Protein Purification

Sa V. Ho

Bio-Products Division, Eastman Kodak Company, Rochester, NY 14652–3605

The development of economical processes for purifying proteins from recombinant sources requires the integration of many isolation and purification methods as well as healthy cross-exchange among molecular biology, fermentation, purification and analytical groups. With regards to purification, two schools of thought seem to have emerged. One approach is to start with a highly specific method early on to achieve the required purification in a single step. While aesthetically appealing, in practice this approach lacks a truly high resolution method that is also economical and scalable. The second approach utilizes a cascade of conventional methods to achieve the required protein purity. Here, what methods to use and in what order represent a major task in the process development effort. Based on the author's industrial experience and process information from the literature, some general guidelines for developing optimal purification processes could be established. Examples showing the applicability of this approach will be discussed.

The development of purification processes for large-scale manufacture of proteins is a very challenging activity. While protein purification itself is already complex, the requirement of "large scale" imposes additional implications such as economy, scalability, and reproducibility, which severely constrain what can and have to be done. The focus in this paper is not on optimizing (or advocating) any particular purification method, for which one could consult experts in the field or draw on the wealth of literature available. Rather, we will try to tackle the challenging task of how to convert a fermentation broth or crude solution into the purified product that satisfies all the requirements (cost, purity, efficacy, etc.). Only recently have several excellent publications appeared addressing various aspects of process development for large-scale protein purification (1-5).

0097–6156/90/0427–0014$06.25/0

STRATEGY DISCUSSION

Purification process development, especially for rDNA-derived products, is a multifaceted activity that requires close participation of many scientific disciplines as illustrated in Figure 1. The significance and the inter-relationship of these elements are the focus of this discussion.

PURIFICATION PROCESS. Unlike fermentation development which basically involves optimizing the operating parameters in a more or less standard vessel, there is not a single technique or equipment that is capable of delivering purified protein in its final form directly from the fermentation broth. So the first major problem confronting a purification development person is, among the fairly large number of techniques and equipment available, not only what techniques should be used, but also in what order (sequence) as well as how each one is operated. These three aspects are interrelated and are determined by the following two main factors.

The first one is the nature of the starting solution, which can come from many different sources (microbes, tissue culture, synthesis), with the product in different forms (soluble or as inclusion bodies) and at different locations (cytoplasm, periplasm, in the broth). Each situation imposes different constraints and challenges to the development effort.

The second key factor relates to the product. Aspects such as purity, form, impurity profile, etc. (product specification) and cost strongly dictate what an acceptable purification process would be like. For both cost and purity requirement the spectrum spans from human therapeutics such as t-PA, insulin, hGH (high cost, ultrapurity) to animal growth hormones (medium cost and high purity) to industrial enzymes where the cost is low but purity is not so critical.

THE ROLE OF FERMENTATION. Fermentation conditions and protocols determine the quantity as well as the quality of the starting material (concentration, conformation, purity, impurity profile), which greatly impacts downstream processing. The optimization (operating conditions, raw materials used, etc.), therefore, should not be based solely on performance at the fermentation stage. Due to the complex nature of the broth, higher titer measured by HPLC or SDS-PAGE, for example, may not be directly related to the final amount of purified, active product obtained. A peak on the HPLC or a band on the gel may contain more than one component, arising from small differences in amino acid residues or different conformations.

A less well-known impact of fermentation on purification is that subtle modifications of the products resulting in a mixture of closely-related compounds that can be extremely difficult to separate may be resolved at the fermentation stage. An excellent example of this is the work done by Amgen scientists on norleucine misincorporation in interleukin-2 (6,7). They found that the incorporation of norleucine instead of methionine in the product, which results in product heterogeneity with unpredictable immunogenic consequences, could be minimized or even eliminated by

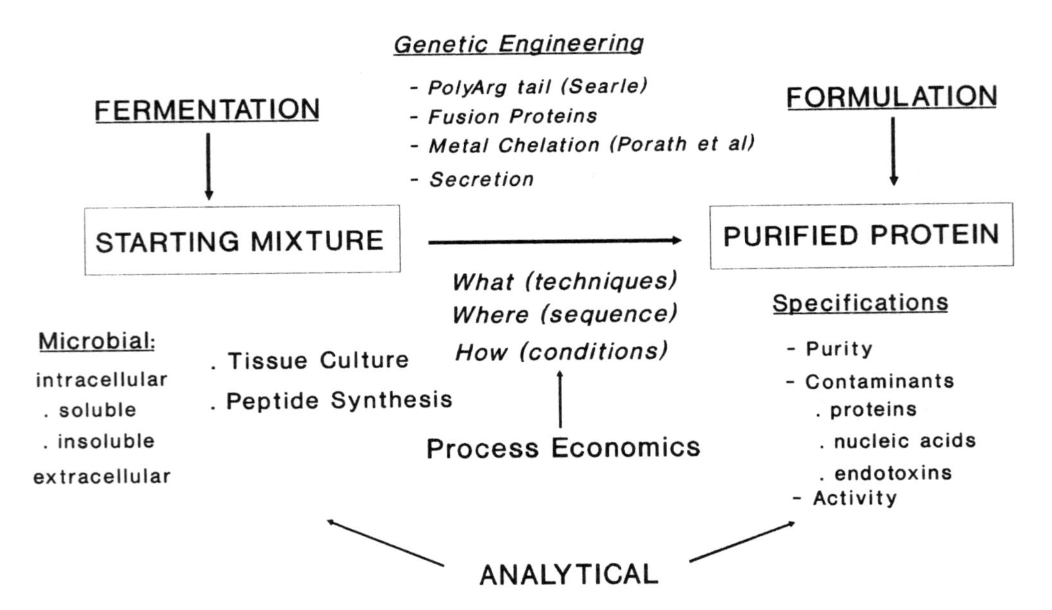

Figure 1. The multifaceted nature of protein purification process development.

simply adding leucine and/or methionine at low levels to the fermentation medium.

FORMULATION CONSTRAINTS. Formulation could impose further constraints on product characteristics since the product in its desired deliverable form (stability, activity, release rate, color, etc.) is truly the final goal. It is possible that a preferred purification process purely from a processing standpoint may produce a less desirable product from a formulation standpoint (such as stability, impurity profile, delivery rate). So a purified product should be rapidly carried through formulation and testing to ensure that it is acceptable before a purification process is locked in.

THE IMPACT OF GENETIC ENGINEERING. The advent of genetic engineering has probably exerted the biggest impact on purification practice, introducing new challenges as well as offering new solutions. It is possible to not only achieve high production of a protein product but also modify it for improved separation and/or stability (e.g., polyarginine tail (8), metal affinity site, fusion proteins) or for determining its eventual residence (cytoplasmic, periplasmic or extracellular). In addition, for some proteins the number of cysteine residues could be changed for improved refold efficiency (9).

THE ROLE OF ANALYTICAL. While not obvious, analytical plays a very critical role in process development for rDNA-derived products. Providing a quantitative assay for the product is only a small part of analytical development. The ability to identify contaminants that not only are present at very low concentrations but also differ very slightly from the main product (e.g. due to misincorporation, process-related chemical modification or aggregation) is absolutely essential in the overall effort of developing an optimal purification process.

In summary, the optimization of a purification process requires close interactions among all of the above functions, careful planning as well as allowing room for iterations. This is, however, an ideal situation. In reality, the pressure of early market introduction and slow regulatory clearance usually forces one to lock in with an "inferior" process, which could be frustrating but necessary. Just being aware of all these issues is in itself an important element of process development. Then early planning in combination with a sound strategy based on experience is probably the best one can do under the circumstances.

STRATEGY DEVELOPMENT

The question here is how to convert an impure protein solution such as a fermentation broth into a purified product that meets all the requirements. Is there a methodology that would offer some guidelines?

There seems to be two general approaches. In one, a highly specific method such as immunoaffinity is utilized early in the process to achieve the maximum degree of purification (e.g.

alpha-interferon, Hoffmann La Roche). This approach is not generic, could be costly, and, if antibodies are used as ligands, may have other operating complications such as significant loss of binding capacity upon immobilization, and leaching of the antibodies into the product stream. Antibodies may also bind to the denatured, unfolded, mispaired, and aggregated forms of the product and its analogs. The other approach is to use a series of "traditional" methods in a concerted way to achieve the desired goal (purity, yield, cost, scalability). The latter approach appears to be quite effective and has been used commercially for purifying insulin (E. Lilly) and human growth hormone (Genentech) and is probably used in development stage for other protein products such as fibroblast interferon, interleukin-1 and animal growth hormones.

Actually the division between the two approaches can be arbitrary. Since no single technique could deliver the final purified product directly from the fermentation broth or crude extract, highly specific methods such as immunoaffinity still require the support of other methods to accomplish the task, especially in light of some of its "unexpected" drawbacks discussed above. Thus, in either case process optimization is still required. The key difference in the two approaches is in the initial focus. In general, unless a highly specific and unique affinity method has already been identified, we recommend that the approach developed below be used.

Bonnerjea et al (2) analyzed 100 publications on lab-scale protein purification procedures. They plotted the results (Figure 2) showing the number of steps used in the purification scheme as a function of the frequency a method is used in each step. A pattern seems to emerge. Homogenization is the most frequently used first step, probably because most protein products are intracellular, hence the need to break the cells open. It is interesting that precipitation is popular as the first purification step, followed by more resolving methods such as ion-exchange and affinity methods, and finished with gel filtration. While these exact methods are certainly not generally applicable, they seem to suggest a fairly sensible sequence of events. Going beyond these specific techniques and combining our experience with others in the literature regarding purifying proteins from recombinant sources, we propose the general purification scheme shown in Figure 3. These are blocks of activities that may require more than one step.

The essence of the proposed scheme is the deliberate breaking down of the development activity for protein purification into two separate blocks: gross purification and high-resolution purification. In gross purification, the focus is to utilize a simple yet effective step(s) to significantly clean up the solution in such a way that it flows naturally into the next block where high resolution methods are used for the final purification. This way, it is the gross purification that determines what and how high-resolution methods should be used. There are several advantages with this approach. For one, there always exist a large number of impurities in the crude solution that have extreme properties (highly charged, either very large or very small, highly hydrophobic, etc.) and that could be removed easily in a simple step if the appropriate method is used. Only after this treatment is it

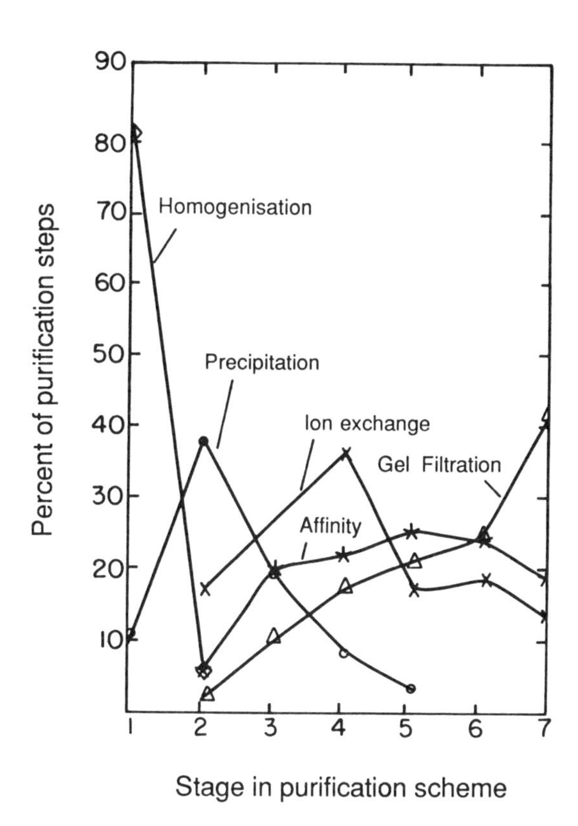

Figure 2. Analysis of the purification methods used at successive steps in the purification schemes (Reproduced with permission from Ref. 2 Copyright 1986 Nature Publishing Company.)

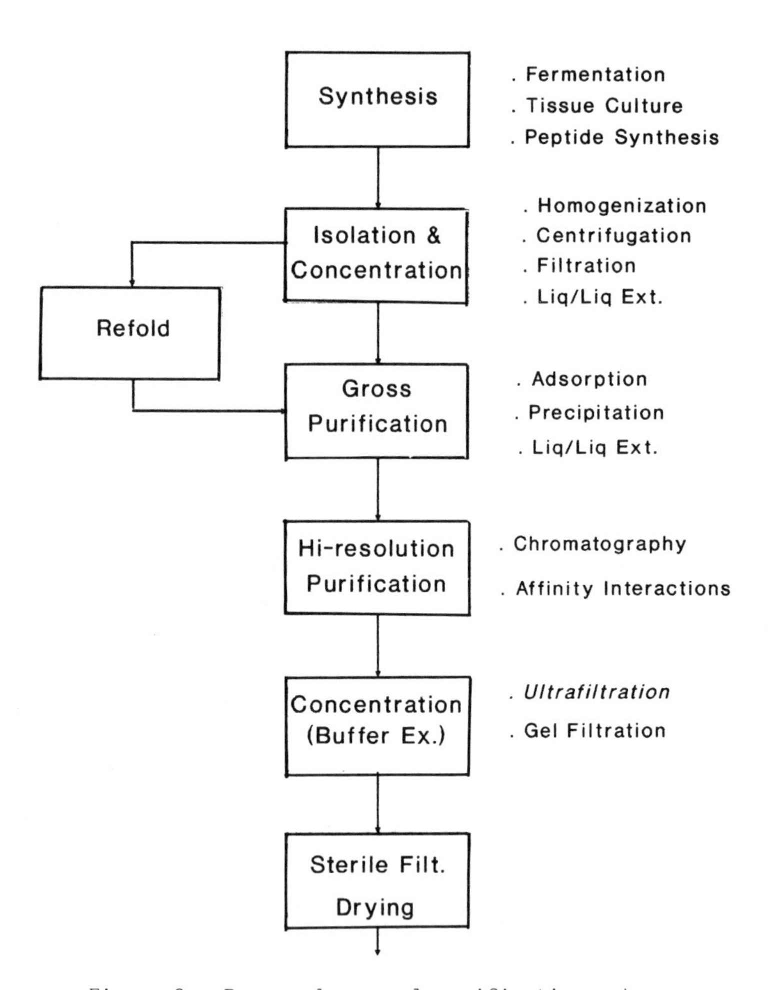

Figure 3. Proposed general purification scheme.

clear what contaminants are left. These tend to have similar properties to the product of interest. Focusing on separation of similar compounds, free from other interfering contaminants, will result in higher capacity, better resolution and longer life for the more complex and more costly high resolution step.

The above scheme automatically organizes the large number of isolation/purification techniques into a few manageable categories as shown in Figure 3 along with the flow diagram and expanded in Tables I-IV. These block activities are considered in detail below.

Table I. Isolation Methods

a) Cell Rupture
- o Homogenization
- o Extraction (Bio/Chemical Treatment)

b) Solids/Liquid Separation
- o Centrifugation
- o Filtration
- o Cross-Flow UF/MF
- o Aqueous Two-Phase Partitioning

Table II. Refold

a) Dissolution
- o Denaturants (urea, guanidine.HCl, SDS, etc.) and/or extreme pH

b) Oxidation
- o Complex, yield loss due to aggregration
- o Key parameters: pH, T, time
  con. of denaturants
  additives
- o Genetic Engineering:
  3 S-H ⟶ 2 S-H (beta-interferon)

Table III. Bulk Purification Methods

- o Adsorption: charged, hydrophobic, affinity adsorbents
- o Precipitation
  - pH
  - Temperature
  - Salts
  - Polymers (Neutral/Charged)
  - Organic Solvents
  * Affinity?
- o Liquid/Liquid Extraction
  - Organic/Aqueous
  - Aqueous/Aqueous

Table IV. High Resolution Methods

o	Chromatography
	- Ion-Exchange (charge)
	- Hydrophobic Interactions
	- Reverse Phase
	- Size-Exclusion (size & shape)
	- Chromatofocusing (pI)
	Affinity Interactions
	- Antibody - - Antigen
	- Hormone - - Receptor
	- Enzyme - - Substrate/Analog/Inhibitor
	- Metal ion - - Ligand
	- Dye - - Ligand
	Characteristics
	- can be costly
	- powerful for Dilute/Low Purity solutions

ISOLATION / RECOVERY. The goal here is to concentrate the starting solution (e.g., fermentation broth), preferably with some degree of purification. This may or may not involve cell breakage since the products may be secreted. Homogenization has been found effective for intracellular products, especially inclusion bodies, even though extraction by bio/chemical means is a feasible option. An advantage with inclusion bodies is that they appear to be quite sturdy and can withstand rather severe operating conditions. For soluble products (enzymes, peptides, etc.), however, caution has to be taken to avoid denaturation and proteolytic clippings. Here speed and solution conditions for minimizing proteolytic activity (temperature, pH, ionic strength, inhibitors, etc.) are of the essence. Product secretion by genetic engineering means is an attractive approach due to the potential process simplification (10), provided that comparable productivity can be achieved.

Depending upon the nature of the product a number of liquid/solid separation methods can be used. For separation of cell debris and products in solid form (e.g., inclusion bodies, precipitated products) centrifugation has emerged as the method of choice in which size and density differences between cell debris and protein particles are exploited (11). The drawbacks here are high capital and operating costs. For some situations cross-flow filtration (UF/MF) may be a feasible alternative.

Aqueous two-phase extraction, in particular PEG/salt systems, has been successfully applied to concentrating cells and/or cell debris in one phase and extracting products into the other phase (12-15). The advantage here is that not only liquid/solid separation but also concentration and partial purification are accomplished at the same time. Compared with other methods, a factor that needs to be considered with aqueous two-phase extraction is the addition of reagents (polymers & salts) to the system, which may complicate downstream processing if they have to be removed at the end. Partly because of this very issue, large-scale applications of aqueous two-phase systems have been limited to purification of industrial enzymes where activity rather than absolute purity is needed .

For secreted products or intracellular products released by permeabilization, cell/liquid separation followed by product concentration/diafiltration can be conveniently carried out by cross-flow microfiltration or ultrafiltration.

REFOLD. This step is necessary if the protein is in an inactive form (denatured, reduced, etc.). It has been found that overexpression of foreign proteins in bacterial systems often results in formation of inclusion bodies. The protein exists in a reduced, polymeric state which necessitates dissolution and renaturation. This step is quite critical since it not only affects the overall yield but through the nature of the impurities generated will also dictate the subsequent purification train. For secreted products, which tend to be in their active form, refold is not normally necessary.

In the case of inclusion bodies, the refold step typically consists of first dissolving the solids in a strong chaotrope such as guadinine hydrochloride, sodium thiocyanate or urea, followed by renaturing/oxidative refolding process. This is a complex process with major yield loss due to aggregation (16). Key parameters for minimizing aggregation or maximizing the refold yield are pH, temperature, time, and concentration of denaturants. Additives such as detergents have also been used (17). The principle at work here seems to be the drive towards an optimal hydrophobic/hydrophilic balance exerted on the protein by its environment: sufficient hydrophobic force is needed to cause the protein to refold but the same force also leads to aggregation.

A great deal of proprietary information related to practical refolding of proteins has been generated in the private sectors. Contrary to theoretical beliefs, a number of proteins have been found to refold quite efficiently (RF efficiency $>$70%) at high concentrations (g/L level) using fairly simple procedures. It appears that since protein refolding is not well understood and since each protein may be unique, generalization to the point of excluding experimental investigation may not be warranted at this time.

GROSS PURIFICATION. The main purpose here is to remove as many impurities as possible in a simple step or step(s). For this, batch adsorption and precipitation seem to be most effective. Batch adsorption is particularly effective for dealing with dilute solutions by selecting an adsorbent that will bind the product. Since high selectivity is not critical at this stage, an adsorbent with high capacity and some specificity for a particular product is not too difficult to find. Common adsorbents available commercially are either charged or hydrophobic. Affinity adsorbents using metal chelate or dye ligands are very effective for group specific interactions and should be exploited when appropriate. With well-thought out applications, this approach could accomplish clarification, concentration, and partial purification in practically one integrated step, as illustrated in one of the case studies discussed later. If the solution is free of particulates, adsorption in column mode can be done, which allows some leverage for improved purification during elution.

Precipitation can be a very powerful method for gross purification. Protein precipitation can be brought about by various means such as changes in pH or temperature of the solution, by addition of salts (e.g. ammonium sulfate), water-soluble polymers (e.g. polyethylene glycol, polyethylene imine, polyacrylic acid), inorganic flocculants (silica or alumina, bioprocessing aids from Rohm and Haas, cell debris remover from Whatman), or organic solvents (e.g. alcohols). An excellent review of these precipitation methods is given by Bell et al (18). Salting out in which molar concentrations of salts such as ammonium sulfate are used is a very common method for protein precipitation. This approach, however, has many drawbacks: low selectivity, high sensitivity to operating conditions, and down stream complications (salt removal & disposal). Simpler and more effective methods such as change in pH or temperature, use of polyelectrolytes alone or in combination with neutral polymers (19,20) should not be overlooked.

There are two ways to use precipitation. One convenient approach is to precipitate most of the impurities leaving the product in solution for further processing. For this, flocculation is very effective. The precipitate is a network of cell debris, extraneous proteins, colored contaminants, and, with anion exchange flocculants such as polyethyleneimine (21), nucleic acids. Instead of precipitating contaminants, in some cases it may be easier to precipitate the product. This, however, would necessitate solid recovery and redissolution, which means additional processing steps and yield reduction.

While the above methods appear common and tend not to be highly selective, they can be very effective if used appropriately, especially for rDNA products, because of the following two reasons. First, most practical rDNA processes achieve fairly high expression (10-30% of total cellular proteins), so highly sophisticated techniques are not usually required to increase the purity to 70-90%. Second, the proteins of interest being foreign to the bacterial systems are likely to have very different properties (pI, hydrophobicity, size, heat stability, etc.) that should be exploited for simple yet effective purification. This will be illustrated in one of the case studies.

Another effective method for gross purification is liquid-liquid extraction, especially aqueous two-phase systems (12-15). These, however, have limited applications so far because of a number of reasons: relatively new method, high polymer cost (for PEG-dextran systems), and the need to remove phase-forming reagents (polymers, salts) from the products.

HIGH RESOLUTION PURIFICATION. Various forms of chromatography (Table IV) are predominantly used for the final purification of proteins to homogeneity (22). Ion-exchange chromatography is widely used due to its versatile applicability for proteins, high capacity and resolution. Based on literature information as well as our experience, ion-exchange or reverse phase/hydrophobic interactions chromatography can have remarkable resolution despite its supposedly nonspecific mode of interactions (charge or hydrophobicity). The nature of the interaction between macromolecules (with their three-dimensional structures) and surfaces is such that unexpected specificity may result from the dynamics of multisite interactions (23). With the large number of parameters available for manipulation in chromatographic separation -- type of resin, degree of loading, washing conditions, elution strategy (step, gradient, displacement or combination; type of buffer or solvent) -- resolution of molecules with minor differences in amino acid residues or between monomer and oligomers could be achievable. Very good review articles and books can be found on the use of chromatography for purification (24-26). Naveh, in an excellent review on scale-up strategies for protein purification (4), offered some practical considerations for utilizing chromatography in a purification scheme. He discussed the importance of maintaining consistent feed solution (e.g. product concentration and impurity profile) due to overloading conditions; the significance of dynamic capacity as related to linear flowrate, solution pH and ionic strength; and the role of matrix packing with respect to column pressure drop through the chromatographic cycle.

Chromatography using affinity interactions is an important method but one has to be very careful in the choice of ligands. Antibodies are generally not a practical choice due to high cost, low capacity, and fairly poor selectivity for closely-related molecules (monomer/dimer, denatured, wrongly-folded). Receptors appear to a better choice with regards to these aspects, as found by Hoffmann La Roche in the purification of interferon. Metal ions and dyes could be very selective for some proteins, especially if they are modified for enhanced interactions. A unique aspect of affinity methods is that, due to the strong interaction involved, they are a powerful tool for concentration/purification of dilute solutions. As such, they could be viewed as concentration & gross-purification methods, which need to be followed with, for example, chromatography for the final purification. Even though this order of usage may seem strange, it makes sense if one recognizes that, due to localized interactions and the nature of single-stage bind/release, most affinity methods can not usually separate similar molecules.

FINAL CONCENTRATION. At this point in the process, the usual requirement is product concentration with or without buffer exchange (either for stability and/or in preparation for the formulation step) prior to the drying step. Both concentration and buffer exchange can be conveniently carried out in one step using ultrafiltration.

For products that are not stable in a cross-flow filtration environment (high circulating rate, long processing time) or that have a strong tendency to foul membranes, the alternative is gel filtration. This is a milder method for buffer exchange and removal of salts or low molecular weight contaminants; it is, however, not suitable for handling large volumes.

SOME PRACTICAL CONSIDERATIONS

RECOMMENDED INITIAL ACTIVITIES. Our experience shows that the first critical step in the process development is to establish a reliable assay for the product. Without this, it would be very difficult to assess the performance of various methods and separation schemes. Next, one should also characterize the starting solution itself -- in addition to the purified product, which is commonly done. The purpose here is to look for main differences between the product and contaminants, which will serve as the basis for devising a sound purification process. Gel electrophoresis is a quick way to assess the size (MW) and purity distribution of proteins in the mixture. Protein net charge and hydrophobicity can be studied with adsorbents (anion/cation, hydrophobic) or aqueous two-phase systems. The isoelectric points (pI) of proteins in solution can be determined by isoelectric focusing.

In addition, solution characteristics such as stability as a function of temperature and time, handling, etc., should be carefully noted and taken into account in the process development. This not only will minimize processing complications at the large-volume stage but sometimes may also offer clues for unique purification approaches.

SOME GUIDELINES FOR PURIFICATION PROCESS DEVELOPMENT

Approaches for selection of early steps in the purification train:

- reduce process volume early on
- eliminate components of extreme nature: particulates, small solutes, large aggregates, nucleic acids, etc.

Appropriate methods here are adsorption (hydrophobic or ion -exchange), precipitation/flocculation, ultrafiltration, and affinity adsorption.

Integration of Purification Steps:

- steps should be complementary to one another both in degree of purification as well as in process flow to achieve the final goals. It should always be kept in mind that optimization (yield, purity & cost) is done for the whole process not for any single step. This is to avoid being trapped into unduly maximizing a step which may not make much difference in the whole scheme.
- minimize the number of solvents and buffers used. While this may sound trivial and is not that critical at the bench scale, unnecessary solvent or buffer exchanges will be costly and time consuming on a large-scale. Also, avoid, if possible, buffers that are expensive, complicated to prepare, or difficult to pH (Tris buffers)

Ease of Operation: The key point here is the simpler the process and the more straightforward the conditions, the less likely for it to fail on a large-scale.

CASE STUDIES

Table V lists the possible sources of protein/peptide products. Only a few representative real examples are discussed here to demonstrate the applicability of the outlined strategies.

MICROBIAL, INTRACELLULAR, INSOLUBLE (E.G., INCLUSION BODIES). An interesting case in this category is the purification process for fibroblast beta-interferon presented by Hershenson of Cetus Corp (9). The interferon was made in the form of inclusion bodies. The cells were first homogenized to release the product, followed by centrifugation to bring down the inclusion bodies leaving cell debris in the supernatant. The recovered inclusion bodies were then dissolved and the protein refolded. In the cloning of interferon, the odd cysteine residue that does not participate in the disulfide formation was deliberately replaced with serine to minimize misparing in the refolding process. This was done to improve the refold efficiency. After the completion of the refold step, there are two types of contaminants present: *E. coli*-derived (proteins, endotoxins, nucleic acids), and product-like (dimer/aggregates of interferon and its modified forms). This type of mixture is fairly typical for rDNA products made by *E. coli* in the form of inclusion bodies. An undisclosed "pretreatment" step (gross purification ?) was carried out next to prepare a cleaned-up load solution for the final purification step in which reverse phase chromatography was used to separate interferon from its other monomeric forms as well as from its polymers.

For the purification of animal growth hormones, which follows more or less the same scheme discussed above for beta-interferon, we discovered a very powerful precipitation method for the gross purification step (19-20). Basically a neutral polymer and a charged polymer, both water-soluble, were used simultaneously to bring about precipitation of almost all of the contaminants (Figures 4 & 5). The role of the charged polymer is to form charged complex with the contaminants. The neutral polymer enhances the precipitation of the complex as well as improves the charged polymer specificity. Both purity and yield in excess of 90% could be achieved in a single precipitation step. Chromatography could then be used for the final purification.

MICROBIAL, EXTRACELLULAR, SOLUBLE. An example here is the purification of a peptide (molecular weight about 300) made by a fungal fermentation. The broth contained a lot of fine particles as well as stringy, slimy substances. The peptide concentration and its purity in the broth were very low (Figure 6). Thus the solution needs to be clarified and, as the proposed strategy indicates, significantly concentrated, preferably with partial purification, before a high-resolution purification method is used. By screening the recommended isolation and gross purification methods (Table I & III) we were able to quickly develop a very efficient, easily scalable process in which a simple adsorption/extraction step was used to achieve substantial concentration and gross purification.

Shown in Figure 7, the broth was first subjected to a coarse filtration step mainly to remove large particles and stringy

TABLE V. SOURCES OF PROTEINS AND PEPTIDES

1) MICROBIAL, INTRACELLULAR, INSOLUBLE

e.g. insulin, animal growth hormones, beta-interferon AP3-Rec A fusion protein, interleukin-2

Characteristics: inclusion or refractile bodies
product in reduced, aggregated forms
high product concentration

2) MICROBIAL, INTRACELLULAR, SOLUBLE

e.g. L-aspartase, human growth hormone, tumor necrosis factor

Characteristics: products susceptible to proteolytic degradation and highly contaminated with soluble cellular components upon cell lysis.

3) MICROBIAL, EXTRACELLULAR (SOLUBLE)

e.g. IGF-1, detergent enzymes, rennin

4) TISSUE CULTURE (EXTRACELLULAR & SOLUBLE)

e.g. t-PA, monoclonal antibodies, interleukin-4

Characteristics: very low product concentration
major contaminants: serum proteins or protein additives (BSA, etc.)

5) PEPTIDE SYNTHESIS

e.g. AP3

Characteristics: no microbial proteins / endotoxins
high product concentration
close analogs as major impurities

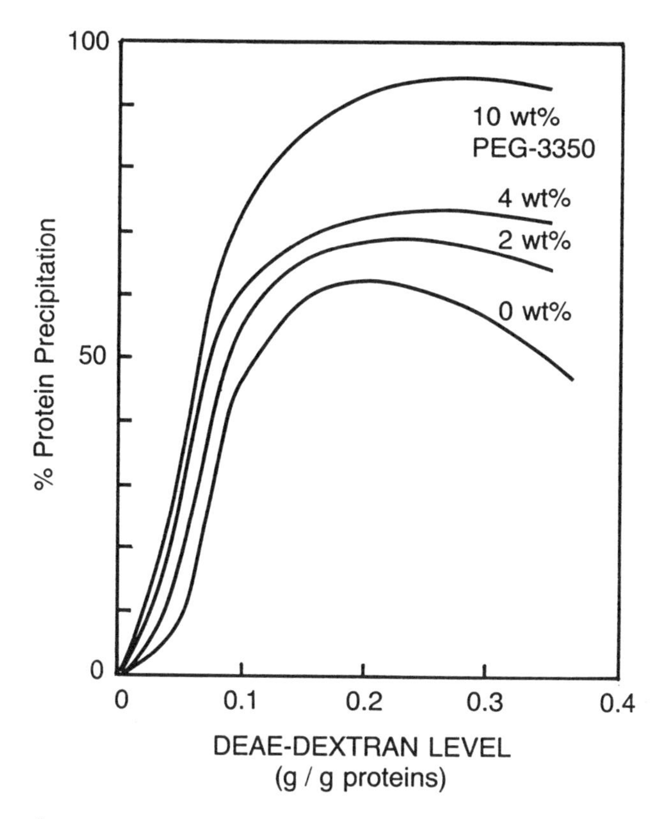

Figure 4. The use of PEG and DEAE-dextran for protein precipitation from a fermentation broth containing an animal growth hormone.

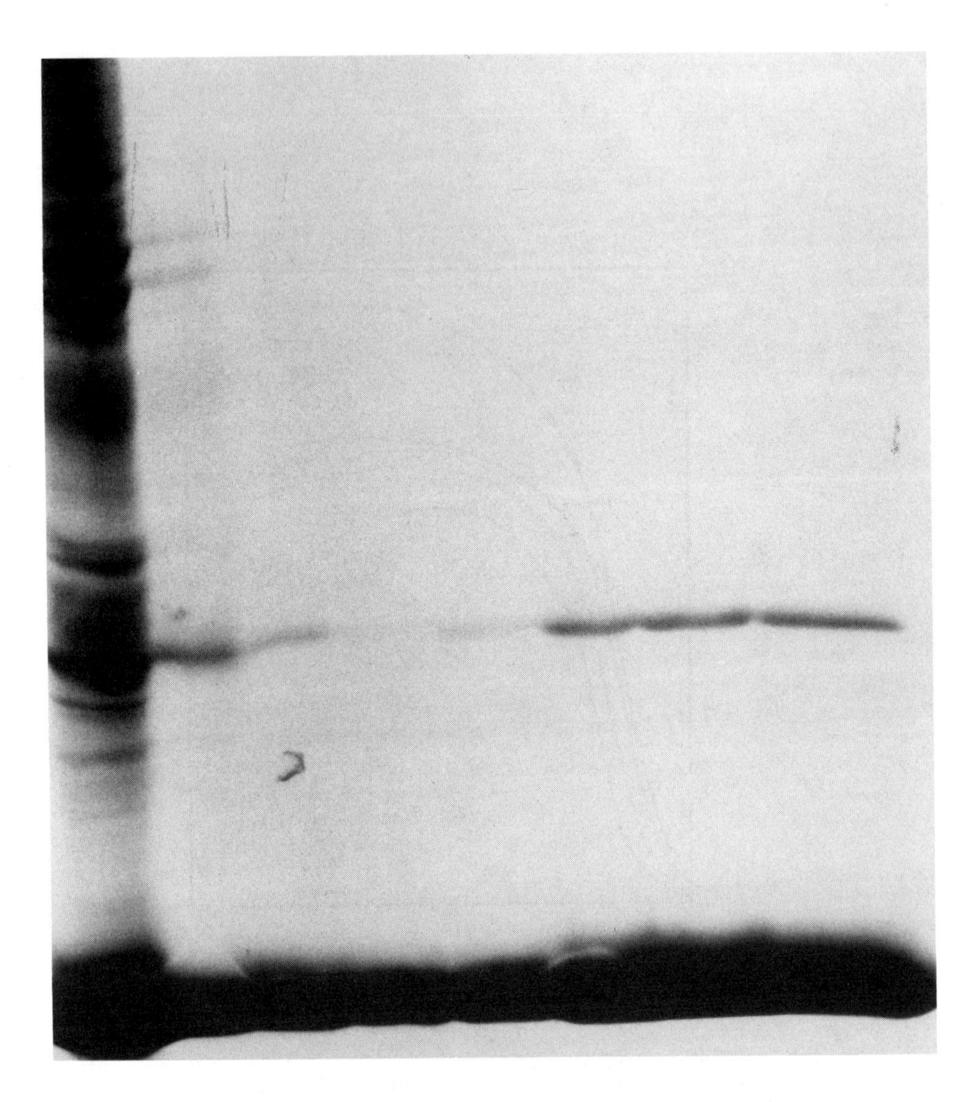

Figure 5. SDS-PAGE of E. coli crude extract (Lane 1) and after precipitation with various combinations of neutral and charged polymers (Lane 2-8). Dark bands at bottom are bovine growth hormone.

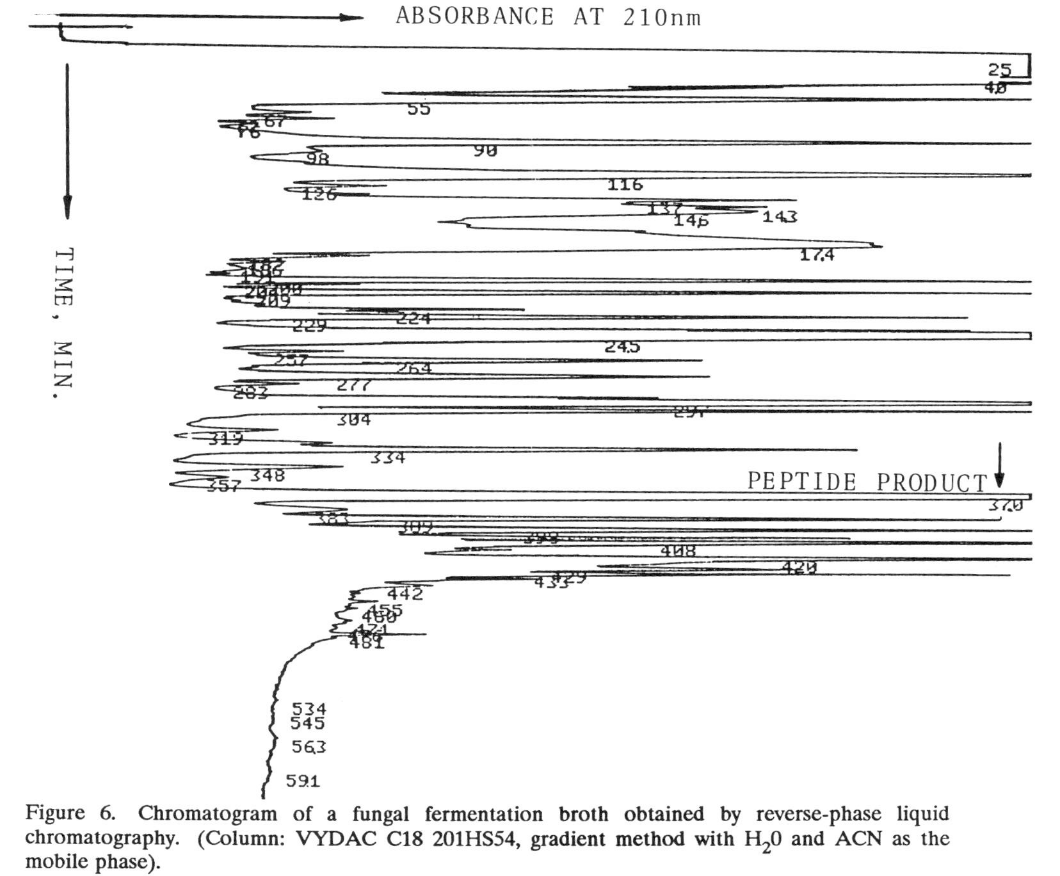

Figure 6. Chromatogram of a fungal fermentation broth obtained by reverse-phase liquid chromatography. (Column: VYDAC C18 201HS54, gradient method with H_2O and ACN as the mobile phase).

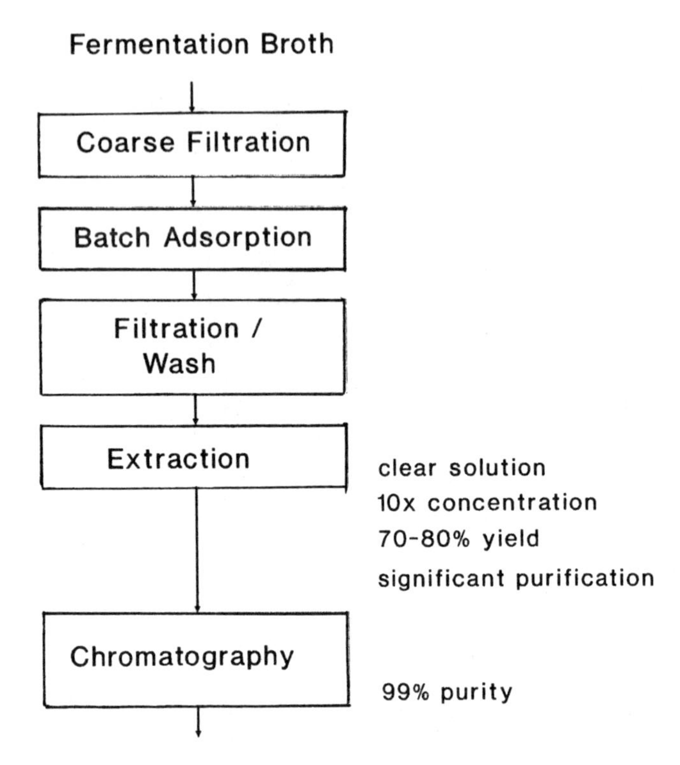

Figure 7. Purification process developed for the fungal peptide.

materials. A hydrophobic resin (XAD series from Rohm & Hass) was then added directly to the cloudy filtrate to adsorb the peptide as well as some other components. The subsequent simple filtration/wash step removed all the fine particulates in solution by exploiting the large size difference between the resin and the particulates. Extraction of the resin resulted in a clear solution that was 10-fold concentrated with 70-80% yield and substantial peptide purification (Figure 8). The final purification using chromatography to get to 99% purity was greatly simplified owing to the relatively pure and concentrated load solution.

TISSUE CULTURE (EXTRACELLULAR, SOLUBLE). For purifying proteins made by tissue culture (e.g., monoclonal antibodies, t-PA) Scott and coworkers at Invitron (27) suggested several schemes illustrated in Figure 9 for conditioned media. Their recommended scheme (#2) is more or less consistent with our proposed strategy. A key characteristic of tissue culture is that product levels tend to be quite low. Concentration is thus a necessary first step, for which ultrafiltration comes in handy. For very dilute solutions, however, affinity adsorption may have to be used also for the concentration step to minimize product loss due to nonspecific binding and long processing time associated with ultrafiltration. For media containing serum proteins, precipitation could be an effective and simple gross purification method for removing these proteins before moving on to chromatography.

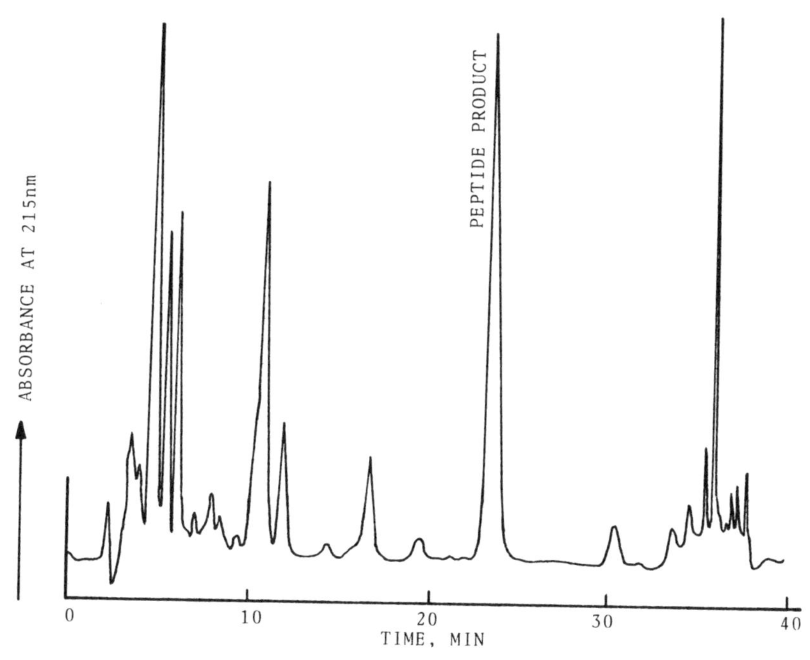

Figure 8. Chromatogram of the solution obtained after the extraction step, as outlined in Figure 7 (isocratic method, 40% ACN in H2O, using the same column as in Figure 6).

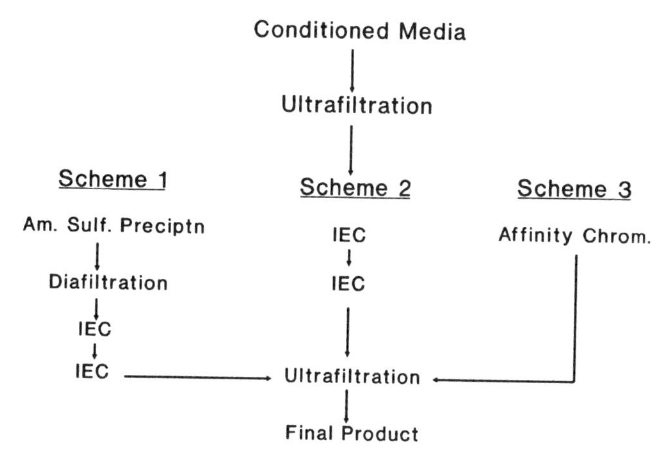

Figure 9. Purification schemes for proteins obtained from tissue cultures. (Adapted from Ref. 27.)

LITERATURE CITED

1. Fish, N. M.; Lilly, M. D. Bio/Technology 1984, July, 623-627.
2. Bonnerjea, J.; Oh, S.; Hoare, M.; Dunnill, P. Bio/Technology 1986, 4, 954-958.
3. Wheelwright, S. M. Bio/Technology 1986, 5, 789-793.
4. Naveh, D. BioPharm 1987, Sept., 34-41.
5. A Strategy For Protein Purification, Pharmacia pamphlet, Vol 13.6.
6. Tsai, L. B.; Lu, H. S.; Kenney, W. C.; Curless, C. C.; Klein, M. L.; Lai, P-H.; Fenton, D. M.; Altrock, B. W.; Mann, M. B. Biochem. Biophys. Research Comm. 1988, 156(2), 733-739.
7. Lu, H. S.; Tsai, L. B.; Kenny, W. C.; Lai P-H. Biochem. Biophys. Research Comm. 1988, 156(2), 807-813.
8. Saaaenfeld, H. M.; Brewer, S. J. Bio/Technology 1984, 2, 76-81.
9. Hershenson, S. in Conference on Frontier in BioProcessing 1986, Boulder, Colorado.
10. Joshephson, S.; Bishop, R. DN&P 1988, 1(5), 271-275.
11. Builder, S. et al., European Patent 190 391, 1983.
12. Albertsson, P. A. "Partition Of Cell Particles And Macromolecules," 3rd edition, John Wiley & Sons, Inc., New York, 1986.
13. Hustedt, H.; Kroner, K. H.; Stach, W.; Kula, M. Biotech. Bioeng. 1978, 20, 1989-2005.
14. Kula, M. R.; Kroner, K. H.; Hustedt, H. Advances in Biochem. Eng. and Biotech. (Fiechter, A., ed.), 1982, 24, 73-118.
15. Datar, R.; Rosen, C.-G. J. Biotechnology 1986, 3, 207-219.
16. Creighton, T. E. Prog. Biophys. Molec. Biol. 1978, 33, 231-297.
17. Tandon, S.; Horowitz, P. M. J. Biol. Chem. 1987, 202(10), 4486-4491.
18. Bell, D. J.; Hoare, M.; Dunnill, P. In Advances in Biochemical Engineering/Biotechnology," ed. Fiechter; New York: Springer-Verlag, 1982, pg 1.
19. Ho, S. V., U.S. Patent 4 645 829, 1987.
20. Ho, S. V. paper presented at AICHE National Meeting 1987, November, New York, NY.
21. Jendrisak, J. In Protein Purification: Micro to Macro," ed. R. Burgess; Alan R. Liss, Inc., 1987.
22. Sofer, G. K. Bio/Technology 1986, 4, 712-715.
23. Regnier, F. E.; Mazsaroff, I. Biotechnology Progress 1987, 3 (1), 22-26.
24. Wankat, P. C. In Large Scale Adsorption and Chromatography CRC Press, Boca Raton, Florida, 1986, pp 55-91.
25. Chase, H. A. In Ion-Exchange Technology; Naden, D.; Streat, M., Eds; Ellis Horwood, Chichester, U.K., 1984, pp 400-406.
26. Chase, H. A. In Discovery and Isolation of Microbial Products; Verrall, M. S., Ed.; Ellis Horwood, Chichester, U. K., 1988, pp 129-147.
27. Scott, R. W.; Duffy, S. A.; Moellering, B. J.; Prior, C. Biotechnology Progress 1987, 3, 49-56.

RECEIVED December 28, 1989

Chapter 3

Separations in Biotechnology

The Key Role of Adsorptive Separations

E. N. Lightfoot

Department of Chemical Engineering, University of Wisconsin, Madison, WI 53706

The key roles of separations processes in biotechnology are identified and briefly reviewed. Discussion begins with very general economic and physico-chemical considerations, to provide overall perspective, and to identify particularly challenging problems in equipment and process design. Bases are then provided for systematic improvement of processes for the large-scale recovery of proteins and other high value pharmaceuticals, high-priority areas where there are pressing needs for faster and more reliable scale-up techniques.

It is shown that adsorptive processes are particularly attractive for the key processes of concentration and purification, both for their high selectivity and flexibility, and because so much information is available about them from experience in biochemical and analytical laboratories.

The final section of the paper deals with the possible advantages of abandoning chromatography completely in the manufacturing process, in favor of potentially more effective operating modes. It will be shown that some alternative techniques are so much more effective that they may compete with chromatography in significant applications, even though they are not yet fully developed.

We begin by examining the significance of Figure 1 which shows the relation of selling price for a wide variety of biologically-produced products to product concentration in the feed to the separation train. This figure, based on data collected in our laboratory (*3*), is consistent with the more comprehensive correlation of Nystrom (*1*) and provides useful insight from several points of view:

1. The strong correlation between price and feed concentration suggests that separations costs are important throughout biotechnology, from the classic commodity chemicals through established antibiotics, and on to therapeutic

0097–6156/90/0427–0035$06.00/0

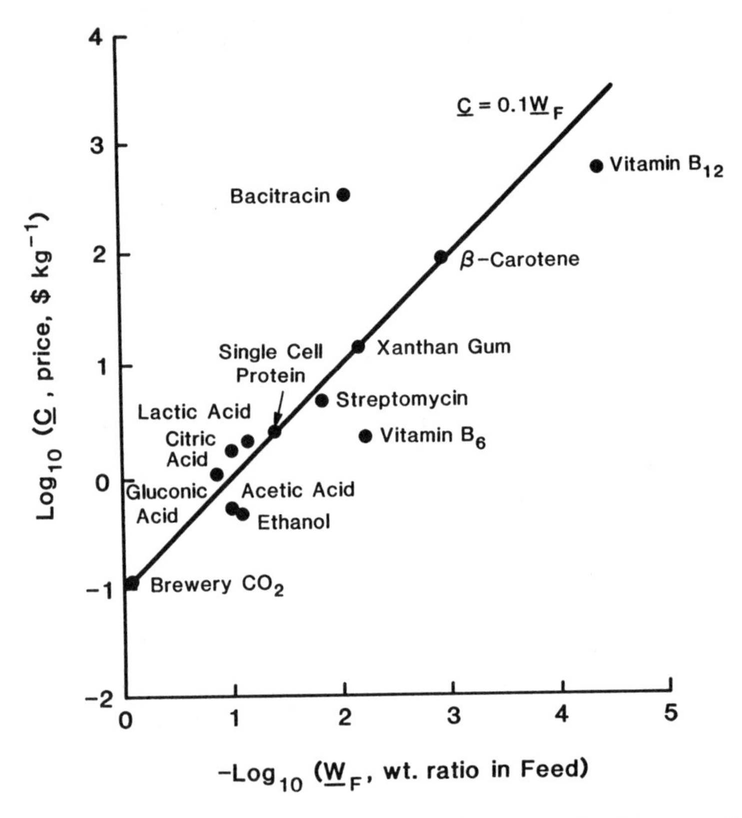

Fig. 1. Selling Price of Biological Products vs. Concentration in Feed to the Separation Train.

proteins. This suggestion is borne out by more thorough examination of cost data (2): separations, or downstream processing, accounts for between 50 and 90% of total production costs, with the higher percentages generally observed for the most dilute feeds.

2. The specific form and generality of the correlation suggest that materials handling tends to dominate the cost of downstream processing. The straight line superimposed on the data shown is of the form:

$$C \alpha W$$

where C = selling price, US$/kg product, and
W = mass ratio of inerts to product in the feed to the separation train.

This relation suggests that for all of the diverse chemical species represented in the figure *processing cost is unrelated to the amount of potentially valuable material being processed.* In other words the cost of processing depends only upon the amount of undesired material that must be processed during separation. This suggestion is borne out to a surprising degree by closer examination, and it will form the basis for the next section of this paper.

3. The very large range of feed concentrations shown demonstrates the heterogeneity of "biotechnology" and suggests that different classes of biologicals may have more in common with selected non-biological products than with each other. This suggestion is supported by the author's experience, and three particular classes of products are of particular importance at the time of writing:

   i. The classic commodity chemicals such as alcohols and carboxylic acids are promising both for their commercial importance and because they can be produced from waste products of the food industry which currently present vexing disposal problems. Lactic acid is a representative example which is currently the subject of considerable industrial attention. However, biological processing suffers from heavy petrochemical competition and is burdened by high capital costs and the dispersion of feed sugars as dilute streams from small isolated food processing plants. From a separation standpoint it is critical to obtain rapid volume reduction and to maximize separator productivity, and many comments of the next section are pertinent. These problems are not at all biological in any intrinsic sense, and they are shared by many non-biological processes.

   ii. The traditional pharmaceutical fermentations, for example the production of penicillin, are mature processes, and they suffer from the general ailments of their non-biological counterparts: obsolescent plants, poorly motivated management and low budgets for process improvement. Established manufacturers are losing ground to more aggressive newcomers and are in many cases abandoning potentially profitable product lines. Technological problems here are to some extent subordinate to considerations of economics and "human factors engineering".

iii. The so-called "new biotechnology" is to a large degree based on processes and equipment scaled up with little modification from the biochemical, or even the analytical, laboratory. There is for example a heavy emphasis on chromatography, which we shall find later in this article to be an inherently expensive procedure. There is a great need here for both equipment and process development, and these problems are aggravated by the strict regulations imposed on this industry: processes must be fixed very early to obtain clearance from government agencies. We shall return to these problems in the last section of the paper, and again it should be noted that these problems are not intrinsically biological. There is for example an analogous situation in the production of synthetic organic drugs, most of which can no longer be purified by crystallization. Chromatographic processes are also being used for these, and not always with great success.

With all of this heterogeneity, both economic and technological, we shall find in the next section, that separations design can be guided by much the same underlying principles, and we now turn to identification of these.

In all cases separations costs are large, and they frequently dominate overall process economics. At the same time the specific needs of the biological processing industry have on balance received less attention than is desirable, and this presents the major challenge to which the remainder of this paper is addressed: to speed the development of biologically oriented separation processes and reduce processing cost.

FUNCTIONAL ORGANIZATION OF SEPARATIONS PROCESSES

To meet the challenges of separations design in the specific field of biotechnoloy it is highly desirable to have a coherent view of separations technology in general. Such a view is just developing at the present time, but useful guidelines are available for our purposes (*3*); we shall briefly review these here.

We begin by suggesting that the functions of separation processes can be usefully organized in terms of three limiting activities:

- Concentration, in which a desired product is removed from large amounts of unrelated impurities.
- Fractionation, in which comparable amounts of related species are separated from each other.
- Purification, in which small amounts of undesired materials are removed from a nearly pure product of interest.

Concentration tends to dominate from a process cost standpoint, for reasons already suggested above: separations costs are primarily related to materials handling, and the mass of the process stream is greatest before concentration has been achieved. Purification is critical to product acceptability and safety. Fractionation is frequently an essential step in processing, but is normally less expensive than concentration and less critical than purification.

Even these very general comments are surprisingly useful to the designer:

- Innovation is particularly important for developing economic concentration processes, and these need not be so highly developed as those for purification: there will be opportunities to remove small amounts of impurities at a later stage. Concentration processes therefore provide useful opportunities for equipment development.

- Purification is too critical from a safety standpoint to take unnecessary chances, and processing costs here tend to be relatively low, because of the small masses of material that need to be handled. It is therefore wise to concentrate attention on the most highly refined and reliable equipment available.

- Fractionation is in an intermediate position, and it is most important that generally suitable equipment be available here. We shall see that this is not always the case at present.

We now review briefly key aspects of process and equipment design in terms of this organization.

Process Aspects

It may readily be shown that the key parameters characterizing recovery from dilute biological feeds are concentration and thermodynamic activity of the product in the process streams, and overall yield. Since yield is simply to be maximized it is useful to plot process trajectories on a phase plot of activity vs. concentration as in Figure 2. The dominance of material handling costs in the processing of dilute solutions suggests that rapid volume reduction is desirable and hence that the trajectory FEP in the figure is preferable to FIP. Here F and P refer to feed and product states, respectively while I and E are intermediate states.

As a specific example we consider recovery of a protein and compare FIP (salting out with Na_2S04 followed by precipitation or extraction) and FEP (adsorption on an ion exchanger followed by elution into a salt solution). The first of these processes requires little specific knowledge, and it is familiar. However, it wastes large amounts of salting agent and creates a severe disposal problem.

A second example is the extraction of a dilute solution in fermentation broth with supercritical carbon dioxide, as indicated in Figure 3. Here use of an intermediate adsorption step greatly decreases both the carryover of water to the next stage and the cost of saturating large volumes of water with CO_2. In addition it avoids contamination of the fermentation broth, and is a reminder that fermentation and extraction processes should be developed jointly since there is significant interaction between them.

Equipment Selection

We now consider some commonly used process equipment from the standpoint of suitability for concentration, fractionation and purification.

Filtration and Centrifugation. Especially for biologicals, filtration and centrifugation are common early steps in the processing of dilute solutions. Typical applications include the removal of unwanted particulates, the recovery of desirable particulates and the removal of water and other solvents. Precoat filtration of

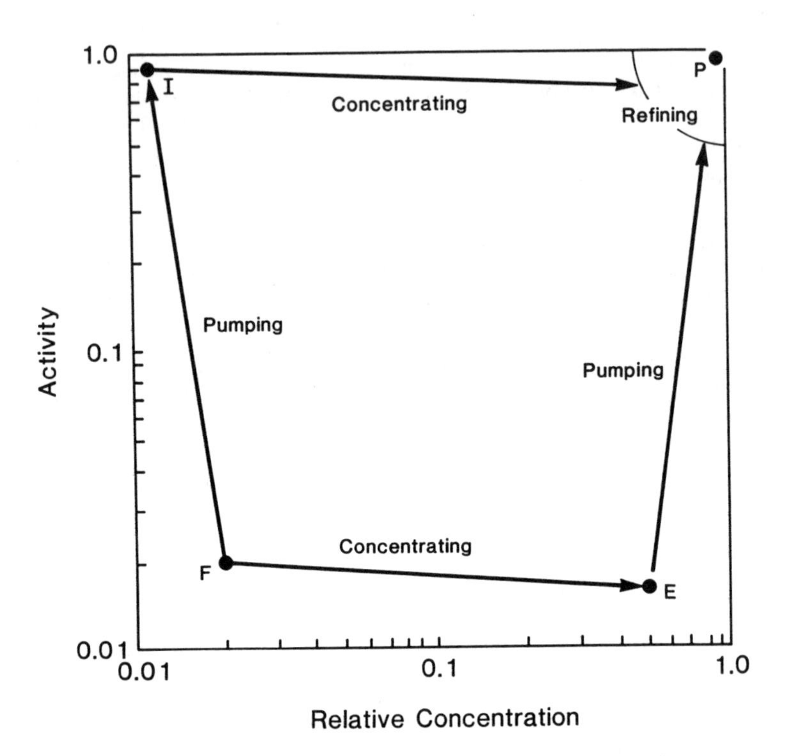

Fig. 2. Recovery Trajectory in the Activity-concentration Phase Plane.

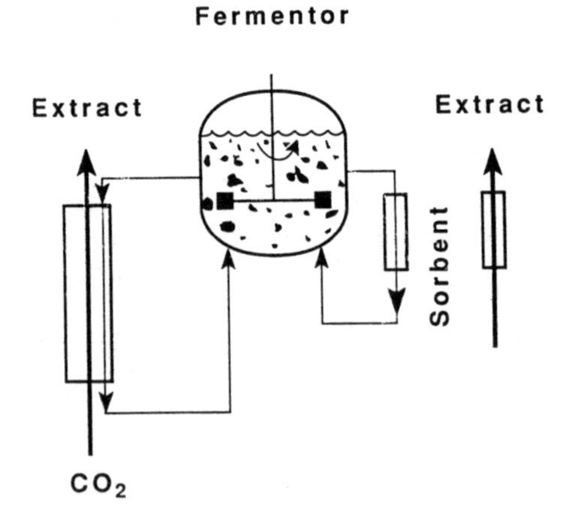

Fig. 3. Comparison of Two Recovery Processes.

mycelia from antibiotic broths is a particularly impressive historical example. Centrifugal separators are very widely used, and membranes for micro-filtration, ultrafiltration and reverse osmosis are currently getting a great deal of attention.

For all of these processes, cost is related only to the volume of broth or other suspending liquid and to the concentration and mechanical characteristics of mycelium or other solids present. It is not appreciably affected by the amount of potential product present, whether in the solution or suspended solids, if these are a minor constituent. Moreover, for individual classes of products at least, for example those produced from genetically altered *E. coli*, filtration characteristics are insensitive to the specific fermentation, at least to the degree of correlation shown in the above diagrams.

Filtration and centrifugation costs should then be proportional to the ratio of inerts to potential product, AND independent of the degree of purification later desired. Filtration costs do then behave consistently with the Sherwood plot and the explanation given for its cost/concentration relation given above. It may also be noted that these processes are *transport limiited* in the full sense of this term. Their cost per unit mass of product thus increases rapidly with product dilution, and there is an increasing tendency to avoid this processing step. We return to this point in connection with adsorption processes.

Liquid Extraction. Liquid extraction is another favorite early processing step, and, at first sight, it seems much more responsive to the amount of valuable material present: it is in principle possible to use highly selective solvents and thus to obtain rapid volume reduction proportionate to concentration of desired solutes. In practice, however, this is not normally true.

Available liquid extractors perform well only if the ratio of extract and raffinate streams is kept within relatively narrow limits. As a result examination of process data shows that volume reductions on extraction from dilute solution are typically no greater than five to ten fold. This is primarily because interfacial area decreases as the ratio of larger to smaller stream rate increases, and in practice liquid extraction performance tends to be dominated by fluid mechanical considerations.

Liquid extractors are thus only suitable as concentration devices in rare situations, and their natural role is for fractionation. Unfortunately there are no readily available high-resolution extractors operable on the relatively small process scale typical of biological conditions. Among the exceptions are the Ito centrifugal devices (*4*) and the Brenner rotating barrel extractor (*5*).

Liquid extractors may also prove useful in purification operations, as suggested in the final section of this paper, where more conventional techniques such as crystallization are not feasible.

Evaporation and Distillation. Evaporation is clearly the thermal analog of filtration, with the liquid-vapor interface taking the place of the filter septum and heat replacing mechanical energy. Hence similar comments apply, irrespective of refinements such as vapor recompression or multiple effect operation.

The picture for fractional distillation is a bit more complex, but once again we find that, for sufficiently low feed concentration, costs should scale with mass of inerts rather than that of desired solutes. To demonstrate this we refer to Figure 4, which is a McCabe-Thiele diagram for rectification of a binary mixture in which the desired constituent is the lower boiling but present in very low concentration. More particularly it is assumed here that at the intersection of the "q-line" and the equilibrium line Henry's law applies in the form

$$y = a\ x \qquad (1)$$

where y and x are vapor and liquid-phase mole fractions, and a is the relative volatility at infinite dilution.

For simplicity we make the reasonable assumption that column operation is effectively determined by the minimum reflux ratio, noting that it is common to operate at 1.05 to 1.5 times minimum reflux. Now it is easily shown, by the usual material balance relationships, that, if the overhead is largely the desired solute

$$(V/P)_{min} = [1/(a-1)](1/x_F) \qquad (2)$$

where V and P are respectively the upflow vapor and product removal rates, and x_F is the feed mole fraction of the desired solute. It can, moreover, be shown that this conclusion is independent of the feed condition, i.e. the slope of the "q-line".

Moreover, since (l/x_F) is very nearly the ratio of unwanted solvent to desired solute, and since P is about equal to the amount of desired solute in the feed to the column, we find

$$V \sim S \qquad (3)$$

where S is the mass flow rate of unwanted solvent in the feed.

We find then that vapor upflow rate, hence both capital and operating costs of the distillation, depend primarily upon the relative volatility and mass flow of solvent and, to a first approximation are independent of the mass of solute. This is the situation required for our definition of recovery from dilute solution to be valid. Distillation then also tends to be unattractive as a concentration step, and it also is more suitable for fractionation. It again becomes unattractive for purification.

It is perhaps now time to put our simplistic description of concentration economics to a test, and to do this we present actual cost data of Busche (*8*) for recovery of acetic acid from water by distillation on a Sherwood plot, in Figure 5, along with a line of unit slope for reference. It may be seen that the predicted behavior is being approached for very dilute solutions, but that these are below the present economic recovery range of about 5 to 10%. Dilute by our standards in this system means below about 1%, and this is too low for economic recovery by any known means. It would be interesting to examine the Busche calculations to obtain insight into the nature of the cost-concentration relation shown here, but this is outside the scope of our present discussion.

Precipitation. We now consider direct precipitation of the desired material from solution, a technique which includes both the sulfate precipitation favored by biochemists and crystallization.

We see immediately that sulfate precipitation and any crystallization technique requiring a change in bulk solution thermodynamics give a cost picture which tends to be dominated by the amount of inerts present. However, filtration cost is related to the amount of product obtained, and our cost/concentration relations begin to become more favorable.

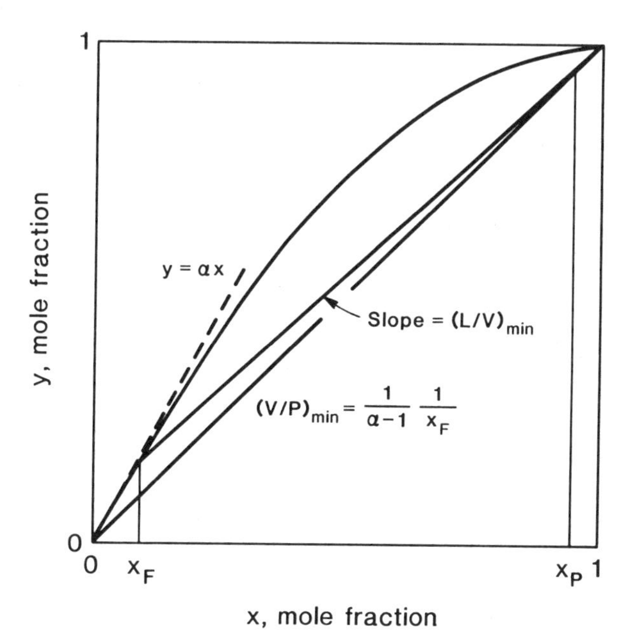

Fig. 4. Distillation of Dilute Feeds.

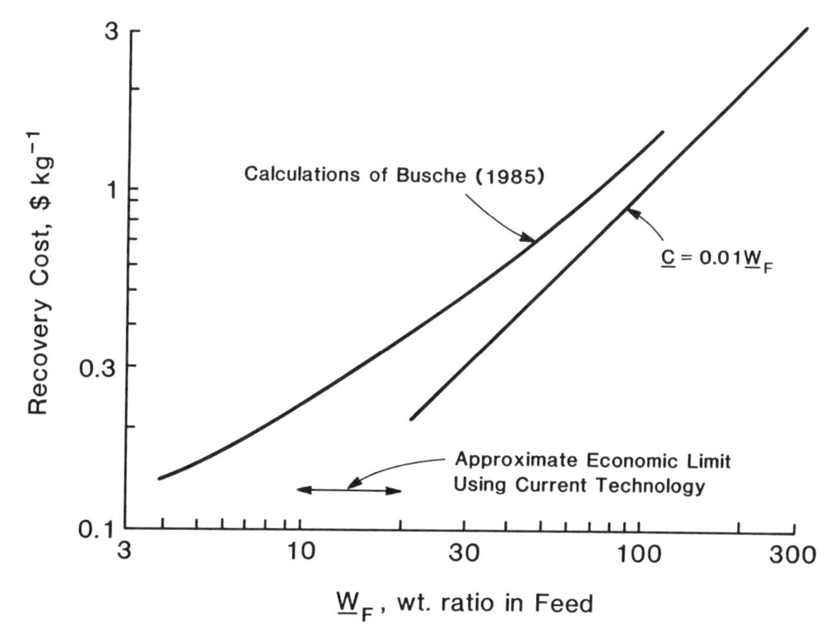

Fig. 5. Costs of Acetic Acid Recovery by Distillation.

Moreover, if we can precipitate the desired solute by reversible reaction of a stoichiometric type, the cost picture becomes much more favorable. Examples include streptomycin and riboflavin recovery (*3*). In fact, as has long been known, this type of precipitation is the favorite among experienced process designers for recovery from dilute solution. We thus find here our first exception to Sherwood-plot economics.

Sorption Processes. Another favorite of designers working with dilute solutions is sorption of solute in fixed beds. Here the restriction on stream-rate ratios does not exist, and it is possible to find highly selective sorbents. Adsorption processes are also very attractive for purification processes, where their typically high selectivity is important, but where their typically low capacity is not a serious drawback. They are generally not considered useful for fractionation, in part because of their low capacity, but also because the chromatographic modes of contacting most widely used are not effective for these purposes. We return to this point later.

Equipment Design

The importance of materials handling costs also provides useful insight into the design of individual separation devices. More specifically, one may view the three primary design criteria to be the maximizing of mass transfer rates while minimizing momentum transfer and providing effective macroscopic distribution of process streams. We concentrate here on mass and momentum transfer, and consider the distribution problem in the final section of the paper.

Since mass and momentum transfer both increase with interfacial area and decrease with the distance over which these quantities must be transported it is necessary to seek differences between these transport processes, and there appear to be only three:

1. The lack of a mass transfer analog to form drag.
2. The greater Schmidt numbers typical of mass transport in liquids.
3. The fact that the momentum transfer load can be controlled independently of the mass to be transferred.

None of these familiar concepts has been fully utilized in the design of mass transfer apparatus to date, and we now consider each of these in turn.

Minimizing Form Drag. We begin by reviewing mass transfer correlations for flow of a single fluid over solid bodies at the high Reynolds numbers characteristic of large-scale processing equipment. Data of this type are conveniently correlated in terms of the Chilton-Colburn j-factor for mass transfer, and the Fanning friction factor f for momentum transfer. Results for a number of common shapes are summarized in Figure 6, and it may be seen from this figure that mass transfer coefficients are shape insensitive: results for a wide variety of shapes can be summarized by the approximation

$$j_M \sim Re^{1/4} \qquad 100 < Re < 100{,}000 \quad (4)$$

It follows that particle shape is not a major design factor for mass transfer. Momentum transfer, however, shows much different behavior, and in general friction factors are orders of magnitude larger than j-factors: most solid bodies are far better momentum transfer than mass transfer devices. There is, however, a major exception in the data shown, for flat plates at zero incidence. Here the Chilton-Colburn analogy is closely approximated, and

$$j_M \sim f/2 \qquad \text{(flat plate)} \qquad (5)$$

The reason for this superior performance is the absence of form drag, which tends to dominate momentum transfer at the high Reynolds numbers characteristic of large process equipment.

It follows from these comparisons that form drag should be minimized in the design of fluid-solid contactors, and similar remarks apply to the more complex situation of two-phase flow over packings. This principle appears to be utilized in the Sulzer packing of Figure 7, where the fluid phases need undergo little change of trajectory once they enter a given packing section. Such packings have the high ratio of mass to momentum transfer predicted from the above simple arguments: high mass transfer efficiency and at the same time high fluid handling capacity. These and other high-performance packings based on the same principles are becoming increasingly popular, and their use may be critical to the development of commodity chemical production from biomass, as capital costs here are high.

Similar remarks apply to the low Reynolds numbers of adsorption columns, and it is easily shown that the classic packed bed of spheroids is also a much better momentum than mass transfer device (*3*). Development of adsorbers with reduced form drag is active at the time of writing, and progress in this area is to be expected.

Increasing Particle Spacing. A second way to reduce momentum relative to mass transfer is to increase the distance between the solid surfaces of a fluid-solid contactor. It is readily shown (see for example Sect. 6.4 of *6*) that friction factors decrease markedly with increasing void fraction whereas mass transfer coefficients are very insensitive (*7*). This has long been recognized in the use of fluidized-bed contactors, but these devices exhibit too much internal mixing to provide effective counterflow contacting. Magnetic stabilization of mobile granular beds (*2*) offers the possibility of markedly decreased back mixing, and these devices are being actively investigated at the time of writing. Like fluid beds they offer the advantage of excellent tolerance of suspended solids.

Decreasing the Momentum Transfer Load. Finally, it should be recognized that the amount of momentum that need be transferred, is proportional to the velocity of the process stream. It can therefore be reduced, at constant bed volume, by using shorter beds of larger cross-section. The ultimate step in this shortening process, for an adsorptive bed, is use of a membrane with adsorptive sites. A variety of such devices is currently on the market, some in the form of hollow-fiber bundles, and these have excellent mass transfer characteristics. They do not give the high resolution of a chromatographic column, but they appear more than adequate for solute concentration, and they can be used without prior removal of solids: the pores in these devices are small compared to the diameters of cells, cell debris and other common solids which are so effective in plugging granular beds.

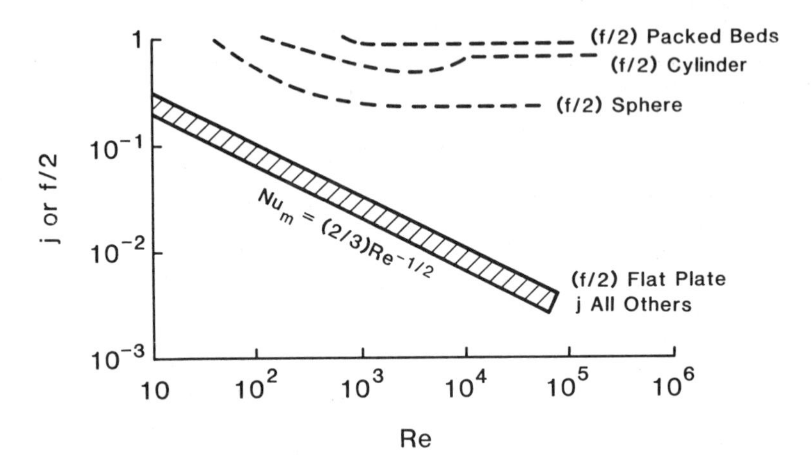

Fig. 6. Effect of Geometry on the Chilton-Colburn Analogy.

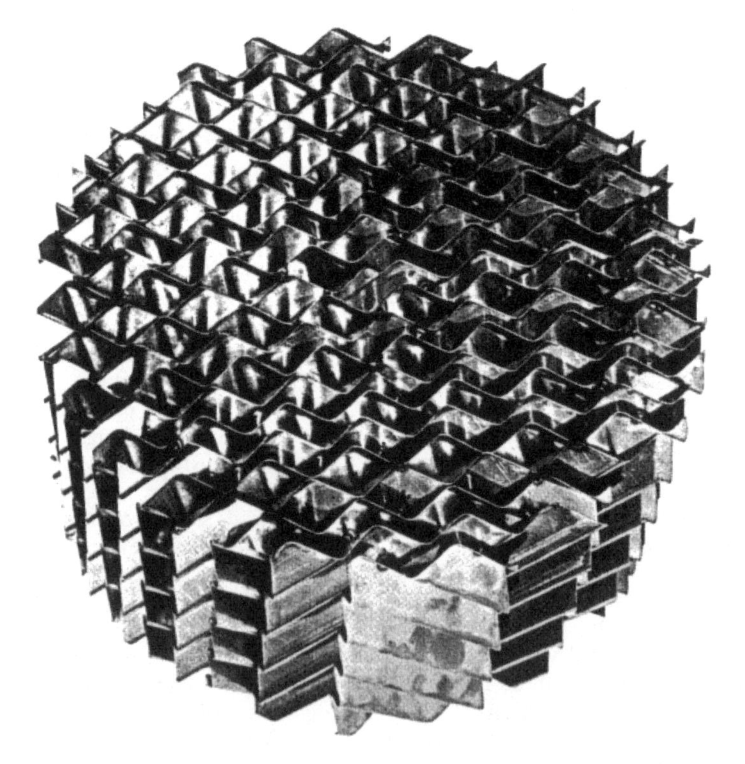

Figure 7. A high-efficiency packing. (Courtesy Koch Industries, Inc.)

DESIGN OF ADSORPTIVE SEPARATIONS

Background

Adsorptive separations presently dominate the "new biotechnology", usually in the form of differential chromatography, in that they are responsible for most separations steps requiring high selectivity. They are also the method of choice for concentration of almost any solute from sufficiently dilute solution, and for many purification steps. They have tended to replace crystallization in the purification of synthetic drugs, because those being produced today often will not crystallize. They show promise for the increasingly important purification of commodity chemicals.

Adsorptive separations should therefore receive particular attention from biotechnologists, and it must be recognized that the bulk of existing data and process experience have been obtained with chromatography. Accordingly it is important to understand chromatography in detail, and an appendix devoted to this subject will be provided to readers of this paper on request to the author. It is, however, too long for general distribution.

Discussion here will be limited to a brief review of contacting modes, important because the author believes that chromatography as such will be effectively eliminated from biotechnology within a decade or so.

We may note in this respect that historically chromatography has organized into three categories (*11*):

Differential Chromatography. Differential chromatography is defined as the elution of a small load of solute from the inlet region of an adsorbent bed. Under the low solute loadings typical of this process each of the solutes spreads into a roughly Gaussian concentration distribution describable in terms of a scaled variance (*9*). At the column outlet this variance is usually written in terms of the "number of theoretical equivalent plates" N, according to

$$\sigma^2 \sim 1/N \qquad (6)$$

where

$N = L/H$

L = column length, and
H = height of a theoretical plate.

At the end of such a process the solute bands are effectively separated as suggested in Figure 8: shown in this figure are the concentration profiles of a desired product (denoted by +) from faster (o) and slower (□) migrating impurities. At the end of such a separation the faster migrating impurity has left the column while the product and slower impurity are still within it. The column length required for such an operation is indicated by the upper horizontal arrow, labeled "elution chromatography".

This process is wasteful of both column and solvent. It may be seen from Fig. 8 that the bulk of the column is free of solute at all stages of the separation, and the fraction of solute-free adsorbent becomes larger the more difficult the separation. In addition such a process is operable only at very low solute loadings. It has been found by numerous investigators that effective plate height

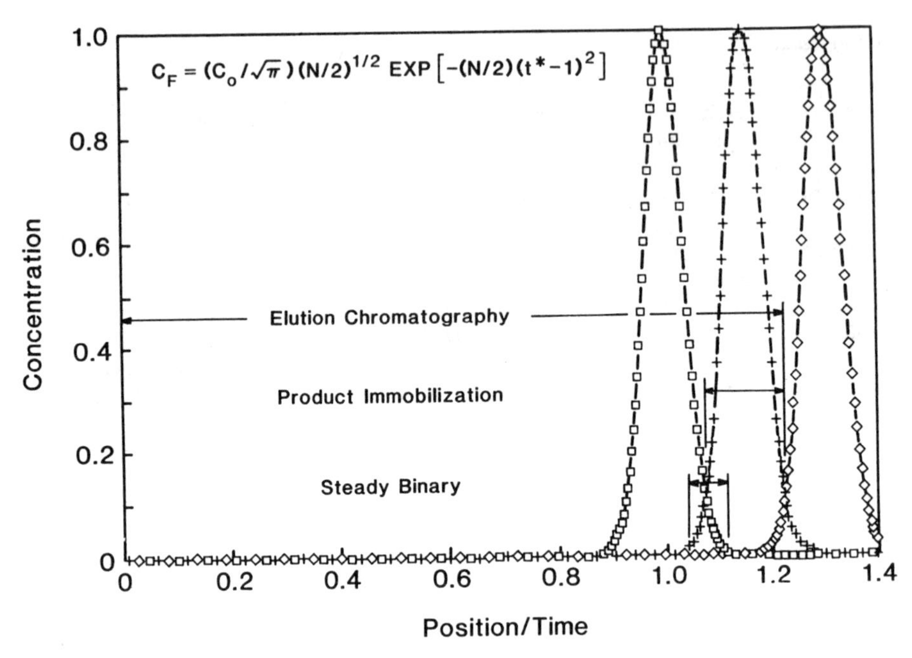

Fig. 8. Comparison of Contacting Modes.

begins to increase rapidly with increasing solute concentration. In the experience of Yamamoto et al. (*14*), for example, the effects of concentration become appreciable at about 1% protein in the feed, and the separtion process becomes inoperable a bit above 5% protein. We are finding that the effects of concentration may vary appreciably from run to run, even in the same column, and that some as yet unidentified hydrodynamic unstability appears responsible for them.

Displacement Chromatography. Displacement chromatography is a modification in which a larger mixed-solute load is fed to the column, and in which these solutes are displaced by a more strongly adsorbed material. As elution proceeds the individual solutes are arranged serially in the order of decreasing adsorptive strength, and they appear one by one in the eluate, separated by relatively short mixed zones, each containing only the adjacent solute pairs in significant concentration. Displacement chromatography is more efficient in use of adsorbent and solvent, but it does not produce the complete separation of differential chromatography.

Frontal Analysis or Batch Saturation. In this simple process mixed solute feeds are pumped through the bed until the desired product begins to appear in the effluent. Frontal analysis cannot achieve a complete separation of any two species, but it is relatively economical in use of adsorbent and solvent. It is particularly effective for concentration, and it can be useful for fractionation and purification when selectivity for the desired species is sufficiently high.

In the limit of extremely high selectivity frontal analysis may be classified as affinity chromatography, but this is not a true chromatographic process either.

Discussion

In spite of its wastefulness differential chromatography has tended to dominate adsorptive processes in biotechnology, and there are probably at least three important reasons for this: the short time span available for process development imposed by regulatory conditions, insufficient quantitative understanding of the chromatographic process, and lack of suitable alternative equipment. At the moment the safest and quickest procedure for the engineer is to simply increase the size of laboratory columns using constant percolation velocity or some equivalent simple criterion for scale up. However, this situation should change as economic constraints become more severe, and our knowledge increases. Many laboratories including our own are making rapid progress in direct measurement of column separation parameters, and many alternate processing systems are coming on the market.

It is the author's belief that only frontal analysis will survive among the existing three processes in the years ahead, for its simplicity as a concentration and preliminary fractionation technique, and in the form of affinity chromatography for final purification. Use of adsorbent membranes such as those already being marketed by Sepracor (*12*) for example should increase. Even frontal analysis should, however, ultimately yield to simpler processes, discussed below, for many applications.

Displacement chromatography may play a strong role in the near term, but it should also give way to more efficient operating modes. The reasons for this may be seen in part from examination of Figure 8, where differential chromatography is contrasted with more efficient processes.

The first of these is the product immobilization mode of Brenner (*5*) in which the solid phase, or its liquid equivalent, is moved against flow of the eluting solvent at a rate keeping the center of mass of the desired product stationary. The product band then steadily spreads about the feed point while the faster solute leaves the system to the right and the slower to the left. The length of column needed for such a process is indicated by the short arrow labeled "product immobilization", and we now find that the entire column is being used at the end of the separation. Moreover, it may be shown that the relative advantage of this mode increases with the difficulty of the separation. It seems likely that this process will see application as soon as suitable means of moving the solid sorbent, or replacing it with a liquid, are available.

The second alternative is a pair of binary separations of the type familiar from counterflow operations such as liquid extraction. It may be seen by simple calculation (publication of the authors in progress) that such a separation, suggested schematically by the lowest arrow in the figure, is far more effective than even product immobilization, and practical experience (*10*) suggests that simple counterflow binary separation is more tolerant of high solute concentrations as well as easier to control than any chromatographic process.

Radical changes of operating mode require either movement of the adsorbent phase, e.g. via use of magnetically stabilized moving beds (*13*), simulation of such movement as via the Sorbex system (*10*), shown in Figure 9, or replacement of the solid by a liquid. This latter in turn can contain a liquid ion exchange

IMPLEMENTING COUNTERFLOW

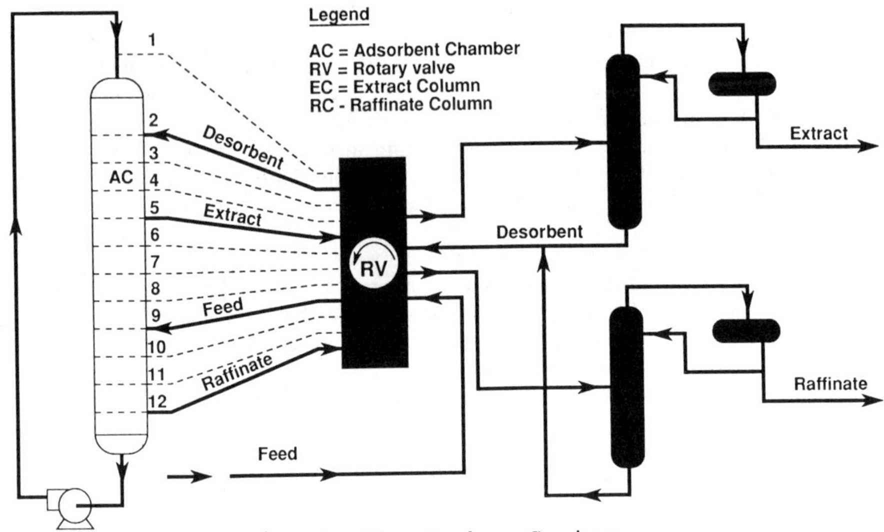

Fig. 9. The Sorbex System.

or other adsorbent analog to provide desired selectivity. Among the more interesting of the latter are ferrofluids in which the dispersed solid is adsorbent as well as ferromagnetic.

The substantial developmental effort for implementation of these alternative processors will require time, but the author is confident of ultimate success. This is in part because the reduction in the number of stages ("plates") required by these alternate schemes is so large as to make up for the relatively primitive state of development.

Bioseparations should prove a very vigorous and fast moving field for some time, and the advances made in this area should improve our general understanding of separations as well.

Literature Cited

1. Figure taken from Nystrom, J. M., "Product Purification and Downstream Processing", 5th Biennial Executive Forum, A. D. Little, Boston, MA. 3-6 June, 1984; Reproduced in J. L. Dwyer, *Biotechnology*. 1, 957 (Nov., 1984).
2. King, C. J., et al., "Separation and Purification: Critical Needs and Opportunities", National Academy Press (1987).
3. Lightfoot, E. N., and M. C. M. Cockrem, "What Are Dilute Solutions?", Sep. Sci. Tech., 22 (2&3), 165-189 (1987).
4. Ito Multilayer Coil Separator-Extractor, P. C. Inc., 11805 Kim Place, Potomac, Md. 20854.
5. Brenner, Max, Institut für Organische Chemie, der Universität Basel, St. Johannsring 19, CH-4056 Basel, Switzerland.
6. Bird, R. B., et al., Transport Phenomena, Wiley (1960).

7. Pfeffer, Robert, "Heat and Mass Transfer in Multiparticle Systems", IEC Fund, 3(4), 380-383 (1964).
8. Busche, R. M., *Biotechnology Prog.* 1 (#3), 165-180 (Sept., 1985).
9. Gibbs, S. J., and E. N. Lightfoot, "Scaling up Gradient Elution Chromatography", IEC Fund., 25, 490-498 (1986).
10. Broughton, D. B., Sep. Sci. Tech., 19, 723 (1985); N. N. Li, Allied Signal Corp., 50 E. Algonquin Rd., Des Plaines, IL 60016-5016.
11. Giddings, J. C., et al., Eds. Advances in Chromatography, Marcel Dekker, 1965 and subsequent.
12. S. L. Matson, Sepracor, Inc., 33 Locke Drive, Marlborough, MA 01752.
13. Rosensweig, R. E., Exxon Corporate Research Science Laboratories, Linden, N. J.; Burns, M. A., Dept. of Chem. Eng., Univ. of Mass., Amherst, MA 01003.
14. Yamamoto, Shuichi, et al., "Scaling up of Medium-performance Gel Filtration Chromatography of Proteins", Journal of Chemical Engineering of Japan, #19(3), 227-231 (1986).

RECEIVED March 12, 1990

Chapter 4

Peptide and Protein Partitioning in Aqueous Two-Phase Systems

Effect of Amino Acid Sequence

Alan D. Diamond, Kun Yu, and James T. Hsu

BioProcessing Institute, Department of Chemical Engineering, Lehigh University, Bethlehem, PA 18015

The effect of amino acid sequence on peptide and protein partitioning in PEG 3400/potassium phosphate/water systems at 20°C is reported. By reversing the amino acid sequence of twenty-three dipeptides, the partition curve, obtained by plotting the natural logarithm of the partition coefficient versus the PEG concentration difference between the phases, is altered depending on the side chain type and location. Utilizing this dipeptide data along with the partition coefficients of twenty amino acids, a method is presented for determining the contribution of an amino acid residue to the partition coefficient of a biomolecule. The method is used to predict the partition coefficient of homogeneous glycine peptides and several tripeptides, and is also used to predict the ratio of partition coefficients of homologous proteins.

Aqueous polymer two phase extraction, such as that utilizing a system composed of polyethylene glycol (PEG), potassium phosphate and water is fast becoming an accepted technique for the purification of biomolecules (1,2). The large scale purification of enzymes has been demonstrated (3,4) and the economic outlook is favorable (5). However, in order to enhance and facilitate their use for protein purification, a means is needed for correlation of biomolecule partitioning.

The partition coefficient of a biomolecule may be influenced by many factors occurring in its surrounding environment and internal structure. The environmental effect, such as the influence of polymer molecular weight (6,7), polymer concentration (8), pH (9), and salts (10,11) has been well studied. The effect of amino acid sequence on peptide partitioning was investigated by Sasakawa and Walter (12) and then by Diamond et. al. (13) where it was shown that when the amino acid sequence of a dipeptide was reversed, the partition coefficient in the systems composed of PEG 3400/potassium phosphate/water was altered depending on the residue type and location. In this paper, the dipeptide partition studies in the PEG/potassium phosphate/water systems are extended and the internal sequence effect quantified by partition of the component amino acids. The amino acid and dipeptide data are then combined to provide a correlation which predicts partitioning based on the contribution to the partition coefficient of the amino acid

0097–6156/90/0427–0052$06.00/0

residues rather than the pure amino acids. The results are applied to the prediction of peptide and protein partition coefficients where the effect of changing the amino acid sequence is observed.

Materials and Methods

Chemicals. PEG of molecular weight 3,400 (Lot 00304EV) was obtained from Aldrich Chemical Company, Milwaukee, WI. Potassium phosphate, both mono- and dibasic, were of A.C.S. reagent grade and also obtained from Aldrich.

The peptides gly-gly, triglycine, tetraglycine, pentaglycine, gly-ala-tyr, gly-tyr-ala, gly-ala, ala-gly, gly-val, val-gly, gly-leu, leu-gly, gly-phe, phe-gly, gly-trp, trp-gly, gly-met, met-gly, gly-pro, pro-gly, gly-ser, ser-gly, ser-leu, leu-ser, gly-tyr, tyr-gly, gly-asp, asp-gly, gly-lys, lys-gly, gly-his, his-gly, ala-trp, trp-ala, ala-val, val-ala, ala-leu, ala-tyr, tyr-ala, ala-asp, asp-ala, val-asp, asp-val, leu-arg, arg-leu, val-lys, lys-val, asp-lys, and lys-asp were purchased from Sigma Chemical Company, St. Louis, MO. The dipeptides gly-ile, ile-gly, and leu-ala were purchased from U. S. Biochemical Corp., Cleveland, OH.

Glycine, along with the following l-amino acids were purchased from Sigma: tryptophan, phenylalanine, tyrosine, isoleucine, leucine, cysteine, methionine, valine, proline, glutamic acid, glutamine, alanine, threonine, aspartic acid, asparagine, serine, arginine, lysine and histidine.

Insulin from equine and porcine pancreas, along with β-lactoglobulins A and B were also purchased from Sigma. It should be pointed out that individual A and B samples were purchased, not a mixture of the two.

Experimental. Five hundred to one thousand milliliter samples of systems b–e of the PEG 3400/potassium phosphate/water phase digram at 20°C were prepared according to the procedure of Albertsson (1). The phase compositions of each of these systems were as follows (1):

b. Bottom Phase: 14.06% phosphate/4.23% PEG, Top Phase: 8.19% phosphate/15.96% PEG

c. Bottom Phase: 15.46% phosphate/2.54% PEG, Top Phase: 7.01% phosphate/19.16% PEG

d. Bottom Phase: 17.41% phosphate/1.30% PEG, Top Phase: 5.56% phosphate/23.90% PEG

e. Bottom Phase: 19.41% phosphate/0.78% PEG, Top Phase: 4.55% phosphate/28.15% PEG

The potassium phosphate utilized in this study consisted of a mono- to dibasic weight ratio of 0.55, thus giving a pH of 7.0 in each of the four systems. The samples were placed in separatory funnels and allowed to settle at 20°C for 24 hours in the laboratory environment. The two phases were collected and then used as stock solutions for the partitioning studies.

5 mL of top and bottom phase were placed together into a 15 mL polypropylene centrifuge tube. 3 mg of either peptide or amino acid, or 10 mg of β-lactoglobulin or cytochrome c was then added, and the tubes tightly sealed. The

biomolecule was then dissolved by gently mixing the contents of the centrifuge tube with a vortex mixer. The systems containing the biomolecule were permitted to settle at 20°C for 24 hours. The phases were then separated. Peptide and β-lactoglobulin concentration was determined by measuring absorbance using a Shimadzu UV-Vis spectrophotometer according to the procedure of Sasakawa and Walter (12,14). Absorbance at 220 nm was utilized for the peptides, while 280 nm was used for the β-lactoglobulins. Amino acid and cytochrome c concentration was determined using the fluorescamine technique (15). The concentrations determined for the top and bottom phases were then used to calculate the partition coefficient, K, which was defined as biomolecule concentration in the top phase divided by the concentration in the bottom phase.

Results and Discussion

Biomolecule Partitioning. Twenty-three dipeptide pairs were partitioned in systems b–e of the PEG 3400/potassium phosphate/water phase diagram at 20°C. The dipeptides that comprise a pair differ from one another by reversal of their amino acid sequence. Six categories of reversed sequence dipeptides were utilized in this study, with each category containing the following sidechain types: 1. uncharged polar and nonpolar sidechains, 2. uncharged polar sidechains, 3. uncharged polar and charged polar sidechains, 4. nonpolar sidechains, 5. nonpolar and charged polar sidechains, and 6. charged polar sidechains.

The dipeptide partition data is presented in Figures 1–7. In each of these figures, the natural logarithm of the partition coefficient has been plotted versus the PEG concentration difference between the top and bottom phases, expressed in %(w/w). This type of plot has been selected since it was demonstrated by Diamond and Hsu (8) to be effective in the correlation of biomolecule partitioning. In Figures 1–3, dipeptide pairs from category 1 are presented. These three plots demonstrate that when glycine, a polar, uncharged amino acid, is at the N-terminal and is attached to an amino acid with a nonpolar side chain, i.e., alanine, valine, leucine, phenylalanine, proline, or methionine, the partition curve is lower than the reverse sequence. Interestingly, it can be seen in Figure 1 that when a CH_2 group is added to the residue with a nonpolar sidechain, the difference between partition curves for the reverse pair increases, with the gly-leu/leu-gly pair showing the greatest difference. Figure 3 reveals that when serine (polar uncharged amino acid) is at the N-terminal and paired with leucine (nonpolar), its partition curve is higher than the reverse case. A similar situation is encountered when tyrosine is combined with alanine. This is opposite of what occurred when glycine was at the N-terminal and paired with a nonpolar residue. Hence, at present, a generalization can not be made as to which dipeptide will have a higher partition curve when polar and nonpolar residues are combined. However, one may conjecture that such factors as side chain size and charge distrubution are of importance.

In Figure 4, dipeptide partitioning from categories 2 (gly-ser/ser-gly, gly-tyr/tyr-gly) and 6 (asp-lys/lys-asp) is presented. These categories each contain dipeptides composed of amino acids with similar sidechains. This figure illustrates that gly-ser, gly-tyr, and lys-asp have lower partition curves than the reverse sequence. In the case of lys-asp/asp-lys, lysine contains a positively charged residue, while aspartic acid is negative. The lower partition curve occurs when the positively charged lysine is at the positive N-terminal, while the negative aspartic acid is at the negative C-terminal.

Figure 5, which contains dipeptides of category 3, reveals that when glycine is paired with a residue containing a charged, polar sidechain, whether positive or

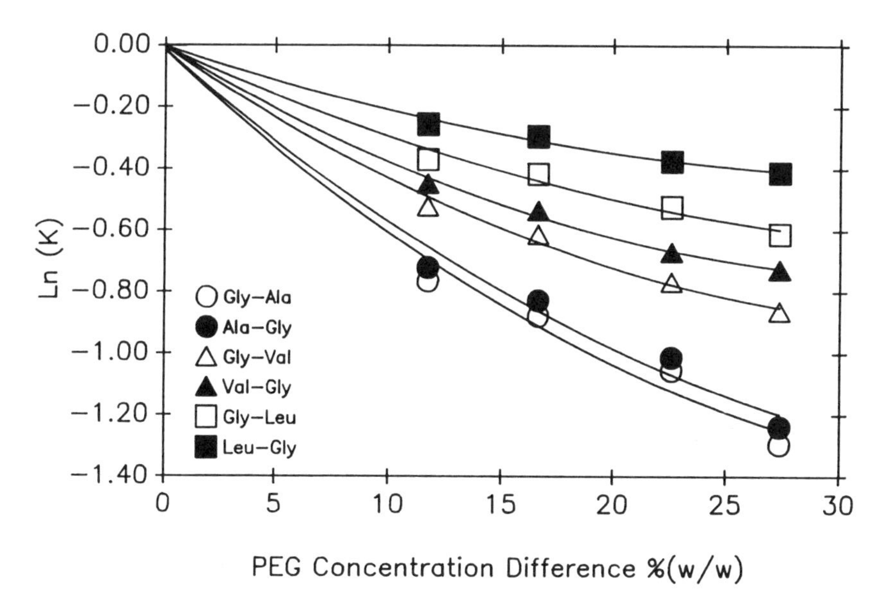

Figure 1. Natural Logarithm of the Partition Coefficient as a Function of the PEG Concentration Difference Between the Phases for Reversed Sequence Dipeptides of Category 1 in the PEG 3400/Potassium Phosphate/Water System at 20°C.

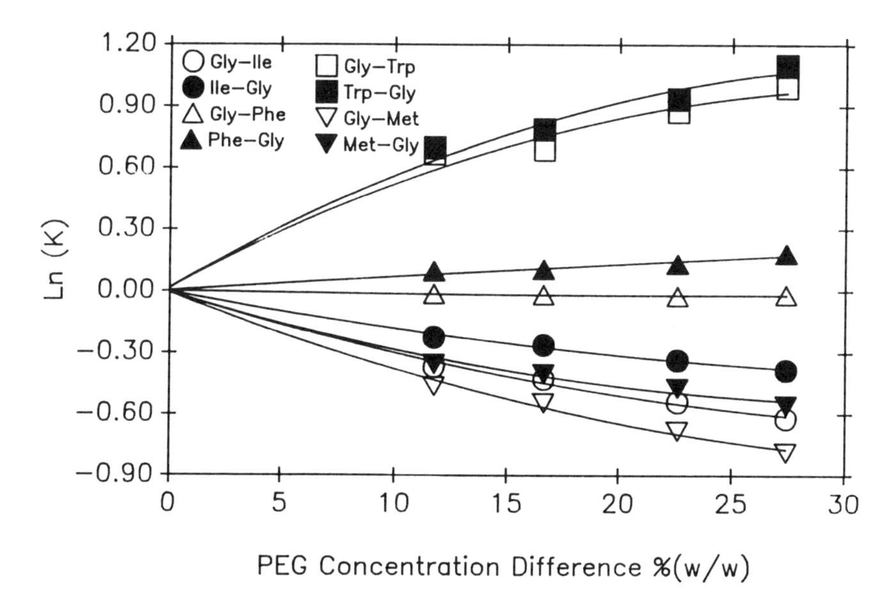

Figure 2. Natural Logarithm of the Partition Coefficient as a Function of the PEG Concentration Difference Between the Phases for Reversed Sequence Dipeptides of Category 1 in the PEG 3400/Potassium Phosphate/Water System at 20°C.

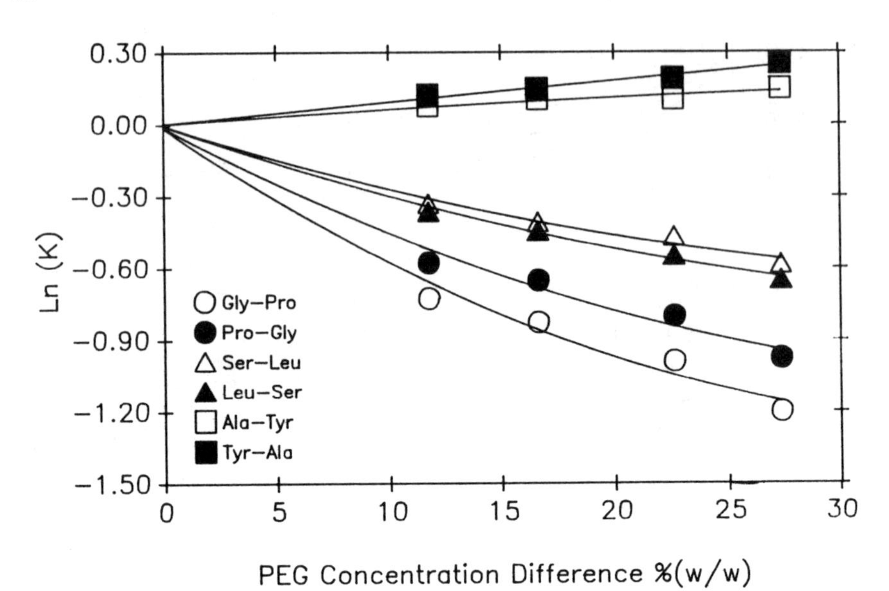

Figure 3. Natural Logarithm of the Partition Coefficient as a Function of the PEG Concentration Difference Between the Phases for Reversed Sequence Dipeptides of Category 1 in the PEG 3400/Potassium Phosphate/Water System at 20°C.

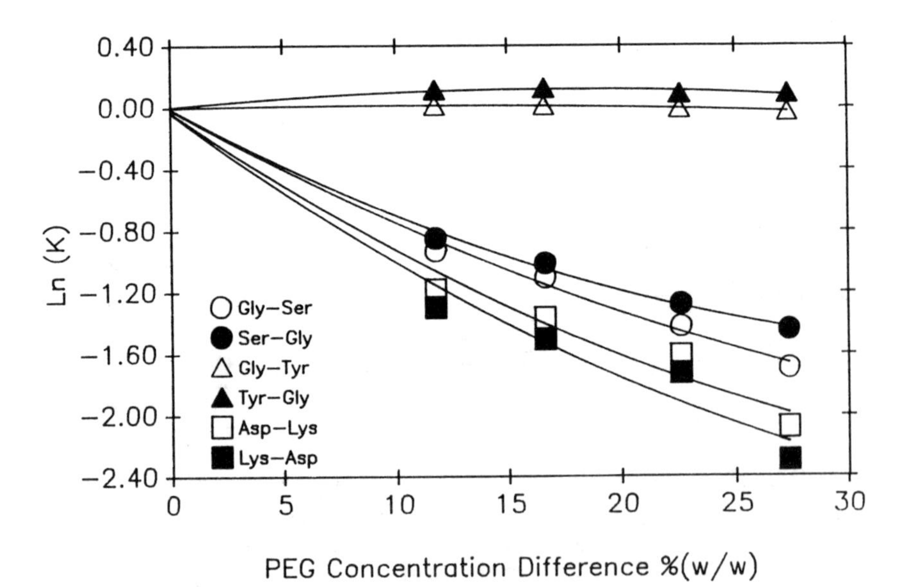

Figure 4. Natural Logarithm of the Partition Coefficient as a Function of the PEG Concentration Difference Between the Phases for Reversed Sequence Dipeptides of Categories 2 and 6 in the PEG 3400/Potassium Phosphate/Water System at 20°C.

negative, the partition curve is lower than the reverse. In Figure 6, when two residues with nonpolar sidechains are combined, i.e., category 4, the partition curve is lower when the smaller of the two residues is at the N-terminal. In Figure 7, when a polar residue such as alanine, valine, or leucine is at the N-terminal and paired with a charged residue, i.e., category 5, the partition curve is lower than the reverse case.

In order to further demonstrate the sensitivity of the PEG 3400/potassium phosphate/water systems, the two proteins, β-lactoglobulin A and B, were partitioned in systems b–e. These two proteins have been selected since there is only a slight difference in their amino acid sequence. Species A has aspartic acid and valine, while species B has glycine and alanine at residues 64 and 118, respectively (16). The proteins, which may be resolved by ion exchange chromatography or electrophoresis, have an isoelectric point difference of only 0.1 pH units (17). The partition results are presented in Figure 8 where the natural logarithm of the partition coefficient is plotted versus the PEG concentration difference. The separation factor, defined as the partition coefficient of species A divided by that of B was found to be 2.0. Although this separation factor is quite high, it should also be noted that the partition coefficients are very low, on the order of 10^{-2}.

The above results demonstrate the effect of amino acid sequence on protein partitioning. By changing only two residues, the partition coefficient was dramatically altered in the PEG/potassium phosphate/water systems. A quantitative explanation for this occurrence will be given later on in the paper.

In order to gain a more fundamental understanding of peptide and protein separation in aqueous two-phase systems, the partition of their building blocks, the amino acids, is undoubtedly required. Twenty amino acids were partitioned in system b of the PEG 3400/potassium phosphate/water phase diagram at 20°C and the results are presented in Table I in the order of decreasing partition coefficient. Although this table by no means represents a hydropathy scale due to the many factors that influence partitioning in the PEG/potassium phosphate/water systems, it is of interest to compare these results with previously established scales, such as the hydropathy index of Kyte and Doolittle (18). This index is presented in Table I. In the case of the PEG/potassium phosphate/water systems, the upper, PEG rich phase may be considered more hydrophobic than the lower, salt rich phase. Kyte and Doolittle's scale has the charged, polar amino acids with the lowest (most negative) hydropathy, followed by the polar and then nonpolar amino acids. Similarly, the charged, polar amino acids have the lowest partition coefficient followed by the polar and then nonpolar amino acids.

Correlation of Biomolecule Partitioning. With knowledge of amino acid partitioning, the next logical step is to utilize the data for correlation and prediction of peptide partitioning. It is of interest to derive the partitioning behavior of peptides from their amino acid sequences. The correlation between individual amino acids and peptides is a revelation of the interaction between amino acids in the sequence and the nature of the peptide bond and may help to decipher the rules of peptide conformation.

The Gibbs free energy of transfer of a molecule between the two phases is determined by its partition coefficient:

$$\Delta G = -RT \text{ Ln } (K) \tag{1}$$

where ΔG is the free energy of transfer, R is the gas law constant T is absolute temperature and K the partition coefficient. Since the experiments were carried out

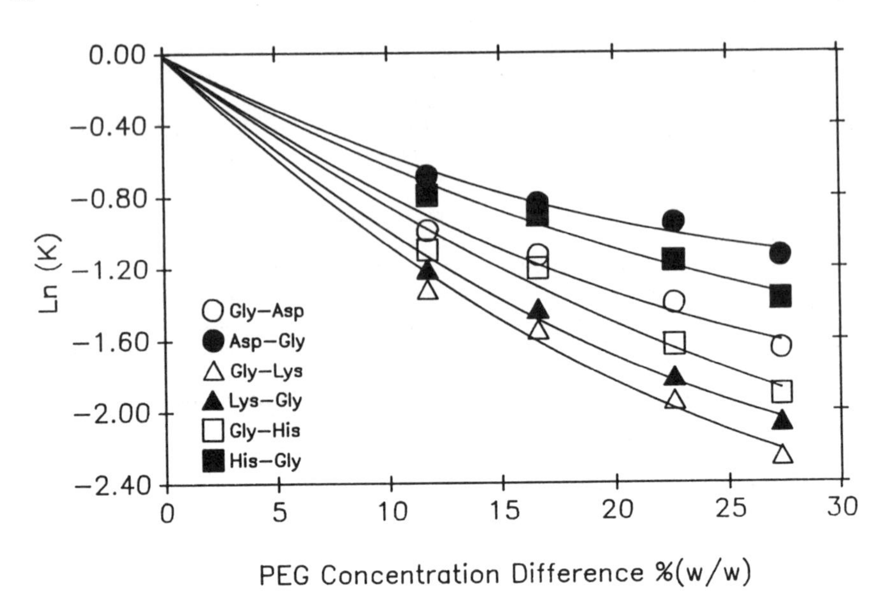

Figure 5. Natural Logarithm of the Partition Coefficient as a Function of the PEG Concentration Difference Between the Phases for Reversed Sequence Dipeptides of Category 3 in the PEG 3400/Potassium Phosphate/Water System at 20°C.

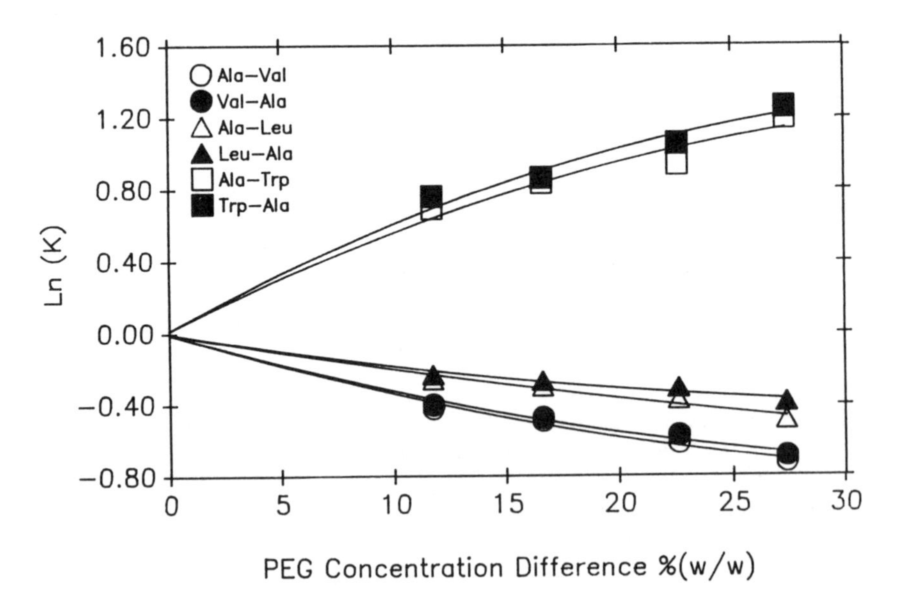

Figure 6. Natural Logarithm of the Partition Coefficient as a Function of the PEG Concentration Difference Between the Phases for Reversed Sequence Dipeptides of Category 4 in the PEG 3400/Potassium Phosphate/Water System at 20°C.

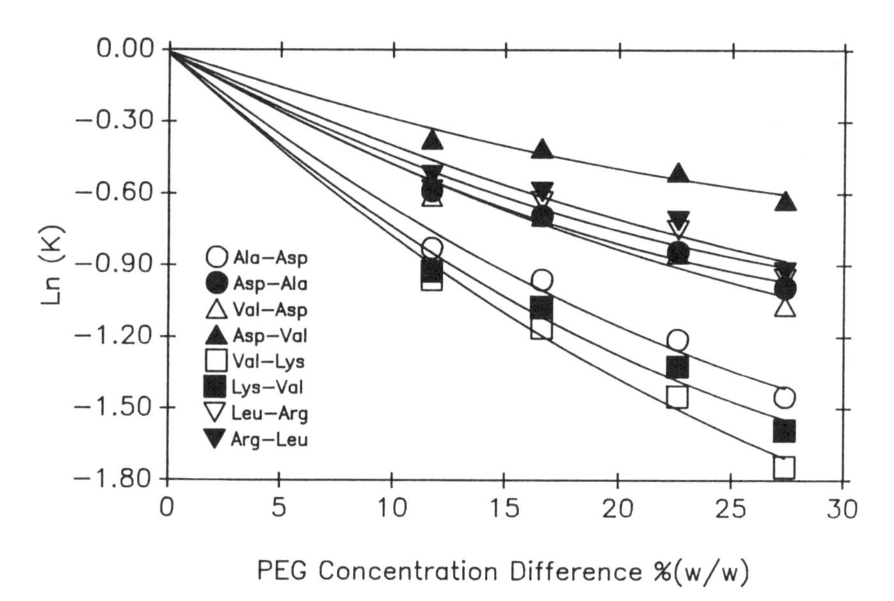

Figure 7. Natural Logarithm of the Partition Coefficient as a Function of the PEG Concentration Difference Between the Phases for Reversed Sequence Dipeptides of Category 5 in the PEG 3400/Potassium Phosphate/Water System at 20°C.

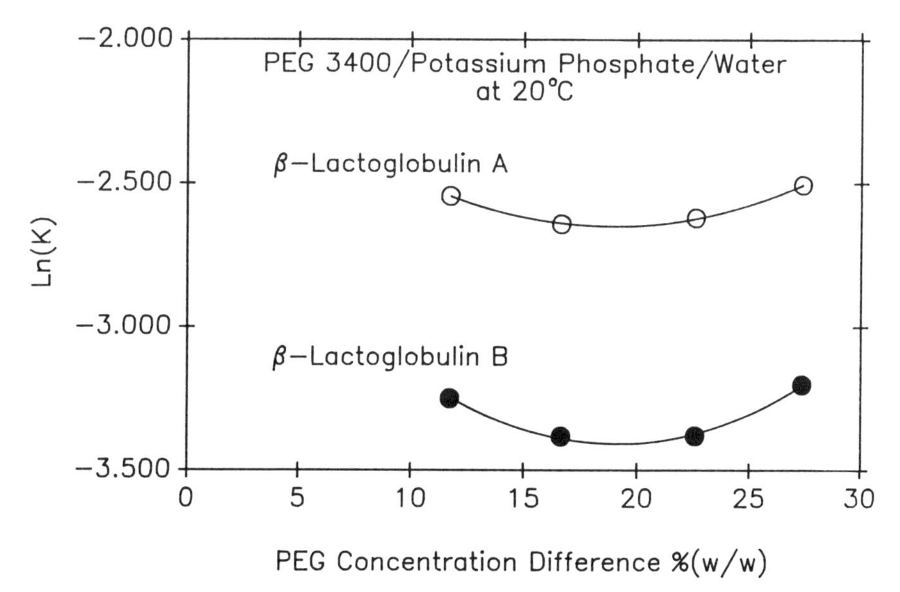

Figure 8. Natural Lorarithm of the Partition Coefficient as a Function of the PEG Concentration Difference for β-Lactglobulins A and B in the PEG 3400/Potassium Phosphate/Water System at 20°C.

Table I. Partitioning Parameters for Amino Acids in System b of the PEG 3400/Potassium Phosphate/Water Phase Diagram at 20°C

Amino Acid	H*	K	E	δ^C	δ^N	ϵ
Tryptophan	−0.9	2.00	0.69	0.27	−1.04	1.46
Phenylalanine	2.8	1.14	0.13	0.31	−0.94	0.77
Tyrosine	−1.3	1.07	0.07	0.31	−1.08	0.84
Isoleucine	4.5	0.86	−0.15	0.35	−0.87	0.37
Leucine	3.8	0.84	−0.17	0.37	−0.92	0.38
Cysteine	2.5	0.78	−0.25	---	---	---
Methionine	1.9	0.71	−0.34	0.28	−0.96	0.34
Valine	4.2	0.70	−0.36	0.38	−0.93	0.19
Proline	−1.6	0.61	−0.49	0.35	−0.86	0.01
Glutamic Acid	−3.5	0.57	−0.56	---	---	---
Glutamine	−3.5	0.56	−0.58	---	---	---
Alanine	1.8	0.54	−0.62	0.40	−0.95	−0.06
Threonine	−0.7	0.53	−0.64	---	---	---
Aspartic Acid	−3.5	0.50	−0.69	0.26	−0.84	−0.11
Glycine	−0.4	0.48	−0.73	0.35	−1.00	−0.08
Asparagine	−3.5	0.50	−0.69	---	---	---
Histidine	−3.2	0.44	−0.82	0.24	−0.81	−0.25
Serine	−0.8	0.43	−0.84	0.25	−1.01	−0.09
Arginine	−4.5	0.37	−0.99	0.27	−0.96	−0.31
Lysine	−3.9	0.32	−1.14	0.33	−0.92	−0.55

* Kyte and Doolittle's hydropathy index (18)

at 20°C, RT remains constant. As a simplification, the quantity E may be defined as

$$E = \frac{-\Delta G}{RT} = \mathrm{Ln}\,(K) \tag{2}$$

Thus, E is proportional to ΔG and retains the sign. It can be regarded as the transfer free energy on a nondimensional scale, or, alternatively, the natural logarithm of the partition coefficient. E values for the twenty amino acids are listed in Table I. In addition to the amino acids, the partition coefficient for the twenty-three dipeptide pairs partitioned in system b will also be utilized in this study. These were obtained from Diamond et. al. (12) as was previously mentioned.

An initial assumption was that the E of a dipeptide equaled that of the N-terminal amino acid plus that of the C-terminal amino acid minus a loss term, denoted by δ:

$$E_{BX} = E_B + E_X - \delta \tag{3}$$

where subscripts B and X refer to two arbitary amino acids, and BX is the dipeptide with B at the N-terminal. A correlation of system b data gives δ between 0.55 and 0.71. however, application of this result to prediction of dipeptide partitioning would produce a maximum error of 15%.

In order to make predictions more accurate, δ is treated as a characteristic of an individual amino acid and is split into two components,

$$\delta_i = \delta_i^{N} + \delta_i^{C} \tag{4}$$

where superscripts N and C refer to the amino and carboxyl terminals, respectively, and subscript i refers to the amino acid in question. For dipeptide BX,

$$E_{BX} = E_B + E_X - \delta_B^{C} - \delta_X^{N} \tag{5}$$

while for dipeptide XB,

$$E_{XB} = E_X + E_B - \delta_X^{C} - \delta_B^{N} \tag{6}$$

where E_B and E_X in Equation 5 are the E values for amino acids B and X, respectively, and $(-\delta_B^{C} - \delta_X^{N})$ in Equation 5 is considered a correction term which takes into account the fact that the first amino acid, in this case B, has lost an oxygen atom from its carboxyl terminal, while the second amino acid, X, has lost two hydrogen atoms from its amino terminal upon formation of the dipeptide bond. Since fifteen different amino acids are present among the twenty-three dipeptide pairs used in the partitioning study, then there are fifteen unknown δ_i^{N} and δ_i^{C} for a total of thirty variables. For each dipeptide, an equation like that for XB or BX may be written. Thus, from the published data, there are fourty-six equations (arrising from the twenty-three dipeptide pairs) with thirty variables. The δ_i^{N} and δ_i^{C} can be solved for by least square fit. The δ_i^{N} and δ_i^{C} thus obtained are recorded in Table I. It should be pointed out that these two parameters were not obtained for all of the amino acids due to the limitations of the dipeptide data.

Now that E_i, δ_i^{N} and δ_i^{C} values have been obtained for individual amino acids, a generalized expression for predicting the E value of a peptide may be written as

$$E = \sum_{j=1}^{n} (E_j - \delta_j^{N} - \delta_j^{C}) + \delta_1^{N} + \delta_n^{C} = \sum_{j=1}^{n} \epsilon_j + \delta_1^{N} + \delta_n^{C} \tag{7}$$

where j is the number of the amino acid in the sequence, counting from the N-terminal, n refers to the last amino acid in the sequence, and $\epsilon_j = E_j - \delta_j^{N} - \delta_j^{C}$. Equation 7 has been written in this form so that it could be extended to the prediction of peptide partition coefficients ($n > 2$), which will be discussed later. The partition coefficient, K, is then evaluated as follows:

$$K = \exp(E) \tag{8}$$

Equation 7 may be applied to the dipeptide gly-ala by letting the subscripts 1 and 2 refer to glycine and alanine, respectively:

$$E_{gly\text{-}ala} = E_1 + E_2 - \delta_1^{C} - \delta_2^{N} \tag{9}$$

The calculated value of $E_{gly\text{-}ala}$ is -0.75. Application of Equation 8 gives a K value of 0.47, which agrees very well with the experimental value of 0.465. The predicted partition coefficients using Equations 7 and 8 for the fourty-six dipeptides are listed in Table II along with the actual K values. The dipeptides yielding the greatest percent error were ala-tyr and tyr-gly with errors of 5.6% and 8.0%, respectively, while for most dipeptides, the error was close to zero.

The parameter $\epsilon_j = E_j - \delta_j^{N} - \delta_j^{C}$ for a particular amino acid is of interest because it represents the Ln (K) of the amino acid residue, i.e., the amino acid

Table II. Comparison of Experimental and Predicted Partition Coefficients for Dipeptides in System b of the PEG 3400/Potassium Phosphate/Water Phase Diagram

Dipeptide	$K_{Exper.}$	$K_{Pred.}$
1. Uncharged Polar and Non Polar Sidechains		
Gly-Ala	0.465	0.47
Ala-Gly	0.486	0.48
Gly-Val	0.593	0.60
Val-Gly	0.638	0.63
Gly-Leu	0.691	0.71
Leu-Gly	0.775	0.76
Gly-Ile	0.691	0.69
Ile-Gly	0.798	0.80
Gly-Phe	0.984	0.98
Phe-Gly	1.10	1.10
Gly-Trp	1.95	1.90
Trp-Gly	2.01	2.01
Gly-Pro	0.484	0.48
Pro-Gly	0.561	0.56
Gly-Met	0.627	0.63
Met-Gly	0.701	0.70
Ser-Leu	0.718	0.70
Leu-Ser	0.695	0.69
Tyr-Ala	1.13	1.10
Ala-Tyr	1.08	1.14
2. Uncharged Polar Sidechains		
Gly-Ser	0.394	0.40
Ser-Gly	0.428	0.44
Gly-Tyr	1.01	1.06
Tyr-Gly	1.12	1.03
3. Uncharged Polar and Charged Polar Sidechains		
Gly-Asp	0.372	0.39
Asp-Gly	0.504	0.51
Gly-Lys	0.276	0.27
Lys-Gly	0.299	0.30
Gly-His	0.334	0.33
His-Gly	0.453	0.45
4. Non Polar Sidechains		
Ala-Val	0.657	0.64
Val-Ala	0.670	0.67
Ala-Leu	0.769	0.77
Leu-Ala	0.792	0.82
Ala-Trp	2.00	2.05
Trp-Ala	2.14	2.15
5. Non Polar and Charged Polar Sidechains		
Ala-Asp	0.436	0.42
Asp-Ala	0.555	0.54
Val-Asp	0.541	0.55
Asp-Val	0.688	0.69
Val-Lys	0.382	0.38
Lys-Val	0.396	0.41
Leu-Arg	0.559	0.56
Arg-Leu	0.593	0.59
6. Charged Polar Sidechains		
Asp-Lys	0.305	0.31
Lys-Asp	0.273	0.27

minus a water molecule. The ϵ values for amino acids are recorded in Table I. It is the amino acid residue, instead of the amino acid itself, which determines the local behavior in a peptide or protein. The ϵ values, as will be demonstrated later in this paper, can be used to compare the partitioning behavior of similar or homologous proteins.

The experimental and predicted partition coefficients of the glycine peptides in system b of the system PEG 3400/potassium phosphate/water phase diagram are listed in Table III. As can be seen, they are fairly consistent and show the same trend of decreasing K with longer chains. In a similar manner, the tripeptides gly-tyr-ala and gly-ala-tyr were partitioned in system b, and the experimental K values were 1.09 and 1.06, respectively. When Equations 7 and 8 are used to predict the K values, the result is 1.09 and 1.00, repectively, which is in good agreement.

Table III. Comparison of Experimental and Predicted Partition Coefficients for Polyglycines in System b of the PEG 3400/Potassium Phosphate/Water Phase Diagram

Peptide	Experimental	Predicted
Diglycine	0.43	0.44
Triglycine	0.41	0.41
Tetraglycine	0.40	0.37
Pentaglycine	0.39	0.34

Examination of Table III reveals that the correlation gives large error as the number of glycine residues is increased, with pentaglycine giving the greatest error. This is most probably due to the fact that the database used for developing Equation 7 contained solely amino acids and dipeptides. In addition, the secondary and tertiary structures begin to become a factors for the larger peptides and proteins. Although these factors will undoubtedly arise when applying the predictive equations to proteins, it is interesting to see how they may predict the ratio of K values for two similar, or homologous proteins, such as β-lactoglobulins A and B, or insulin from different species.

Utilizing Equations 7 and 8, the ratio, ρ, of K values for β-lactoglobulins A and B may be expressed as

$$\rho = \frac{K_A}{K_B} = \exp\left(E_A - E_B\right) \tag{10}$$

where E for β-lactoglobulin (which has 162 residues) may be expressed as

$$E = \sum_{j=1}^{162} \epsilon_j + \delta_1^N + \delta_n^C \tag{11}$$

and E_A and E_B differ at $n = 64$ and $n = 118$ as was previously discussed. Substituting Equation 11 into 10 and simplifying gives

$$\rho = \exp\left[(\epsilon_{asp} + \epsilon_{val}) - (\epsilon_{gly} + \epsilon_{ala})\right] \tag{12}$$

It is interesting to note that only values of ϵ, which represent the effect of an amino acid residue on the partition coefficient, remain in the partition ratio. As was mentioned earlier, it is the value of ϵ which will be used to compare the partition coefficients of similar proteins. When the ϵ values for aspartic acid, valine, glycine, and alanine are substituted into Equation 12 a predicted ratio of 1.25 is obtained. The actual partition coefficients of β-lactoglobulins A and B were found to be 0.079 and 0.039, respectively, giving an experimental ratio of 2.03. The error between experimental and predicted ratios is undoubtedly due to the complex nature of the proteins and the simplicity of the predictive method.

The predictive nature of Equation 7 was again tested when horse and pig insulin were partitioned in system b. The partition coefficient for both species was found to be 21.2 and 19.4, respectively, thus giving a partition ratio, ρ, of horse to pig insulin of 1.09. The two species differ at position 9, in which the horse and pig species have glycine and serine, respectively. In order to predict the ratio of horse to pig insulin, an equation similar to that used for the β-lactoglobulins may be written

$$\rho = \frac{K_{\text{horse}}}{K_{\text{pig}}} = \exp\left(\epsilon_{gly} - \epsilon_{ser}\right) \tag{13}$$

Substituting the appropriate values of ϵ in Equation 13 yields a predicted ratio of 1.00, which agrees extremely well with the experimental value. This prediction, which is only 9% in error, is much improved over that of the β-lactoglobulins due to the fact that glycine and serine, which have similar partitioning behavior in the system, and which both have polar uncharged sidechains, are the only differing residues in the two insulins.

Conclusions

The effect of amino acid sequence on peptide partitioning was investigated by partitioning twenty-three pairs of reversed sequence dipeptides in four systems of the PEG 3400/potassium phosphate/water phase diagram at 20°C. When the sequence was reversed, a plot of Ln (K) of the dipeptide versus the PEG concentration difference between the phases reveals two different partition curves for each dipeptide pair depending on the sidechain type and location. The dipeptide partition data along with the partition coefficients measured for twenty amino acids in the above system were utilized to obtain Equation 7 for the predicition of peptide and protein partition coefficients. The equation, which takes into account the effect of the amino acid residues rather than the pure amino acids, was used to successfully predict the partition of homogeneous glycine peptides containing two to five residues, and several tripeptides. Although the correlation is too simple to predict protein partitioning, it provides a promising start for analyzing the effect of amino acid sequence. The ratio of β-lactoglobulin A and B partition coefficients, along with the ratio of horse to pig insulin partition coefficients were predicted and compared with experimental results. Future work will involve investigation of the effect of protein secondary and tertiary structure on the partition phenomenon so that a more general correlation may be obtained.

Acknowledgments

This work was supported by the National Science Foundation Grant Nos. EET-8708839 and CBT-8702912.

Literature Cited

1. Albertsson, P. -A. Partition of Cell Particles and Macromolecules, 3rd Ed.; John Wiley & Sons: New York, 1986.
2. Walter, H.; Brooks, D. E.; Fisher, D. Partitioning in Aqueous Two-Phase

Systems. Theory, Methods, Uses and Aplications to Biotechnology; Academic Press, Inc.: Orlando, 1985.
3. Kroner, K. H.; Hustedt, H.; Kula, M. -R. Biotech. Bioeng. 1982, 24, 1015.
4. Tjerneld, F.; Johansson, G.; Joelsson, M. Biotech. Bioeng. 1987, 30, 809.
5. Kroner, K. H.; Hustedt, H.; Kula, M. -R. Process Biochem. 1984, 19, 170.
6. Hustedt, H.; Kroner, K. H.; Stach, W.; Kula, M. -R. Biotech. Bioeng. 1978, 20, 1989.
7. Albertsson, P. -A.; Cajarville, A.; Brooks, D. E.; Tjerneld, F. Biochim. Biophys. Acta 1987, 926, 87.
8. Diamond, A. D.; Hsu, J. T. Biotech. Bioeng. 1989, 34, 1000.
9. Albertsson, P. -A.; Sasakawa, S.; Walter, H. Nature (London) 1970, 228, 1329.
10. Johansson, G. Biochim. Biophys. Acta 1970, 221, 387.
11. Albertsson, P. -A. Partition of Cell Particles and Macromolecules, 2nd Ed.; John Wiley & Sons: New York, 1971.
12. Sasakawa, S.; Walter, H. Biochemistry 1974, 13(1), 29.
13. Diamond, A. D.; Lei, X.; Hsu, J. T. Biotech. Techn. 1989, 3(4), 271.
14. Sasakawa, S.; Walter, H. Biochemistry 1972 11(15), 2760.
15. Bohlen, P.; Dairman, W.; Udenfriend, S. Arch. Biochem. Biophys. 1973, 155, 213.
16. Walstra, P.; Jenness, R. Dairy Chemistry and Physics; John Wiley and Sons: New York, 1984.
17. Righetti, P. G.; Caravaggio, T. J. J. Chromatogr. 1976, 127, 1.
18. Kyte, J.; Doolittle, R. F. J. Mol. Biol. 1982, 157, 105.

RECEIVED December 28, 1989

Chapter 5

Protein Separation via Polyelectrolyte Complexation

Mark A. Strege, Paul L. Dubin, Jeffrey S. West, and C. Daniel Flinta

Department of Chemistry, Indiana University–Purdue University, Indianapolis, IN 46205–2810

The complexation of proteins with synthetic strong polycations may lead to selective coacervation. For two proteins, A and B, selectivity may be defined as $S = ([A]_c/[B]_c)([A]_s/[B]_s)^{-1}$, where the subscripts c and s refer to coacervate and solution phases. We have measured S for mixtures of two proteins and the polycation poly(dimethyl-diallylammonium chloride), and found it to be a function of solution pH and protein isoelectric point.

Protein-polyelectrolyte coacervation occurs abruptly at a critical pH which depends strongly on the ionic strength I. Plots of pH_{crit} vs. I may thus be viewed as phase boundaries. Analysis of the effects of protein:polyion stoichiometry on the phase boundary suggests that phase separation may be a consequence of the saturation of binding sites on the polymer with protein. Assuming that electroneutrality of the complex occurs at the point of coacervation, it is possible to estimate the number of proteins bound per polycation at the point of phase separation, $\bar{n}_{crit}$. Comparison of the phase boundaries for several proteins reveals that the net protein surface charge density does not control phase separation, but instead suggests the importance of charge patches on the protein surface.

Size-exclusion chromatography, via the Hummel-Dreyer method, reveals the existence of a primary complex in which the degree of intrapolymer protein binding depends on pH. The average number of proteins bound per polymer as obtained by this technique, $\bar{n}$, is seen to depend upon the free protein concentration. The value of $\bar{n}$ corresponding to polymer saturation is in fair agreement with $\bar{n}_{crit}$ from phase separation studies.

Oppositely charged polyelectrolytes interact to form complexes. Depending primarily on the molecular weights and linear charge densities, these complexes may be amorphous solids (1), liquid coacervates (2,3), gels (4), fibers (4), or soluble aggregates (5-7). One particular case of inter-macroion complex formation involves synthetic polyelectrolytes and globular proteins. The formation of these complexes is generally evidenced by phase separation, where the denser, polymer-rich phase may

0097–6156/90/0427–0066$06.00/0

be a liquid "complex coacervate" (8) or a solid precipitate. Examples of the former have been observed for gelatin and polyphosphate (9), and serum albumin and poly(dimethyldiallylammonium chloride) (10). Systems that exhibit precipitation include hemoglobin and dextran sulfate (11), carboxyhemoglobin and potassium poly(vinyl alcohol sulfate) in the presence of poly(dimethyldiallylammonium chloride) (PDMDAAC) (12), lysozyme and poly(acrylic acid) (13), and RNA polymerase and poly(ethyleneimine) (14).

The ability of polyelectrolytes to remove proteins from solutions represents tremendous potential in the area of protein fractionation. Protein separation may occur at two steps in the protein complexation/recovery process: at the point of complex formation, or at the point of complex redissolution. The second approach has recently been exploited through the incorporation of protein-polymer precipitation stages into a variety of protein purification procedures, wherein the precipitated proteins are selectively recovered from the insoluble complex aggregate via step-wise selective redissolution by pH or ionic strength adjustment (14-16). However, selectivity in the first step in the complexation/recovery process, complex formation, has not been thoroughly studied. Earlier demonstrations of preferential complexation of polyelectrolytes with specific proteins (13,17), suggest the value of further investigations along this line.

The use of polyelectrolyte-precipitation to separate proteins offers several advantages over other protein fractionation techniques. The recovery of proteins through the formation of insoluble complexes with polyelectrolytes appears to be a non-denaturing process, inasmuch as Sternberg and Hershberger reported high recoveries of activities for enzymes precipitated with polyacrylic acid (13), while other workers have reported the non-denaturing fractionation at slightly alkaline pH of intracellular proteins using synthetic polycations (14,16,18). Furthermore, compared to other methods for protein separation, e.g. chromatography, selective precipitation offers great economy with regard to materials and process, and, furthermore, is virtually unlimited in scale. Thus, an elucidation of the principles governing protein selectivity in polyelectrolyte separation would be of considerable applied significance.

Evidence for the existence of a stable, soluble intrapolymer BSA-PDMDAAC complex has been obtained (20). The model for protein-polyelectrolyte coacervation may thus prove to be similar to that proposed by Tsuchida, Abe, and Honma for the coacervation of linear oppositely charged polyelectrolytes (21). The initial (possibly cooperative) binding of proteins via electrostatic interaction leads to soluble complex formation. A subsequent increase in the energy of binding occurs in response to an increase in protein surface charge, and results in the exclusion of water molecules. At the same time the net charge of the complex approaches zero. The complex thus acquires hydrophobic character, leading to aggregation and phase separation . This process is schematically depicted in Figure 1.

Studies of complexation between globular proteins and synthetic strong polycations were undertaken to gain insight into protein-polyelectrolyte coacervation selectivity. In the present work, two approaches were applied toward the study of the mechanism of protein-polyelectrolyte interaction. The first approach involved the investigation of phase separation, and the analysis of the phase boundary plots that show the ionic strength dependence of the critical pH. Since selective binding in soluble complexes is believed to be prerequisite to coacervation selectivity, a second approach involved the characterization of soluble complexes at conditions prior to phase separation. The goal of this second approach is the determination of fundamental parameters, such as protein-polymer binding constants, from which coacervation selectivity can be predicted. Soluble complexes were characterized by size-exclusion chromatography (SEC) to gain insight into the cooperativity of

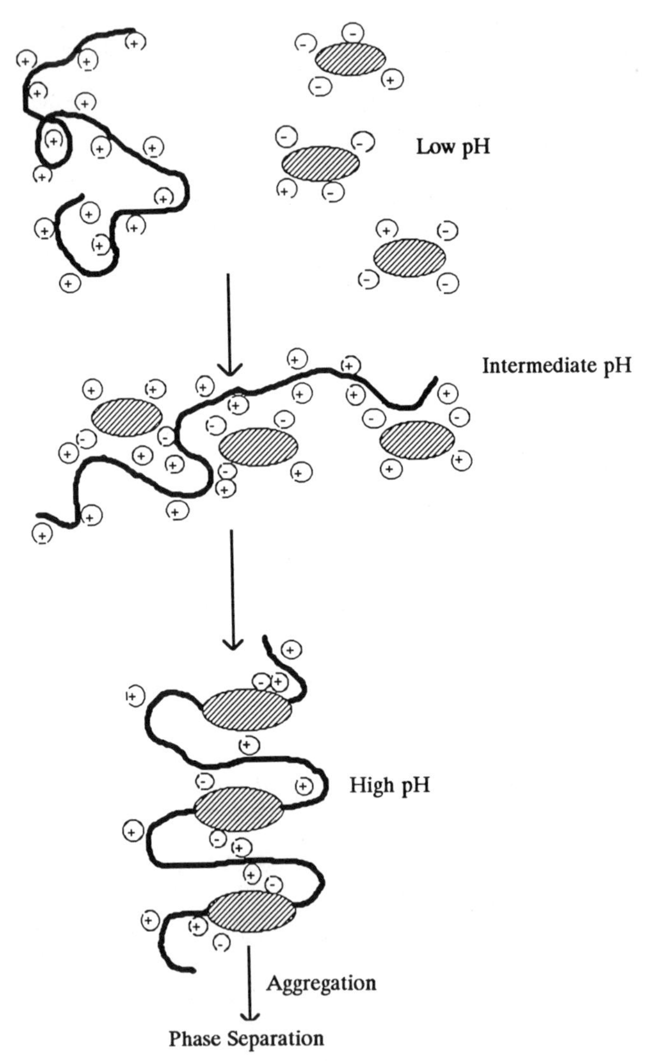

Figure 1. A model of protein-polyelectrolyte coacervation.

binding, the structure of the complex, and the complex stoichiometry, i.e.. the number of protein molecules bound per polyion. In principle, such methods may also yield the number of binding sites per polymer molecule, and the intrinsic association constant.

EXPERIMENTAL

Materials: Poly(dimethyldiallylammonium chloride) (PDMDAAC), a commercial sample "Merquat 100" from Calgon Corp. (Pittsburgh, PA) possessing a nominal molecular weight of $2x10^5$ and a reported polydispersity of $M_w/M_n \approx 10$, was dialyzed and freeze-dried before use. All proteins were obtained from Sigma Chemical Corp.

Turbidimetric Titrations: Solutions were prepared as mixtures of PDMDAAC (0.05 - 1 g/l) and protein (0.25-25 g/l), corresponding to protein/polymer weight ratios (**r**) ranging from 0.25 to 200, at pH 4 - 5, in dilute (0.05-0.3 M) NaCl. The optical probe (2 cm path length) of a Brinkmann PC600 fiber optics probe colorimeter, and a pH electrode connected to an expanded scale pH meter (Orion 811, or Radiometer model 26), were both placed in the solution. Changes in turbidity were monitored as %T, relative to a blank (polymer-free) solution, as the pH was adjusted by the addition of dilute (0.01-0.10 M) NaOH. Critical pH, the pH at which phase separation takes place, was determined using the method described elsewhere (10).

Coacervation Selectivity Measurements: In order to determine the selectivity of synthetic polycations toward complexation with specific proteins, the following experimental procedure was devised. A mixture of PDMDAAC and two proteins, in 0.01 M NaCl, was titrated with NaOH until a desired pH or percent transmittance was reached. The titrated solution was centrifuged (2000 rpm, 10-15 minutes), the supernatant was removed, and the coacervate centrifugate was resuspended in a small volume (2.0-10.0 ml) of acidified (pH 4-5) 0.4 M NaCl buffer. The protein contents of both supernatant and redissolved coacervate were then analyzed via SEC (all protein peaks were baseline resolved).

Size-Exclusion Chromatography: SEC was carried out on an apparatus comprised of a Minipump (Milton Roy), a model 7012 injector (Rheodyne) equipped with a 100 µl sample loop, an R401 differential refractometer (Waters), and a Model 120 UV detector (λ = 280 nm) (Gilson). A Superose-6 column (30cm x 1cm OD) (Pharmacia) was eluted at 0.53 ml/min. Column efficiency, determined with acetone, was at least 12,000 plates/meter.

Injections were performed in mobile phases at pH 8.0, which is below critical pH at this ionic strength. To determine complex stoichiometry, we have used the Hummel-Dreyer (HD) method (22, 23), as recently applied to dextran-hemoglobin complexes (19). HD experiments were carried out employing a 0.25 M NaOAc buffer as mobile phase, and the protein concentration in the mobile phase varied from 0.10 to 8.00 mg/ml. Polymer samples (2.0 mg/ml) were filtered (0.2 µm) before injection.

To decrease the degree of polymer MW polydispersity, PDMDAAC was fractionated via SEC, prior to use in HD experiments. The fractionation of PDMDAAC was carried out using a mobile phase of 0.4 M NaOAc buffer, pH = 7.0, which has been found to sufficiently repress adsorption effects, especially in regard to the polycation (24). 40.0 mg of polymer were applied to a Superose-6 prep. gel column (100 ml gel bed) via a 1.0 ml sample loop. The mobile phase was eluted through the column at a velocity of 2.0 ml/min, and the eluent was monitored using a R401 differential refractometer (Waters). The injected sample was separated into nine fractions, collected at 6.7 ml intervals following the beginning of sample elution, possessing average polydispersities (calculated through the use of a pullulan calibration curve) of 1.5.

RESULTS AND DISCUSSION

Coacervation Selectivity Experiments: Although separative coacervation of proteins by synthetic polyelectrolytes has been reported (13-18), no parameter describing the selective binding has been formally described. Thus, we define protein-polyelectrolyte coacervation selectivity as the comparative tendency of a given protein to be removed, in the presence of a second protein, by an oppositely charged polyelectrolyte. Since the concentrations of proteins present in both the coacervate and soluble phases, after centrifugation, may readily be determined, it is convenient to define the selectivity as

$$S = \frac{[A]_C/[B]_C}{[A]_S/[B]_S} \qquad (1)$$

where $[A]_C$ and $[B]_C$ are the weight concentrations of protein A and protein B in the pelleted coacervate, respectively, and $[A]_S$ and $[B]_S$ are the weight concentrations of the two proteins in the supernatant. For $S>1$, protein A must have an isoelectric point (pI) (the pH at which a protein has zero electrophoretic mobility) equal to or lower than that of protein B.

Complex formation and coacervation occur in response to electrostatic interactions. Since the charge of PDMDAAC is independent of pH, the surface charge density of the protein molecule or some related variable must - along with the ionic strength - govern phase separation. The pI of a protein should therefore reflect its tendency toward complex formation. For example, in a mixture of BSA (pI = 4.8) and lysozyme (pI = 11.0), the former, with a greater net negative surface charge density at any solution pH, is expected to complex more strongly at intermediate pH. At pH = 6.4, analysis of the coacervate of BSA and PDMDAAC in the presence of lysozyme revealed the phase separation of 90% of the BSA and only 4% of the lysozyme initially present in solution , corresponding to $S = 190 \pm 10$. On the other hand, coacervation of a mixture of BSA and ribonuclease (pI = 9.0), at pH = 5.7, revealed infinite selectivity, with no ribonuclease in the coacervate, showing that coacervation selectivity is not simply a function of ΔpI. We propose that the strength of the binding depends not only on the global or net charge, but is also sensitive to the protein surface charge distribution. Further evidence from turbidimetric titrations in support of this suggestion will be discussed later.

In a solution of two proteins possessing different pI, S would be expected to be largest at a pH where the potential created by the surface charges on the more basic protein is less than some critical value. For BSA and ribonuclease at I = 0.01, S is plotted vs. pH in the pH region intermediate to the pI's of the two proteins in Fig. 2. S is infinite at pH 5.7, and rapidly drops as pH approaches 9.0. Thus, at pH ≤ 5.7, it appears that the negative charges distributed on the surface of ribonuclease are not capable of generating a negative potential of sufficient magnitude to bind the polycation.

Turbidimetric Titrations: Since selectivity depends upon coacervation, it is important to study the dependence of phase separation upon solution variables. Turbidimetric titrations were utilized to determine the critical pH of solutions of BSA and Merquat

100 over a range of I and bulk solution protein:polymer ratio (**r**). The results are shown in Figure 3 as plots of critical pH vs. I (phase boundaries) over a range of **r**. In addition, the effect of total solute concentration on critical pH was obtained from Type I titrations at **r** = 0.25 at BSA concentrations from 0.05 g/l to 1.0 g/l, for $0.01<I<0.3$.

At constant I, the critical pH varies inversely with **r**. This effect may be explained by assuming that hydrophobicity or desolvation of the complex is prerequisite to phase separation. We suggest that such desolvation occurs only when some critical fraction of polycation sites is electrically neutralized by binding the requisite number of protein-borne carboxylates. The conditions required for phase separation may then be represented by:

$$Z_T = Z_P + \bar{n} Z_{pr} = 0 \quad (2)$$

where Z_T represents the net charge of the complex (which can be assumed to approach zero at phase separation), Z_P is the net positive charge of the polycation, Z_{pr} is the net negative charge of the protein, and $\bar{n}$ is the average number of proteins bound per polycation. According to the foregoing electroneutrality assumption, the value of $\bar{n}$ obtained from phase separation data is identified as $\bar{n}_{crit}$, the number of proteins required to neutralize the polyion charge. Since the charge of PDMDAAC results from the quaternary nitrogen, both Z_P and the dimensions of the polymer are independent of pH. At high values of **r**, i.e. "excess protein", the polycation sites may become saturated with proteins, so that $\bar{n}$ is large. Then Z_{pr} at critical conditions may be relatively small. At low values of **r**, the number of protein molecules bound per polycation may be decreased, and larger values of Z_{pr} - hence higher critical pH - may be required. It must be recognized that the number of protein molecules bound under any condition, $\bar{n}$, is a function of both an intrinsic binding constant, which is pH dependent, and a mass action equilibrium, which is linked to **r**. It is expected that $\bar{n}$ increases with **r**; then, from eq. 2, $(Z_{pr})_{crit}$ must decrease with **r**. Thus, critical pH decreases with increasing **r**, as observed. This effect appears to reach a limit at **r** = 50, which may correspond to saturation of the polymer with protein molecules.

Values of $\bar{n}_{crit}$ estimated from equation 2 are plotted vs. **r** over a range of I in Figure 4. All plots reveal a plateau at high **r** corresponding to polymer saturation. The value for $\bar{n}_{max}$ at large I, ca. 30, is in good agreement with the maximum number of proteins bound at I = 0.25 deduced from the Hummel-Dreyer analyses (see below). However, much larger values of $\bar{n}_{crit}$ are observed at low I. This effect could be explained in two ways. First, the polymer chains expand at low I in response to intrapolymer charge repulsion, possibly increasing their binding capacity. Second - and more probably - the unscreened coulombic forces are sufficiently large at low I so that proteins bind at pHs near their pI. When proteins with such small Z_{pr} bind, many must bind before the complex can achieve electroneutrality. Consequently, phase separation at low I and low pH corresponds to large $\bar{n}_{crit}$.

Protein pI clearly affects critical pH, so that plots of pH_{crit} vs I are expected to diverge for proteins of different pI. To observe whether any single protein charge parameter might control the coacervation of several proteins, phase boundaries were developed for PDMDAAC and β-lactoglobulin, chicken egg albumin, lysozyme, ribonuclease, and trypsin, all at **r** = 100. For each protein, critical pH was converted to global surface charge density, obtained by dividing the net negative surface charge by the protein surface area, using data from published pH titration curves (25-30) and hydrodynamic radii (31) found in the literature. The plots of surface charge density (Z_{pr}/A) vs. I shown in Figure 5 are observed to diverge, suggesting that the formation of coacervate is governed by a parameter more complex than Z_{pr}/A.

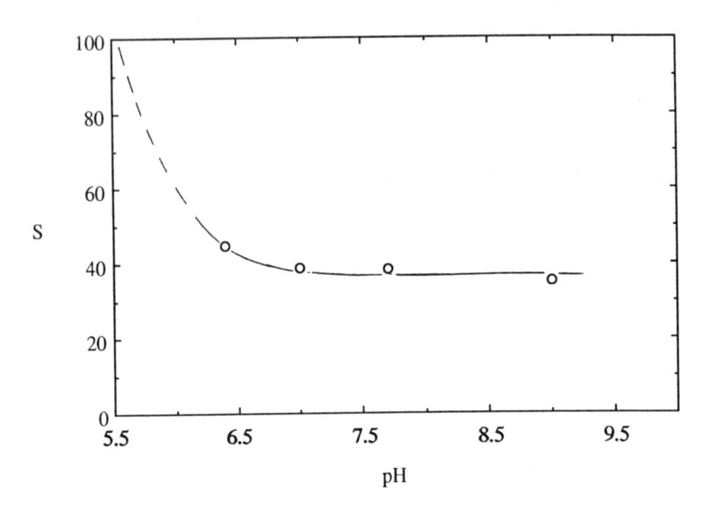

Figure 2. Selectivity of BSA-PDMDAAC coacervation in the presence of ribonuclease as a function of pH. Infinitely large values of S are obtained at pH < 5.8.

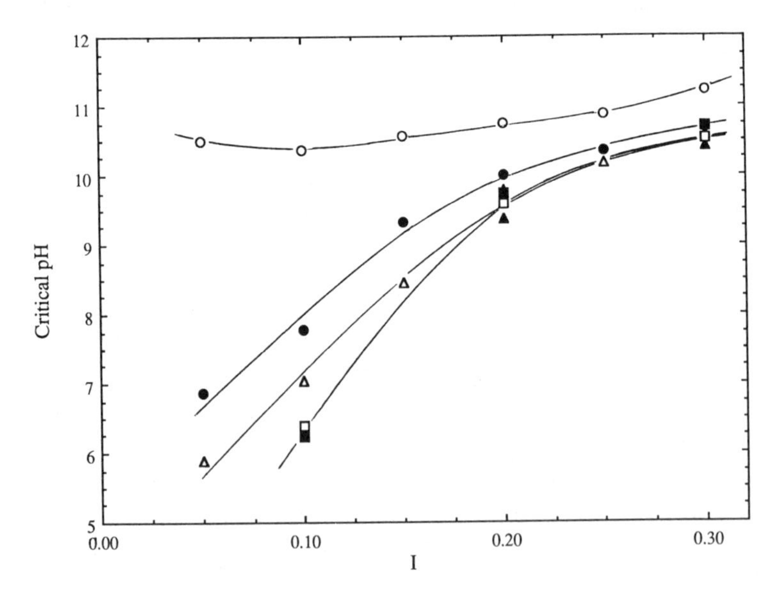

Figure 3: BSA-PDMDAAC phase boundaries over a range of r: (○) r = 0.25; (●) r = 2.5; (△) r = 25; (▲) r = 50; (□) r = 100; (■) r = 200.

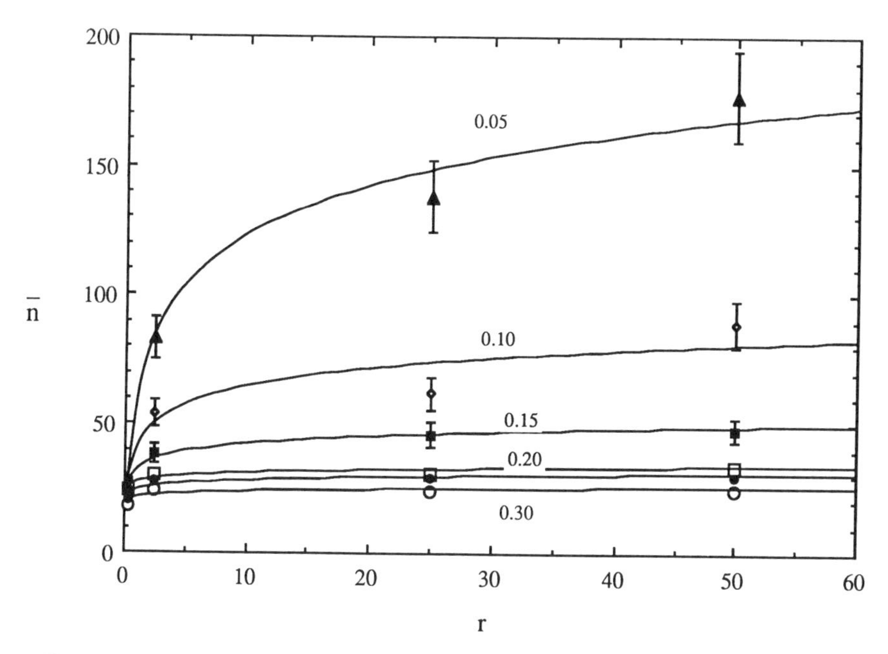

Figure 4. Plots of $\bar{n}$ vs. r at various I: (▲) I = 0.05; (◇) I = 0.10; (■) I = 0.15;(□) I = 0.20; (■) I = 0.25; (○) I = 0.30.

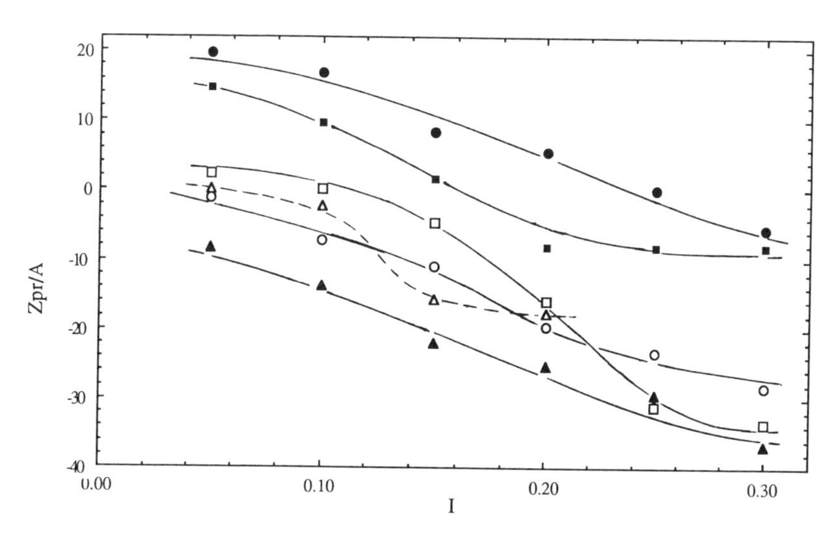

Figure 5. Phase boundaries (**r**=100), plotted as net surface charge density (net charge/nm^2) vs. I (ionic strength): (○) bovine serum albumin; (●) lysozyme; (△) ribonuclease; (▲) chicken egg albumin; (□) β-lactoglobulin; (■) trypsin.

Especially striking is the fact that basic proteins, e.g. lysozyme and trypsin, phase separate with the polycation even while possessing a net positive charge, i.e. below their respective isoelectric points. It is believed that the existence of local negative "charge patches" on the protein surface, rich in acidic groups, are responsible for overcoming the net positive charge and initiating the electrostatic interaction between the basic proteins and the polycations. The variable "patchiness" of the different proteins may also explain the divergence of the plots of Figure 5. Evidence for patch-controlled attractive electrostatic interactions between charged proteins and chromatographic adsorbents has recently been reported elsewhere (32).

Size-Exclusion Chromatography: Information on the binding of BSA to PDMDAAC was sought by an adaptation of the Hummel-Dreyer method (19, 22). In this modified technique, polyelectrolyte is injected onto an SEC column in which the mobile phase is a buffer, adjusted to the desired pH and ionic strength, and containing some appropriate concentration of protein. As an example, a UV (280 nm) chromatogram resulting from the injection of a PDMDAAC fraction (estimated MW = $3x10^6$) into a mobile phase of pH 9.0, 0.25 M sodium acetate buffer, containing 5.0 mg BSA/ml, is displayed in Figure 6a. The initial UV-absorbing peak is observed at a smaller elution volume than that of protein alone, and must correspond to soluble complex. One also sees the loss of protein from the mobile phase, as evidenced by the negative peak at the retention volume of BSA (ca. 16.5 ml). The area under this peak is proportional to the amount of protein bound to the injected polymer sample. At pH = 9.00, I = 0.25, the UV peak corresponding to the complex resembles, in shape and position, the refractive index trace of polymer alone, suggesting that the complexes existing at these conditions may be solvated, intrapolymer species, in accord with previous results (20). At higher pH's the complex peak shifts toward a larger elution volume; this effect probably results from sample adsorption. Such higher pHs produce negatively charged sites on the column packing which can interact with polycation residues, leading to larger elution volumes.

At I = 0.25, the amounts of protein corresponding to the negative peak area were determined from a calibration plot of peak area vs. protein concentration, and were used to compute $\bar{n}$. The results, displayed in Table I, suggest that an increase in the degree of protein binding in the soluble complexes occurs in response to an increase in pH, i.e. as the negative protein surface charge is increased. The values of $\bar{n}$ in Table I are very low compared to the ones obtained from phase separation data, presented in Figure 4. The pH values in Table are considerably lower than the critical pH at the conditions (**r** and I) of Table I, 10.60. The intrinsic protein binding constant at these lower pH's is relatively small, so free protein is in large excess relative to bound protein. The combination of $pH<pH_{crit}$ and small **r** thus account for the low values of $\bar{n}$.

Table I. Average Number of BSA Molecules Bound Per Polymer Chain at Various pH (**r** = 0.04, I = 0.25)

pH	$\bar{n}$
8.90	0.2
9.40	0.3
9.98	1.4

The value of $\bar{n}_{sat}$, the maximum number of proteins per polymer chain that may bind in the presence of excess protein, may be determined by the method applied by Barberousse et al. to hemoglobin-dextran sulfate complexes (19). While chromatographic overlap was minimized by the use of a high MW PDMDAAC fraction, the high absorbance of the mobile phase at 254 nm lead to poor chromatographic peak reproducibility. Hence, as shown in the plot of $\bar{n}$ vs mobile phase BSA concentration in Figure 7, $\bar{n}$ contained a 25% relative error. Nevertheless, one may observe $\bar{n}_{sat} \cong 15$, for this pH and ionic strength.

A more refined analysis of the binding data of Figure 7 is possible through a variety of graphical procedures, such as those originally proposed by Scatchard, Hill and Klotz, respectively (33). These procedures, all based on simple models for macromolecule-ligand interactions, characterize the binding curves in terms of two parameters: the intrinsic ligand binding constant, and the number of binding sites per macromolecule. In our situation, the protein is considered to be the ligand. Unfortunately, the large error in $\bar{n}$ mentioned above rendered meaningless a direct conversion of the data into a Scatchard plot. However, a hand-fitted line drawn assuming non-cooperative binding (33) generated the linear Scatchard plot of Figure 8. This procedure yields values of $K = 7x10^{-5}M^{-1}$ and $\bar{n}_{sat} = 23$. A cooperative fit (33) yielded the Scatchard plot displayed in Figure 9, suggesting $\bar{n}_{sat} = 15$. These results are of the same order of magnitude as the value of $\bar{n}_{crit} = 30$ calculated for coacervation conditions at pH 10.0 using equation 2. Differences in $\bar{n}$ could result from the difference in pH for the two methods.

Hummel-Dreyer chromatograms (UV, 280 nm) obtained following the injection of 2.00 mg/ml polymer (high MW fraction) into mobile phases (pH = 8.0, I = 0.25) containing 5.00 mg/ml of ribonuclease or 3.00 mg/ml lysozyme are displayed in Fig. 6. In contrast to the elution profile of the BSA-PDMDAAC complex, the ribonuclease-PDMDAAC chromatogram does not reveal a negative peak at the elution volume of ribonuclease (ca. 17.5 ml). The baseline in that region of the chromatogram is difficult to interpret, as a portion of the complex trace appears to be overlapping the negative protein peak, suggesting that some adsorption of the former may be occurring. The position of the entire lysozyme-PDMDAAC complex chromatogram suggests adsorption. Complexes formed between polycations and ribonuclease or lysozyme may reveal a greater tendency toward adsorption because, at any pH, these two basic proteins neutralize fewer polycation charged sites than does the more acidic protein, BSA.

The following model for protein-polyelectrolyte coacervation is consistent with our results. The initial step in the formation of a complex coacervate is the binding of protein molecules by the oppositely charged polyion. Polymer segments near the protein will experience a potential arising from opposite charges on the protein surface. Because of screening, a high protein surface charge density is required for complex formation at higher ionic strengths. Protein charge distribution appears to be critical, as local concentrations of negative charge on the protein surface may create the potentials responsible for binding. This local potential is reflected in a pH-dependent intrinsic binding constant. The number of protein molecules bound per polymer chain, $\bar{n}$, depends on both this local potential and also on mass action equilibria, i.e. the protein:polymer weight ratio. When $\bar{n}$ attains a value sufficient to neutralize the polyion charge, the electroneutral complexes can aggregate to yield a coacervate phase. According to this model, selectivity in coacervation is a consequence of selectivity in complexation. Therefore, a knowledge of the pH-dependent binding curve for soluble complexes should lead to predictions of coacervation selectivity.

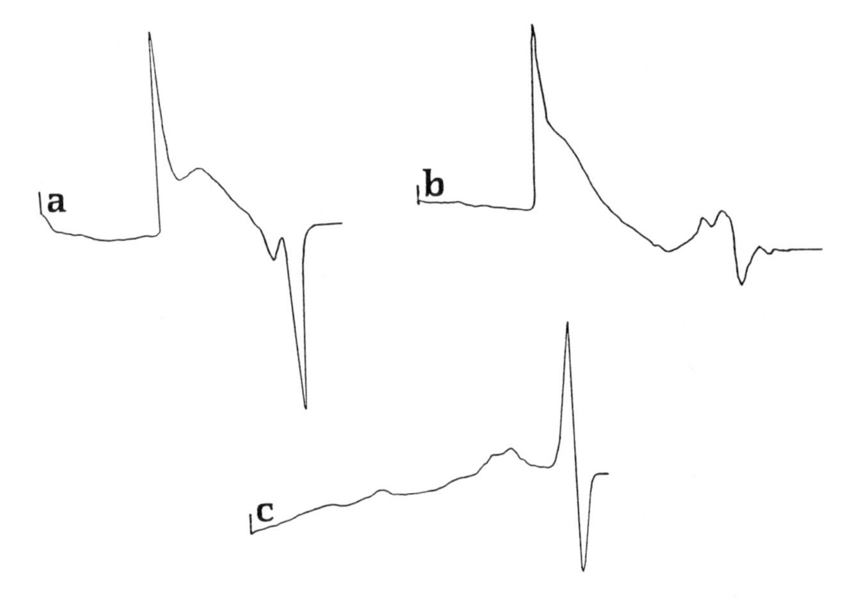

Figure 6. UV (254 nm) chromatograms (Superose 6) of PDMDAAC fraction 3 (MW≈3.2×10^5) using, as mobile phase, 0.25M, pH 9.0, NaOAc buffer containing (a) 5.0 mg/mL BSA, (b) 5.00 mg/mL ribonuclease, and (c) 3.00 mg/mL lysozyme.

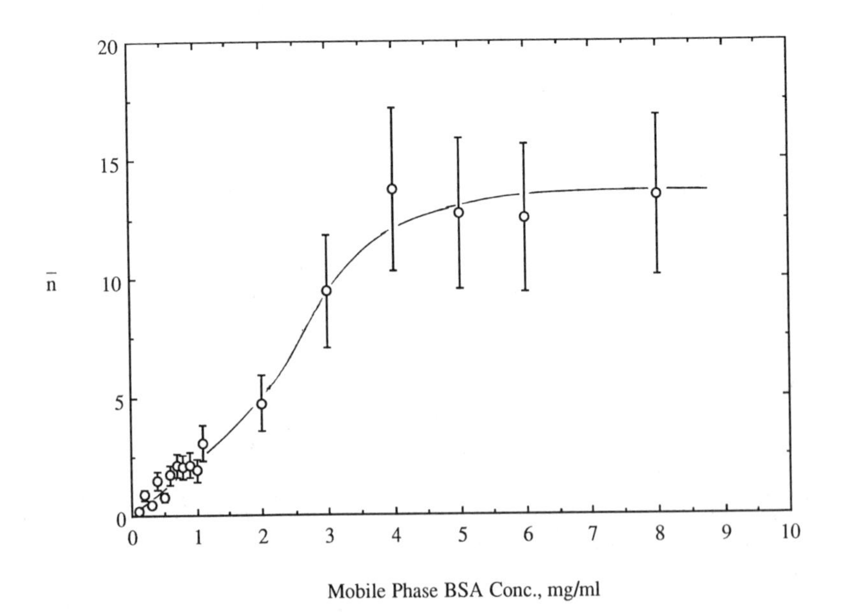

Figure 7. $\bar{n}$ versus mobile phase BSA concentration (I = 0.25, pH 9.0, PDMDAAC concentration = 2.5 mg/ml).

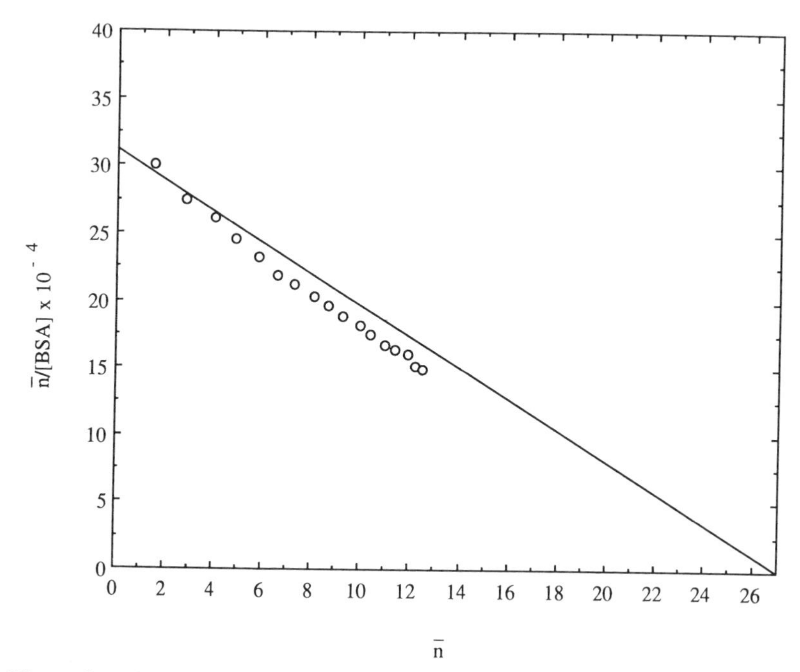

Figure 8. Scatchard plot developed from a non-cooperative fit to the data of Figure 7.

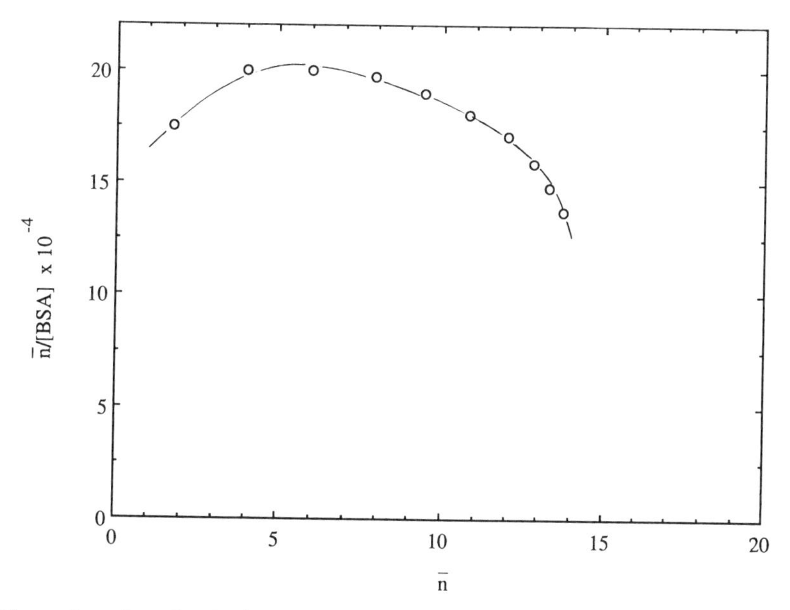

Figure 9. Scatchard plot developed from a cooperative fit to the data of Figure 7.

CONCLUSIONS

Protein-polycation coacervation selectivity was found to be a function of pH and protein isoelectric point. This result is clearly a consequence of the electrostatic nature of the primary process, the formation of a coulombic complex between some sequence of polyion segments and an oppositely charged area on the protein surface.

Phase boundary investigations via turbidimetric titrations provide indirect evidence for polymer saturation and complex electroneutrality at the point of phase separation, and also suggest the existence of protein surface charge patches. Estimations of the average number of proteins bound per polymer at the point of phase separation $\bar{n}_{crit}$ may be made. $\bar{n}_{crit}$ depends on the ionic strength and on the bulk solution protein:polymer weight ratio (**r**).

Hummel-Dreyer SEC investigations confirmed the existence of stable, soluble intrapolymer complexes, previously deduced from dynamic light scattering studies. The number of proteins bound per polymer molecule $\bar{n}$ was found to be a function of pH and free protein concentration. The value of $\bar{n}$ extrapolated to saturation conditions is found in fair agreement with $\bar{n}_{crit}$. An estimate of an equilibrium binding constant (K) was made through the development of a Scatchard plot for BSA-PDMDAAC, but attempts to determine the binding constants of basic proteins were hindered by column adsorption effects.

Implementation of selective polyelectrolyte coacervation will confront additional questions not dealt with yet. First, removal of the polycation may in principle be carried out in several ways, e.g. precipitation at low pH using a strong polyion. Second, it will be necessary to evaluate the retention of enzymatic activity for the recovered protein. Third, while the contribution of the polyelectrolyte to the solution viscosity prior to coacervation is negligible (because of its low concentration), the coacervate itself is quite viscous. These issues are the subjects of continued study.

ACKNOWLEDGMENTS

This work was partially supported by the National Science Foundation (DMR-8507479) and by the Eli Lilly Corporate Research Center.

REFERENCES

1. Michaels, A.S.; Miekka, R.G. J. Phys. Chem. 1961, 65, 1765.
2. Veis, A.; Bodor, E.; Mussel, S. Biopolymers 1967, 5, 37.
3. Polderman, A. Biopolymers 1975, 14, 2181.
4. Tsuchida, E.; Abe, K.; and Honma, M. Macromolecules 1976, 9, 112.
5. Kabanov, V.A.; Zezin, A.B. Macromol. Chem. Suppl. 1984, 6, 259.
6. Dauzenberg, H.; Linow, K.; Philipp, B. Acta Polymerica 1982, 33, 619.
7. Shinoda, K.; Sakai, K.; Hayashi, T.; Nakajima, A. Polymer J. (Japan) 1976, 8, 208.
8. Bungenburg de Jong, H.G. In Colloid Science; Kruyt, H.R., Ed., Elsevier, 1949; Vol. II, Chapter 10.
9. Lenk, T.; Theis, C. In Coulombic Interactions in Macromolecular Systems; Eisenberg, A. and Bailey, F.E., Eds., American Chemical Society: Washington, DC, 1986; Chapter 20.
10. Dubin, P.; Ross, T.D.; Sharma, I.; Yegerlehner, B. In Ordered Media in Chemical Separations; Hinze, W.L., and Armstrong, D.W., Eds., American Chemical Society: Washington, DC, 1987; Chapter 8.
11. Nguyen, T.Q. Makromol. Chem. 1986, 187, 2567.

12. Kokofuta, E.; Shimizu, H.; Nakumura, I. Macromolecules 1981, 14, 1178.
13. Sternberg, M.; Hershberger, D. Biochim. Biophys. Acta 1974, 342, 195.
14. Jendrisak, J.J.; Burgess, R.R. Biochemistry 1975, 14, 4639.
15. Bell, D.J.; Hoare, M.; Dunnill, P. In Advances in Biochemical Engineering/Biotechnology; Fiechter, A., Ed., Springer-Verlag: New York, 1982; p 1.
16. Burgess, R.R.; Jendrisak, J.J. Biochemistry 1975, 14, 4634.
17. Morawetz, H.; Hughes, W.L. J. Phys. Chem. 1952, 56, 64.
18. Jendrisak, J.J. In Protein Purification: Micro to Macro; Burgess, R.R., Ed., Alan R. Liss, Inc., 1987. p. 85.
19. Barberousse, V.; Sacco, D.; Dellacherie, E. J. Chromatogr. 1986, 369, 244.
20. Dubin P.L.; Murrell, J.M. Macromolecules 1988, 21, 2291.
21. Tsuchida, E.; Abe, K.; Honma, M. Macromolecules 1976, 9, 112.
22. Hummel, J.P.; Dreyer, W.J. Biochim. Biophys. Acta 1962, 63, 530.
23. Korpela, T.K.; Himanen, J.P. In Aqueous Size-Exclusion Chromatography; Dubin, P.L., Ed., Elsevier: Amsterdam, 1988; Chapter 13.
24. Strege, M.A.; Dubin, P.L. J. Chromatogr. 1989, 463, 165.
25. Tanford, C.; Swanson, S.A.; Shore, W.S. J. Amer. Chem. Soc. 1955, 77, 6414.
26. Cannan, R.K.; Palmer, A.H.; Kilbrick, A.C. J. Biol. Chem. 1942, 142, 803.
27. Cannan, R.K.; Kilbrick, A.C.; Palmer, A.H. Ann. N.Y. Acad. Sci. 1941, 41, 243.
28. Tanford, C.; Wagner, M.L. J. Amer. Chem. Soc. 1954, 76, 3331.
29. Tanford, C.; Hauenstein, J.D. J. Amer. Chem. Soc. 1956, 78, 5287.
30. Duke, J.A.; Bier, M.; Nord, F.F. Arch. Biochem. Biophys. 1952, 40, 424.
31. Cartha, G.; Bellow, T.; Harker, D. Nature 1967, 213, 826.
32. Lesins, V.; Ruckenstein, E. Colloid Polym. Sci. 1988, 266, 1187.
33. See, for example, Chang, R. Physical Chemistry with Applications to Biological Systems, 2nd Ed., Macmillan: New York, 1981, pp. 247-251.

RECEIVED January 30, 1990

Chapter 6

Mechanisms of Protein Retention in Hydrophobic Interaction Chromatography

Belinda F. Roettger[1,2], Julia A. Myers[1], Michael R. Ladisch[1,2], and Fred E. Regnier[3]

[1]Laboratory of Renewable Resources Engineering; [2]Agricultural Engineering Department; and [3]Biochemistry Department, Purdue University, West Lafayette, IN 47907

Protein retention in hydrophobic interaction chromatography (HIC) depends on surface hydrophobicity of the support and solute and the kosmotropic nature and concentration of the salt used in the mobile phase. Wyman's linkage theory, extended to provide a unifying model of HIC retention, relates protein retention to the preferential interactions of the mobile phase salt with the support and protein. Preferential interactions of ammonium salts with HIC supports were determined by extremely sensitive densimetric measurements. Chromatographic retention of lysozyme was also determined on a column packed with hydrophilic polymeric supports and retention of myoglobin was determined on butyl-derivatized polymeric sorbents. Mobile phases containing ammonium salts of $SO_4^{=}$, $C_2H_3O_2^-$, Cl^-, and I^- at several concentrations were used to probe retention behavior of lysozyme and myoglobin with respect to these supports. Preferential interactions of the salts with the supports and proteins were found to explain adsorption behavior in hydrophobic interaction chromatography, and results in an equation which predicts capacity factor as a function of lyotropic number and salt molality.

Hydrophobic interaction chromatography (HIC) is a purification technique used to separate proteins and other biological molecules on the basis of surface hydrophobicity (*1,2*). Typically, a solute is adsorbed to a mildly hydrophobic stationary phase in an aqueous mobile phase at a high salt concentration and is eluted at a lower salt concentration (*3,4,5*). Because of the mild adsorption and desorption conditions, protein activity is often maintained. The concept of separating biomolecules under HIC conditions was first suggested in the 1940's by Tiselius (*6*), and further developed in the late 1960's and early 1970's (*7,8,9,10,11,12*). More recently, the synthesis of a variety of stationary phases has encouraged a wide range of HIC applications (*13,14,15,16*). Currently, HIC is used for large scale purification of proteins, such as the bacterial enzymes aryl acylamidase, salicylate monooxygenase, glucokinase and glycerokinase (*17*).

Salt composition of the mobile phase markedly affects retention (*18*). Salt anions are ranked in the lyotropic series published by Hofmeister in 1888 (*19*). The order:

0097-6156/90/0427-0080$06.00/0

$$SO_4^{=} > C_2H_3O_2^- > Cl^- > I^- > SCN^-$$

denotes strength in precipitating proteins. The anions which are ranked before chloride are called polar kosmotropes and bind water tightly in the first hydration shell (*20*). These anions are very effective at high concentrations at promoting protein precipitation. Anions ranked after chloride are chaotropes, disrupt water in the first hydration shell, and promote protein solubility. The lyotropic series is also the general order of effectiveness of salt anions in promoting protein stability and protein retention in HIC (*21,22,18,23*).

Theory

Models for HIC adsorption have been recently summarized and discussed elsewhere with emphasis on Wyman's thermodynamic theory of linked functions (*24,25,26*). Adsorption in HIC may be modeled as follows:

$$P + nI \rightleftharpoons C \tag{1}$$

where P is the protein, n is the number of interaction sites (I) on the support, and C is the protein-support complex. Complex formation is favored in high salt concentrations. By applying the thermodynamic theory of linked functions (i.e., Wyman's Linkage Theory) (*27*), the change in adsorption with the change in activity of the salt in the mobile phase, a_s is related by (*24,25,26*)

$$\frac{\partial \ln K}{\partial \ln a_s} = \left[\frac{\partial g_s}{\partial g_C}\right]_{T,\mu} \frac{M_C}{M_s} - \left\{ \left[\frac{\partial g_s}{\partial g_P}\right]_{T,\mu} \frac{M_P}{M_s} + n \left[\frac{\partial g_s}{\partial g_I}\right]_{T,\mu} \frac{M_I}{M_s} \right\} \tag{2}$$

where K is the adsorption equilibrium constant, $[\frac{\partial g_s}{\partial g_i}]_{T,\mu}$ is the change in component i required to maintain constant chemical potential of both water and salt, μ at constant temperature, T, where i denotes complex, protein, or support interaction site, g is molality (g/1000 g water) and M is molecular weight. The term $[\frac{\partial g_s}{\partial g_i}]_{T,\mu}$ is called the "preferential interaction parameter," $\xi_{s,i}$.

The preferential interaction parameter may be described as

$$\xi_{s,i} = B_s - E_s - g_s B_w \tag{3}$$

where B_s is the salt bound to the component (g salt/g component), E_s is the grams of salt excluded from the component due to electrostatic effects (g salt/ g component), g_s is the salt molality (g salt/1000 g water), and B_w is the amount of water bound to the component (g water/g component). Therefore, the preferential interaction parameter is a measure of the net salt exclusion or inclusion in the immediate environment of a component.

Reviewing Equation 2, preferential interactions affect protein retention. If the interactions of the salt with the protein ($\xi_{s,P}$) and the support ($\xi_{s,I}$) are negative, the interaction with the complex is also negative ($\xi_{s,C}$). Since total surface area exposed to the solvent upon complex formation is minimized, we postulate that

$$|\xi_{s,C}|\frac{M_C}{M_s} < |\xi_{s,P}|\frac{M_P}{M_s} + n\,|\xi_{s,I}|\frac{M_I}{M_s} \tag{4}$$

Therefore, if the interactions are negative, $\partial \ln K/\partial \ln a_s$ is positive and complexation is favored. In contrast, when $\xi_{s,C}$,$\xi_{s,P}$ and $\xi_{s,I}$ are positive, $\partial \ln K/\partial \ln a_s$ is negative, promoting desorption. If the interaction parameters of the mobile phase salt with the protein and support have opposite signs, adsorption will depend on the relative magnitudes of the interactions.

To study interaction parameters and protein retention, two methods were employed: densimetric techniques and isocratic elution studies. Experimental procedures are outlined in the following section.

Materials and Methods

Materials Sodium chloride, sodium sulfate and sodium iodide were obtained from Mallinckrodt (Paris, KY). Sodium thiocyanate was obtained from Aldrich (Milwaukee, WI). All salts were analytical reagent grade. Analytical grade ammonium sulfate and analytical grade TRIS (tris(hydroxymethyl)aminomethane) were purchased from Serva Fine Biochemicals (Heidelberg, Germany). Analytical grade ammonium phosphate (monobasic) was purchased from Mallinckrodt, Inc. (Paris, KY). Grade I chicken egg white lysozyme and horse heart myoglobin were purchased from Sigma Co. (St. Louis, MO). Tsk-gel TOYOPEARL HW-65S and Tsk-gel Butyl-TOYOPEARL 650S were gifts from TosoHaas (Philadelphia, PA). Tsk-gel TOYOPEARL HW-65S is a hydrophilic vinyl polymer support with an average particle diameter of 30 μm and pore size of 1000 angstroms. Tsk-gel Butyl-TOYOPEARL 650S is a butyl-derivatized support, using Tsk-gel TOYOPEARL HW-65S as the base sorbent. The butyl support has the same pore size and particle diameter as Tsk-gel TOYOPEARL HW-65S.

Support Preparation The Tsk-gel TOYOPEARL supports were prepared for use by removing fines and particulates. Details on the procedure are given in reference (*28*).

Chromatography Techniques

Instrumentation: Eluent was pumped into the column via the LDC/Milton Roy minipump (Milton Roy, Riviera Beach, FL) and the SSI Lo-Pulse pulse damper from Scientific Systems (State College, PA). The 50 cm x 8 mm i.d. stainless steel column was packed with Tsk-gel TOYOPEARL HW-65S by attaching the column to a 250-ml packing bulb and packing at an average flowrate of 3.5 ml/min of support slurried in an aqueous solution of 0.5 M $(NH_4)_2SO_4$ and 20 mM TRIS. The 50 cm x 8 mm i.d. stainless steel Tsk-gel Butyl-TOYOPEARL 650S column was similarly packed. A model 7125 Rheodyne Injector (Cotati, CA) was used to inject the sample. The detectors were a differential refractometer manufactured by Waters Chromatography Division (Milford, MA), and an Isco (Lincoln, NE) model 228 absorbance detector. The recorder was a Linear 1200 dual pen chart recorder (Linear Instruments Corp., Reno, NV).

Chromatography: Isocratic lysozyme retention studies were conducted for both columns. Mobile phases were the ammonium salts of acetate, chloride, iodide, phosphate, and sulfate in 20 mM TRIS at concentrations ranging from 0.0 M to 1.4 M ammonium sulfate. Column temperature was ambient (21° C) and pH = 8. The flow rate was 1.0 ml/min (pump setting = 15) which corresponds to a linear velocity of 2.0 cm/min based on the empty column cross-sectional area. The sample injection volume was 20 µl, and the protein concentration in the sample was 3 mg/ml. The absorbance detector was set at a wavelength of 280 nm with the linear recorder setting of 1 V and chart speed of 0.25 cm per minute.

Densimetric Techniques Preferential interaction parameters were measured using the densimetric techniques of Lee, Gekko, and Timasheff (*29*). All measurements were made at 20.00 ± 0.02° C using a Mettler DA-210 density meter (Mettler Instruments Co., Hightstown, NJ). Temperature was controlled with a Lauda RMS-20 water bath. Interaction parameters for Tsk-gel TOYOPEARL HW-65S with 0.5 M, 1.0 M and 1.5 M ammonium sulfate and with 1.5 M ammonium chloride were determined. A 20 mM TRIS buffer was used for all trials (pH = 8.2). Details of the densimetric techniques used are given in reference (*28*).

Results and Discussion

Densimetric Studies Densimetric studies were used to determine the interaction parameters of ammonium sulfate with TOYOPEARL HW-65S. Confidence limits of 95% are reported for all values. At 0.5 M ξ_s was -0.00977 ± 0.0109 and decreased to -0.0553 ± 0.009 at 1.0 M. The value of the interaction parameter decreased further to -0.0760 ± 0.0109 when the ammonium sulfate concentration was increased to 1.5 M. The interaction parameters of ammonium sulfate with the support are negative at each concentration studied. Since ξ_s is negative, adsorption is favored, which has been shown to be true for a variety of proteins on HIC supports.

The interaction parameter also becomes more negative with increasing salt concentration. This trend is nonlinear and tends to plateau at the highest concentrations. Similar behavior was shown for CsCl and NaCl by Reisler, Haik and Eisenberg (*30*), and for Na_2SO_4 by Arakawa and Timasheff (*31*).

Measurements were also taken to determine the preferential interaction parameter of ammonium chloride with TOYOPEARL HW-65S. The value of ξ_s was -0.0177 ± 0.011 at 1.5 M NH_4Cl. Compared to ξ_s for ammonium sulfate at 1.5 M (-0.0760), the value for ammonium chloride is much less negative. This trend is expected based on the lyotropic series, since the chloride anion is less effective than the sulfate anion in promoting HIC retention.

Experimental results in the literature show that preferential interaction parameters with proteins follow similar trends to those found for supports (see Figure 1) (*31,32*). The interaction parameters of polar kosmotropes with Bovine Serum Albumin are negative and linearly follow the anion lyotropic ranking (*33*). The $\xi_{s,P}$ for the chaotrope, KSCN, was slightly positive. Initial results with supports indicate that interaction parameters of salts with supports vary with anionic lyotropic ranking, as well.

The preferential interaction parameters have important effects in HIC retention as predicted by Wyman's linkage theory. Since ξ_s of polar kosmotropes with supports and proteins are negative, adsorption is favored in mobile phases consisting of salts such as ammonium sulfate and sodium acetate (*2,34*). On the other hand, ξ_s of chaotropes with supports and proteins are positive and desorption is favored. In practice, chaotropes such as guanidine HCl have been used to desorb proteins from HIC supports (*35*).

Finally, based on our results with supports and data from the literature with proteins, $\xi_{s,P}$ and $\xi_{s,I}$, follow the lyotropic series. Therefore, at a given concentration, chromatographic retention should show similar behavior. Chromatography retention studies were conducted in a variety of mobile phases to assess lyotropic effects.

Chromatography Studies Lysozyme retention was measured on TOYOPEARL HW-65S in mobile phases of ammonium salts of sulfate, acetate, chloride and iodide in a 20 mM TRIS buffer. Representative peaks at 1.3 M salt are illustrated in Figure 2 and results are given in Table I. As shown in Figure 3, retention increases sharply with increasing ammonium sulfate concentration. Retention increases in ammonium acetate and ammonium chloride mobile phases were less pronounced. Retention in ammonium iodide was virtually unchanged as the concentration of the salt in the mobile phase increased.

Table I. Retention of Lysozyme on Tsk-gel TOYOPEARL HW-65S

Salt	Concentration (M)	Replicates	Retention Volume (ml) (95% Confidence Limits)
$(NH_4)_2SO_4$	0.5	3	19.18 ± 0.15
	0.6	3	19.22 ± 0.15
	0.8	3	20.13 ± 0.15
	1.0	3	21.84 ± 0.15
	1.1	3	23.39 ± 0.15
	1.2	3	27.84 ± 0.15
	1.3	3	31.15 ± 0.15
$NH_4C_2H_3O_2$	0.5	3	19.97 ± 0.15
	1.0	3	20.19 ± 0.15
	1.1	3	20.80 ± 0.15
	1.2	2	21.84 ± 0.19
	1.3	3	22.05 ± 0.15
NH_4Cl	0.5	3	19.20 ± 0.15
	0.8	3	20.35 ± 0.15
	1.0	3	19.97 ± 0.15
	1.1	3	20.05 ± 0.15
	1.2	3	19.58 ± 0.15
	1.3	3	20.58 ± 0.15
NH_4I	0.5	3	19.39 ± 0.15
	1.0	2	19.60 ± 0.19
	1.3	3	19.40 ± 0.15
	3.0	2	19.10 ± 0.19

The capacity factor, k′, of lysozyme, is calculated by:

$$k' = \frac{V_R - V_o}{V_o} \qquad (5)$$

where V_R is retention volume of the protein and V_o is the retention volume of a non-interacting solute. By plotting capacity factor of lysozyme at a given concentration versus the lyotropic number, the dependence of retention on the lyotropic number is shown in Figure 4. This dependence is predicted by the trends shown for preferential interaction parameters,

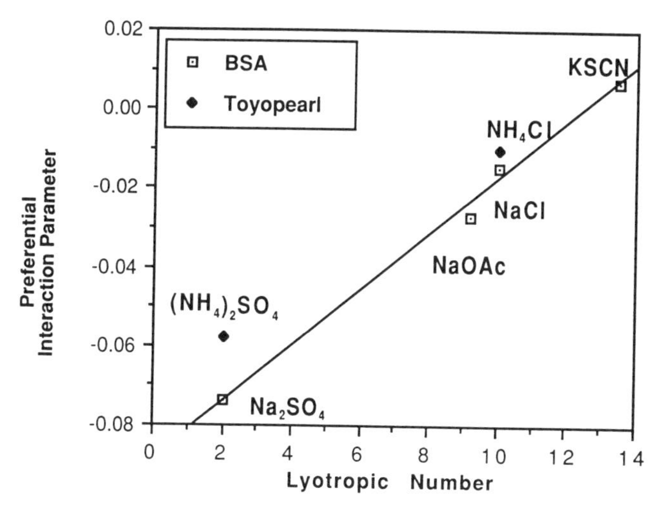

Figure 1. Preferential interaction parameter versus lyotropic number. Data from references [31] and [32].

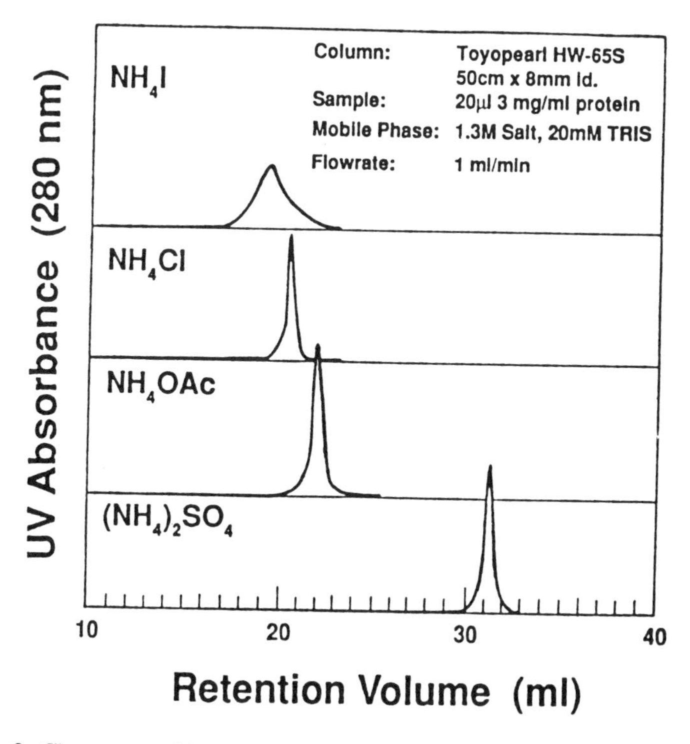

Figure 2. Chromatographic retention of lysozyme on Toyopearl HW-65S in 1.3 M ammonium salt mobile phases.

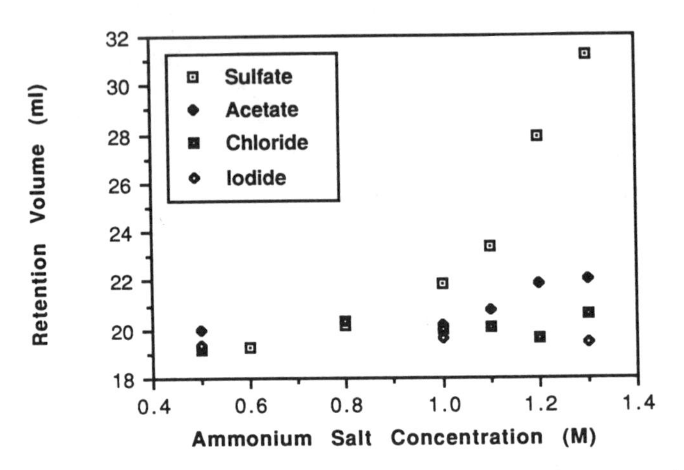

Figure 3. Lysozyme retention on Toyopearl HW-65S versus salt concentration.

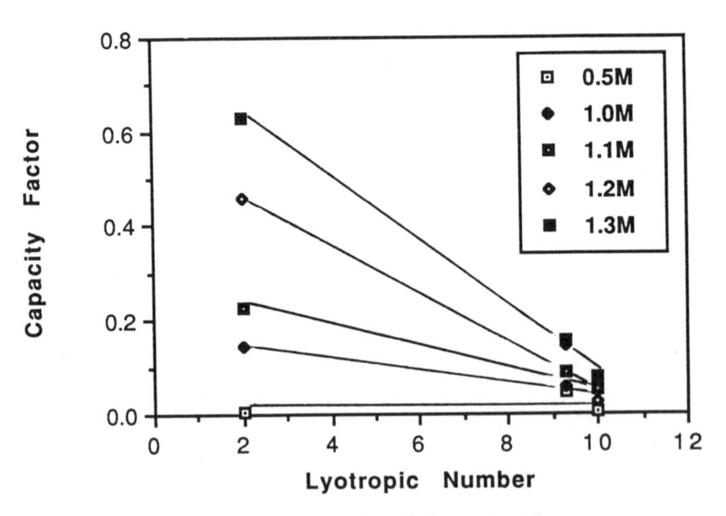

Figure.4. Lysozyme capacity factor versus lyotropic number.

and is shown for concentrations ranging from 1.0 to 1.3 M. Each of the isocratic lines shown in Figure 4 approximately approach a common intersection and the slope of the line becomes more negative with increasing concentration. The change in slope m, can be modeled by:

$$m = n_1 (c)^{n_2} \quad (6)$$

where c is salt molarity, (M), and n_1 and n_2 are constants. The model assumes that the slope of k′ versus the lyotropic number, N_x approaches zero as the concentration of salt in the eluting buffer decreases.

To approximate the constants n_1 and n_2, slopes of k′ vs N_x were determined at 1.0 M and 1.3 M salt. Using k′ = 0.6309 for lysozyme in 1.3 M ammonium sulfate and k′ = 0.0775 in 1.3 M ammonium chloride, m (1.3 M) = -0.0692. At 1.0 M, k′ = 0.1435 in ammonium sulfate, k′ = 0.0456 in ammonium chloride and m (1.0 M) = -0.01224. Substituting the values at 1.0 M and 1.3 M into Equation 6 and solving for a and n yields: n_1 = -0.01224 and n = 6.603. Therefore, the slope of k′ vs N_x varies as m = -0.01224 $M^{6.603}$.

To find the common intercept, the intersection of the isocratic lines at 1.0 and 1.3 M was determined to be at N_x = 10.557 and k′ = 0.03876. The capacity factor may now be determined as a function of salt molarity and lyotropic number as:

$$k' = [-0.01224\, c^{6.603}]\, [N_x - 10.57] + 0.03876 \quad (7)$$

Table II shows a comparison of actual versus calculated values of k′ for lysozyme on TOYOPEARL HW-65S.

Table II. Calculated and Experimental Values of k′ and V_R of Lysozyme In Ammonium Salt Mobile Phases

Salt	Concentration (M)	k′	Estimated k′	V_R (ml)	Estimated V_R (ml)
$(NH_4)_2SO_4$	0.5	0.004	0.040	19.18	19.86
	0.6	0.006	0.042	19.22	19.91
	0.8	0.054	0.063	20.13	20.30
	1.0	0.143	0.143	21.84	21.84
	1.1	0.225	0.458	23.39	23.59
	1.2	0.458	0.388	27.84	26.51
	1.3	0.631	0.631	31.15	31.15
$NH_4C_2H_3O_2$	0.5	0.046	0.039	19.97	19.84
	1.0	0.057	0.054	20.19	20.14
	1.1	0.089	0.068	20.80	20.40
	1.2	0.143	0.090	21.84	20.83
	1.3	0.154	0.126	22.05	21.52
NH_4Cl	0.5	0.005	0.039	19.20	19.84
	0.8	0.065	0.040	20.35	19.87
	1.0	0.046	0.046	19.97	19.97
	1.1	0.050	0.052	20.05	20.08
	1.2	0.025	0.061	19.58	20.27
	1.3	0.077	0.077	20.58	20.58

Investigations were also carried out to study retention of myoglobin on Butyl-TOYOPEARL 650S in the ammonium mobile phases. Initial studies with lysozyme were conducted, but lysozyme adsorbed very strongly to the butyl support, and myoglobin, which has a lower surface hydrophobicity was used instead. Results are summarized in Table III. Retention trends were similar, and as illustrated in Figures 5 and 6, myoglobin was retained quite strongly at moderate concentrations of ammonium sulfate. Slight retention increases were found in high concentrations of ammonium phosphate, ammonium acetate and ammonium chloride. Virtually no changes in retention were obtained for ammonium iodide. Plots of capacity factor versus lyotropic number again showed dependence of retention on the anionic lyotropic ranking (Figure 7). By following the procedure used to develop Equations 6-7, similar expressions were obtained for myoglobin on butyl-TOYOPEARL. Retention behavior may be estimated as a function of salt concentration and anionic lyotropic number. Calculated and experimental values are summarized in Table IV, with good agreement.

Table III. Retention of Myoglobin on Tsk-gel Butyl-TOYOPEARL 650S

Salt	Concentration (M)	Replicates	Retention Volume (ml) (95% Confidence Limits)
$(NH_4)_2SO_4$	0.1	3	19.03 ± 0.15
	0.4	3	18.76 ± 0.15
	0.5	3	19.32 ± 0.15
	0.6	3	19.32 ± 0.15
	0.8	3	20.80 ± 0.15
	1.0	3	23.30 ± 0.15
	1.1	3	25.93 ± 0.15
	1.2	3	32.26 ± 0.15
	1.3	2	44.16 ± 0.19
$NH_4H_2PO_4$	0.5	2	19.22 ± 0.19
	1.0	3	20.20 ± 0.15
	1.2	3	22.18 ± 0.15
	1.3	3	25.30 ± 0.15
$NH_4C_2H_3O_2$	0.5	3	19.04 ± 0.15
	1.0	3	19.22 ± 0.15
	1.2	3	19.37 ± 0.15
	1.3	3	19.36 ± 0.15
	1.4	3	19.74 ± 0.15
NH_4Cl	0.5	3	18.66 ± 0.15
	1.0	3	18.66 ± 0.15
	1.2	3	18.72 ± 0.15
	1.3	3	18.82 ± 0.15
	1.4	3	18.90 ± 0.15
NH_4I	0.5	3	18.61 ± 0.15
	1.0	3	18.50 ± 0.15
	1.4	4	18.54 ± 0.13
	3.0	2	18.13 ± 0.19

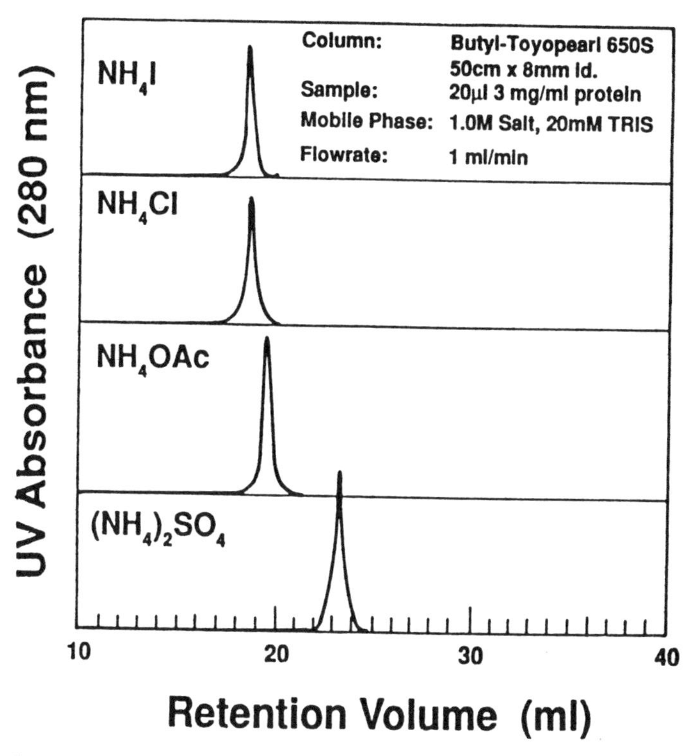

Figure 5. Chromatographic retention of myoglobin on Butyl-Toyopearl 650S in 1.0 M ammonium salt mobile phases.

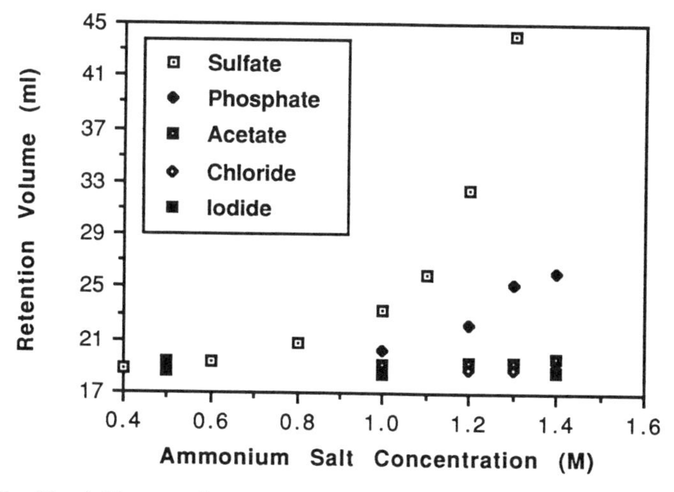

Figure. 6. Myoglobin retention on Butyl-Toyopearl 650S versus salt concentration. (Reproduced with permission from Reference 28. Copyright 1989 American Institute of Chemical Engineers.)

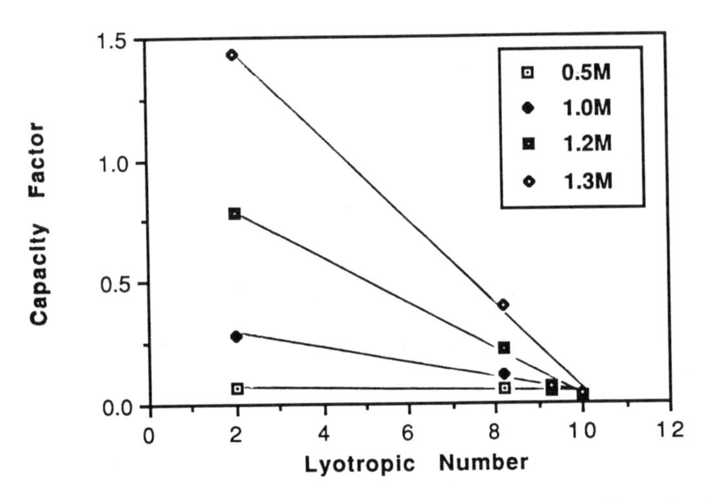

Figure 7. Myoglobin capacity factor versus lyotropic number. (Reproduced with permission from Reference 28. Copyright 1989 American Institute of Chemical Engineers.)

Table IV.
Calculated and Experimental Values of k′ and V_R of Myoglobin In Ammonium Salt Mobile Phases

Salt	Concentration (M)	k′	Estimated k′	V_R (ml)	Estimated V_R (ml)
$(NH_4)_2SO_4$	0.1	0.050	0.027	19.03	18.62
	0.4	0.035	0.028	18.76	18.64
	0.5	0.066	0.030	19.32	18.68
	0.6	0.066	0.037	19.32	18.80
	0.8	0.147	0.088	20.80	19.73
	1.0	0.285	0.285	23.30	23.30
	1.1	0.430	0.505	25.93	27.29
	1.2	0.779	0.866	32.26	33.83
	1.3	1.436	1.436	44.16	44.16
$NH_4H_2PO_4$	0.5	0.060	0.028	19.22	18.64
	1.0	0.114	0.087	20.20	19.70
	1.2	0.223	0.221	22.18	22.14
	1.3	0.395	0.353	25.30	24.52
$NH_4C_2H_3O_2$	0.5	0.050	0.028	19.04	18.63
	1.0	0.060	0.052	19.22	19.07
	1.2	0.068	0.108	19.37	20.08
	1.3	0.068	0.162	19.36	21.07
	1.4	0.089	0.245	19.74	22.57
NH_4Cl	0.5	0.029	0.027	18.66	18.62
	1.0	0.029	0.029	18.66	18.66
	1.2	0.033	0.034	18.72	18.74
	1.3	0.038	0.038	18.82	18.82
	1.4	0.042	0.045	18.90	18.94

Conclusions

HIC adsorption behavior reflects preferential interaction parameters as predicted by Wyman's linkage theory. Preferential interactions of polar kosmotropes with supports and proteins are negative, promoting HIC retention. Preferential interaction parameters of salts with a common cation are linearly dependent on the anionic ranking of the lyotropic series at a given salt concentration. As a result, the protein capacity factor in HIC is a linear and decreasing function of the lyotropic number at a given mobile phase salt concentration. The dependence of k′ on the lyotropic ranking was observed on both hydrophilic and hydrophobic stationary phases. Therefore, the interaction parameter correlates the salt type and concentration to protein adsorption. The impact of this work on separations design is that protein retention can be anticipated, given the salt type and concentration used in the eluent.

Acknowledgments

The material in this work was supported by the National Science Foundation, Grant No. ECE-8613167. We thank TosoHaas (Philadelphia, PA), for donating Tsk-gel TOYOPEARL HW-65S and Tsk-gel Butyl-TOYOPEARL 650S; Dr. Istvan Mazsaroff for many scholarly discussions; and Rick Hendrickson and Cheryl Benko for technical assistance in setting up the chromatography runs.

Literature Cited

1. Barth, H.; Barber, W.; Lochmuller, C.; R. Majors, R.; F. Regnier, F., *Anal. Chem.* 1988, *60*, 387R-429R.
2. G. Osthoff, G.; Louw, A. I.; Visser, L., *Anal. Biochem.* 1987, *164*, 315-319.
3. Fausnaugh, J.; Regnier, F., *J. Chromatogr.* 1986, *359*, 131-146.
4. Porath, J., *Biotechnol. Prog.* 1987 *3*, 14-21.
5. Cacace, M.; Sada, A., *J. Chromatogr.* 1986, *376*, 103-109.
6. Tiselius, A., *Chem. Engr. News* 1949, *27*, 1041-1044.
7. Gillam, I.; Millward, S.; Blew, D.; Tigerstrom, M. V.; Wimmer, E.; Tener, G., *Biochemistry* 1967, *6*, 3043.
8. Weiss, H.; Bucher, T., *Eur. J. Biochem.* 1970, *17*, 561.
9. Er-el, Z.; Zaidenzaig, Y.; Shaltiel, S., *Biochem. Biophys. Res. Commun.* 1972, *49*, 383.
10. Hofstee, B., *Anal. Biochem.* 1973, *52* 430-448.
11. Yon, R., *Biochem. J.* 1972, *126*, 765-767.
12. Hjerten, S., *J. Chromatogr.* 1973, *87*, 325-331.
13. Goheen, S.; Engelhorn, S., *J. Chromatogr.* 1984, *317*, 55-65.
14. Melander, W.; Corradini, D.; Horváth, Cs., *J. Chromatogr.* 1984, *317*, 67-85.
15. Schmuck, M.; Gooding, K.; Gooding, D., *Liq. Chrom.* 1985, *3*, 9.
16. El Rassi, Z.; Horváth, Cs., *J. Liq. Chromatogr.* 1986, *9*, 3245-3268.
17. Hammond, P. M.; Sherwood, R. F.; Atkinson, T.; Scawen, M. D., *Chim. Oggi.* 1987, 157-59.
18. Melander, W.; Horváth, Cs., *Arch. Biochem. Biophys.* 1977, *183*, 200-215.
19. Hofmeister, F., *Arch. Exp. Pathol. Pharmakol.* 1888, *24*, 247-263.
20. Washabaugh, M.; Collins, K., *J. Biol. Chem.* 1986, *261*, 12477-12485.
21. Ahmad, F.; Bigelow, C., *J. Protein Chem.* 1986, *5*, 355-367.
22. Arakawa, T., *Arch. Biochem. Biophys.* 1986, *248*, 101-105.
23. Potter, R. L.; Lewis, R. V., *High-Perform. Liq. Chromatogr.* 1986, *4*, 1-44.
24. Wu, S.; Benedek, K.; Karger, B., *J. Chromatogr.* 1986, *359*, 3-17.
25. Wu, S.; Figueroa, A.; Karger, B., *J. Chromatogr.* 1986, *371*, 3-28.
26. Mazsaroff, I.; L. Varady, L.; G. Mouchawar, G.; F. Regnier, F., (Submitted to *J. Chromatogr.*) 1989
27. Wyman, J., *Adv. Prot. Chem.* 1964, *19*, 223-286.
28. Roettger, B.; Myers, J.; Ladisch, M.; Regnier, F., *Biotech. Prog.* 1989, *5*, 79-88.
29. Lee, J.; Gekko, K.; Timasheff, S., *Methods Enzym.* 1979, *61*, 26-49.
30. Reisler, E.; Haik, Y.; Eisenberg, H., *Biochem.* 1977, *16*, 197-203.
31. Arakawa, T.; Timasheff, S., *Biochem.* 1984, *23*, 5912-5923.
32. Arakawa, T.; Timasheff, S., *Biochem.* 1982, *21*, 6545-6552.
33. Voet, A., *Chem. Rev.* 1937, *20* 169-179.
34. Berkowitz, S. A., *Anal. Biochem.* 1987, *164*, 254-260.
35. Iny, D.; Sofer, J.; Pinsky, A., *J. Chromatogr.* 1986, *360*, 437-442.

RECEIVED December 28, 1989

Chapter 7

Anion Exchange Stationary Phase for β-Galactosidase, Bovine Serum Albumin, and Insulin

Separation and Sorption Characteristics

Michael R. Ladisch, Richard L. Hendrickson, and Karen L. Kohlmann

Laboratory of Renewable Resources Engineering, Potter Engineering Center, Purdue University, West Lafayette, IN 47907

Anion exchangers have widespread utility in protein purification where protein fractionation is achieved using salt gradients. The literature reports that the retention time of various proteins can be correlated with their respective distribution coefficients at a given ionic strength. Experiments in our laboratory show that bovine insulin, and β-galactosidase are readily separated. These results were achieved with a gradient of 0 to 0.5 M NaCl using a polymeric macroporous stationary phase derivatized with DEAE groups. Fundamental sorption characteristics of bovine serum albumin (BSA) and β-galactosidase showed loadings on the order of 100 and 220 mg protein/g sorbent, respectively. The loading of BSA on the support at various salt concentrations was studied as well as the loading of BSA on support which had been preloaded with β-galactosidase. The qualitative interpretation of the BSA/β-gal loading data with the Craig distribution model indicates that BSA retention could be affected by high β-gal if chromatographic operations were carried out at overload conditions.

Anion exchange media are widely used in the chromatography of proteins. Examples include: fractionation of immunoglobulins from human serum; isolation of factors VIII, IX, and X during processing from blood plasma; and purification of numerous enzymes (*1*). The separations costs of many biotechnological products are a significant fraction of the production costs, with chromatography often being the most expensive unit operation for a given separation sequence (*2*). Consequently, significant research efforts are being directed to the development and scale-up of improved chromatography systems. There appears to be a need for approaches which combine existing theory with measurable separation characteristics of proteins in a manner which is readily applicable to designing and optimizing large scale chromatography columns.

We report studies on Toyopearl DEAE 650M which is an anion exchanger consisting of a vinyl polymer. The structure is highly substituted with hydroxyl groups to make it hydrophilic and thereby minimizes non-specific protein adsorption. This material, derivatized with diethyl amino ethyl (DEAE) groups, gives an anion exchanger with an exchange capacity of 0.10 ± 0.02 meq/ml of (wet) stationary phase. This material has previously been demonstrated for the ion exchange chromatography of β-galactosidase (β-gal) (*3*), ovalbumin,

0097–6156/90/0427–0093$06.00/0

lactoglobulins, myoglobin, ribonuclease, and cytochrome C (*4*). We show separation capabilities of this stationary phase may also be applicable to fractionation of bovine insulin from β-galactosidase, bovine insulin A and B chains, as well as bovine serum albumin (BSA).

The interesting separation characteristics of this material indicated that further studies on equilibrium characteristics were worthwhile. Consequently, we report the equilibrium capacities of the Toyopearl DEAE 650M with respect to BSA and β-gal. The DEAE 650M was chosen since it is relatively hydrophilic, and has large pores with an exclusion limit corresponding to 5 million. BSA and β-gal were chosen as model adsorbates due to size (60,000 and 520,000, respectively) and availability as well as being representative of larger proteins which might be encountered in a bioseparations process. In addition, when insulin is produced as a fusion protein, a linkage protein can be β-gal. Characteristics which were probed included retention of enzyme activity upon chromatography, the effect of the salt gradient on protein elution, and comparison of batch equilibrium results to protein loading obtained in column runs.

Scale-up of a protein purification is concerned with obtaining purity at high throughput and loading. If the protein to be isolated is in solution with other adsorbable components at relatively high concentrations, overload phenomena may be encountered in attempting to process large amounts of protein during a single run. This is sometimes referred to as "mass-overload" and can result in different retention behavior and peak bandwidths than what would be encountered at small sample concentration conditions. According to some researchers, there is "considerable debate among experts as to whether analytical scale separations can provide useful information for the design of corresponding overload separations (*5*)." It is conceivable that commercial scale operation could quickly push a separation to overload conditions. For example, the development of a microbial system which facilitates recombinant protein secretion could result in low levels of a product in a fermentation broth having relatively high overall protein concentration. Separation of the product from other extracellular proteins could change as column loading increases or decreases, since changes in adsorbed protein could affect the retention time of the product peak. Since most separations are developed from research initiated and carried out at an analytical or bench scale, a modeling approach is needed which uses analytical or bench scale results to anticipate changes in separation behavior upon scale-up. In this context, the Craig distribution model appears to be a simple but appropriate starting point (*5-7*), since it gives a useful representation of the distribution of two solutes where the presence of one may affect the other.

Craig Distribution Model

Overload conditions in liquid chromatography results from operating in the nonlinear region of the competitive adsorption isotherms of at least one of the components involved. For a binary system consisting of components A and B, a Craig distribution model can be used to conceptualize overload conditions (*7*). This approach assumes that local equilibrium is attained at each plate of the column, and that the solute is distributed between two phases in a countercurrent manner. A basic approach to modeling overload conditions can be formulated in terms of the observed distribution coefficients, K_A and K_B for components A and B, respectively:

$$K_A = \frac{A_{stat}}{A_{mobile}} = \frac{[C_A]_{stat} V_{stat}}{[C_A]_{mobile} V_{mobile}} \tag{1}$$

$$K_B = \frac{B_{stat}}{B_{mobile}} = \frac{[C_B]_{stat} V_{stat}}{[C_B]_{mobile} V_{mobile}} \tag{2}$$

The subscripts "stat" and "mobile" refer to the stationary and mobile phases. A and B refer to the total mass of the components, while C_A and C_B denote concentrations, and V_{stat} and V_{mobile} would be the volumes associated with the stationary and mobile phases, respectively. The observed values of the distribution coefficients can be expressed in terms of the ideal distribution coefficients K_A^o and K_B^o, corrected for effects of one component on another using the parameter functions f (A,B) and g (A,B). The resulting equations are:

$$K_A = K_A^o \; f(A,B) \quad (3)$$

$$K_B = K_B^o \; g(A,B) \quad (4)$$

The functions f (A,B), g (A,B) must follow

$$\frac{\partial f}{\partial A} > 0, \; \frac{\partial f}{\partial B} \leq 0, \; \frac{\partial g}{\partial A} \leq 0, \; \frac{\partial g}{\partial B} > 0$$

for competitive sorption. Many functional forms obey these criteria, e.g., the Langmuir isotherm, the Fritz-Schluender isotherm, etc. The total mass A_{tot} or B_{tot}, is given by:

$$A_{tot} = A_{stat} + A_{mobile} \quad (5)$$

$$= [C_A]_{stat} V_{stat} + [C_A]_{mobile} V_{mobile} \quad (6)$$

and

$$B_{tot} = B_{stat} + B_{mobile} \quad (7)$$

$$= [C_B]_{stat} V_{stat} + [C_B]_{mobile} V_{mobile} \quad (8)$$

In the special case that $V_{stat} = V_{mobile} = 1$, Equations 1, 2, 3, 4, 6, and 8 can be expressed in terms of values for concentration, i.e.,

$$K_A = \frac{[C_A]_{stat}}{[C_A]_{mobile}} \quad (9)$$

$$K_B = \frac{[C_B]_{stat}}{[C_B]_{mobile}} \quad (10)$$

$$K_A = K_A^o \; f(A,B) \quad (11)$$

$$K_B = K_B^o \; g(A,B) \quad (12)$$

$$A_{tot} = [C_A]_{stat} + [C_A]_{mobile} \quad (13)$$

$$B_{tot} = [C_B]_{stat} + [C_B]_{mobile} \quad (14)$$

The values of K_A^o, K_B^o, f (A,B) and g (A,B) are obtained from equilibrium data. Solution of Equations 9 through 14 in an iterative calculation procedure for each stage, where components A and/or B are present, results in profiles useful for simulating chromatograms.

A key difference in calculating chromatograms for non-overload chromatography as opposed to overload chromatography is that the distribution coefficient K is assumed to be constant for non-overload conditions, while K changes during the course of the separation in overload conditions. The utility of the plate model is illustrated by a hypothetical example where component B elutes more rapidly than component A. If component B affects retention of A but A does not affect retention of B there would be minimal interference effects. In comparison, if A affects elution of B, then the retention and relative concentration of B can change quite significantly depending on the value of the interaction embodied in g (A,B). If A and B both affect each other, significant peak skewing and altered peak retention can occur.

Materials and Methods

Proteins. The proteins used in this work were β-galactosidase from *Escherichia coli* (Grade VIII, 80% protein), bovine serum albumin (BSA - Fraction 5 powder, 98% to 99% albumin), and bovine insulin (from bovine pancreas, zinc content of 0.5%), insulin chain A (oxidized from bovine insulin, ammonium salt purity >80%, contains less than 1% chain B), and chain B (oxidized from bovine insulin, free acid, purity >80%, contains less than 1% chain A); all were from Sigma. Bovine serum albumin has a molecular weight of 66,000, while that of monomeric insulin is 6100. A trace amount of EDTA was added to the insulin solution to help dissolve the insulin. Insulin can exist as a monomer, dimer, or hexamer (*8*). The structures of bovine and human insulin differ on the A chain by the amino acids Ala and Val at positions 8 and 10 (for bovine insulin), vs. Thr and Ile (for human insulin) at positions 8 and 10; and the terminal amino acid in the B chain with Ala (bovine) replacing Thr (human)(*9*).

Protein Assays. Protein assays were carried out using the bicinchoninic method (BCA) (*10*) following the procedure of Sigma (TPRO-562). The BCA method consists of preparing a protein reagent by adding one part of a 4% copper sulfate ($5H_2O$) solution to 50 parts (vol/vol) bicinchoninic acid solution. This protein reagent was freshly prepared each day. Standard curves were based on triplicate analysis of each of 7 BSA standards. The assay consisted of adding 2 ml of protein reagent to 100 µl of sample, incubating the sample for 30 min at 37°C, cooling to room temperature for 10 minutes, and reading absorbance in a spectrophotometer at 562 nm using micro-cuvettes. The readings were corrected against a blank with 2 ml protein reagent and 100 µl buffer. Samples of unknown protein concentration were run at several dilutions. The dilutions were then adjusted for a repeat set of analyses to bring the sample in the linear range of the standard curves. Analyses were carried out in duplicate. Protein concentrations were also measured by uv absorbance at 280 nm for appropriately diluted samples. The BCA assay was standardized using bovine serum albumin, and the protein determination by absorbance at 280 nm was standardized using β-gal and BSA standards.

Electrophoresis. Electrophoresis of the protein samples was run to provide a second analytical method (other than chromatography) to detect different proteins in the sample. The standard procedure was to cast polyacrylamide-SDS gels, load the protein with tracking dye onto the gel, run for ca. 30 to 45 minutes at about 600 volts, remove the gel, stain with Coomassie blue dye, and then destain. Since SDS was used, multimeric proteins would be dissociated into their basic subunits. The standards (from Sigma) used as electrophoresis marker proteins were: α-lactalbumin (MW = 14,200); soybean trypsin inhibitor (MW = 20,000); trypsinogen (MW = 24,000); bovine carbonic anhydrase (MW = 29,000); rabbit muscle glyceraldehyde-3-phosphate dehydrogenase (MW = 36,000); egg albumin (MW = 45,000); and bovine albumin (MW = 66,000). Electrophoresis of β-galactosidase from *E. coli* found it to be relatively pure and of high molecular weight.

The enzyme from *E. coli* is a tetramer having a molecular weight of 520,000 (*11*). An excellent review of the properties of β-galactosidase from microbial sources is given by Richard et al. (*12*). The optimum activity appears to be in the range of pH 7.5 (*12*). In comparison, β-galactosidase from bovine testes has a molecular weight of 68,000 and a pH optimum of 4.3 (*13*). Enzyme activities from both sources are readily measured using o- or p-nitrophenyl-β-galactopyranoside (*13,14*).

Activity Assays. β-Galactosidase activity was analyzed using orthonitrophenyl-β-D-galactopyranoside (ONPG) as the substrate. One enzyme unit will hydrolyze 1.0 micromole per minute of ONPG to o-nitrophenyl and galactose at pH 7.3 and 37° C. The presence of enzyme activity results in hydrolysis of the substrate to give nitrophenyl chromogen which is readily detected by absorbance at 410 nm.

Stationary Phase. The stationary phase being studied is Toyopearl DEAE 650M (from TosoHaas, Philadelphia, PA). The material was prepared for packing by washing with DI water in a 1000 ml graduate cylinder. After swirling, the resin was allowed to settle for 1.5 hours. The liquid above the resin as well as trace amounts of fines remaining suspended in the liquid were decanted; this procedure was repeated. Next, the resin was washed with the 0.2 molar and 0.5 molar sodium chloride, respectively, and decanted as described above. The resin was then washed with three liters of DI water. For each of these wash-settle cycles, the resin was allowed to settle to the 260 ml mark before decanting.

Wet particle size of the DEAE 650M was measured in 5% NaCl using a Zeiss light microscope/image analysis system. Based on replicate analysis of 389 particles, the average particle size ($\pm$ 1 σ) was found to be 60 $\pm$ 10 microns with a particle size range of 42 to 89 microns.

Column Packing. Standard column packing procedures were used to pack the DEAE 650M into a 316 SS column having dimensions of 1.09 x 20 cm long with a measured volume of 18.7 ml. Packing was carried out at ambient temperature, with the stationary phase previously equilibrated in 16.7 mM Trizma buffer (from Sigma) at pH 7.3. The flow rate during packing was 8.7 ml/min which gave a system pressure of about 25 psig. The column was packed at this flow rate for 2 hours, and then capped with standard 10 micron end fittings.

Chromatography. The chromatography system consisted of a dual chamber gradient former (Kontes) with the mixing chamber resting on a stir plate. A 16-160 ml/hr pump (Milton Roy) pumps the mobile phase through a pulse dampening system consisting of a 50 cm 3/8" o.d. SS column half filled with liquid followed by a Lo-Pulse pulse dampener (SSI Instruments). Upon exiting the Lo-Pulse pulse dampener the flow passes through a tee to which a 0-600 psi pressure gauge (WIKA) and an adjustable pressure relief valve (Nupro) is plumbed and finally attached to the sample injector (Model 7125, Rheodyne) and chromatography column. Detection of eluting components was by a model VUV-10 (Varian) detector set at 280 nm, followed by a model 402 differential refractometer (Waters). Detector output signals are monitored by a dual channel chart recorder (Linear).

Gradient Chromatography Conditions. Liquid chromatography for protein separations was carried out using a gradient of 0 M to about 0.5 M sodium chloride. The mobile phase consisted of a buffer 16.7 mM Trizma (pH 7.3) containing the described amount of salt. The gradient was formed using Kontes gradient chambers. At the end of a run, a step change to the initial conditions was effected. Protein was not detected after elution of the last (β-gal) protein peak. This indicates that there was little or no irreversible binding of proteins during the separation process. This is further supported by the fact that the separations were reproducible over the period of this study which used the same column.

The total (α), extra-particle (ε_b), and intra-particle (ε_p) void fractions were determined using 20 µl injections of 1 M NaCl and 2 mg/ml of DNA and measuring retention times. The resulting void fractions obtained were 0.72 (= α), 0.44 (= ε_b), and 0.28 (= $\alpha - \varepsilon_b = \varepsilon_p$). The DNA (Type III from salmon testes) was used to estimate extraparticle void fraction since it exhibits minimal interaction with the stationary phase and has a large molecular weight (approximately 2 million). DEAE 650M has a molecular weight cut-off which has been reported to be as high as 5 million (15). Consequently, a very large excluded solute is needed to obtain an estimate of the extraparticle void fraction. We suspect the DNA partially penetrated the stationary phase, and therefore, gave a somewhat higher than expected void fraction.

Equilibrium Measurements (Batch). Equilibrium studies were carried out for BSA sorption on DEAE 650M using screw-cap culture tubes. The procedure entailed weighing fresh (previously unused) DEAE 650M stationary phase (200 mg, air dried) and placing the stationary phase into 12 screw cap culture tubes. Next, 5 ml of the tris buffer at an appropriate salt concentration was added, and the contents of the tubes were mixed. The caps were put on the tubes and taped in place. The tubes were then placed in a shaking incubator (New Brunswick Scientific, model G24) at 30°C and allowed to equilibrate for 2 hr. After equilibration, 5 ml of a BSA solution (10 mg/ml in tris buffer) was added, and the contents of the tubes were again mixed and capped. Blanks were prepared by the addition of buffer at each salt concentration used in place of the BSA. The tubes were then put back into the incubator for 30 min. After incubation, 1 ml portions were removed and microcentrifuged for 1 min in a microcentrifuge (Beckman, model B). The time zero tubes were prepared by adding 5 ml of the BSA solution to 5 ml of buffer (no contact of the protein with the resin). The absorbance of the supernatant was read by direct injection of 500 µl into the spectrophotometer.

Adsorption of BSA by DEAE 650 M-Micro Scale. Because of the expensive nature of the proteins in the equilibrium studies, it was necessary to scale down amounts used. We began this with the scale down of the adsorption of BSA by the DEAE 650 M support. Ten mg of DEAE 650 M was weighed and placed in a capped microcentrifuge tube (volume of tube 1.5 ml). The tris buffer was added (0.5 ml, 16.7 mM, pH 7.2); the tube was capped and taped horizontally in the shaking incubator to equilibrate (30°C) for 2 h. Following equilibration, 0.5 ml of a 5 mg protein/ml buffer BSA solution was added, and the tubes placed back into the incubator for 30 min. In preliminary experiments it was found that 30 min was sufficient for maximum adsorption of the BSA by the support. Tubes were run in duplicate, and blanks were tubes which had buffer in place of the BSA addition.

Loading of BSA by DEAE 650 M (Support Preloaded with β-Galactosidase). It had been determined in preliminary experiments that approximately 220 mg of β-galactosidase (β-gal) would load per g air dried weight of the DEAE 650 M support. We wished to determine the sorption of BSA by support which had been preloaded with β-gal approximately to this degree. It was difficult to prepare β-gal in solutions of greater than 1 mg protein/ml buffer due to solubility problems. Therefore, the β-gal was loaded onto the support gradually in increments. To accomplish this, 10 mg of air dried DEAE 650 M was weighed and placed in microcentrifuge tubes. The tris buffer (0.5 ml) was added to each tube, and the tubes were placed in the shaking incubator to equilibrate for 2 h. A β-gal solution was prepared at a concentration of 0.8 mg protein/ml tris buffer. After the equilibration of the support, 0.5 ml of the β-gal was added, and the tubes placed in the shaking incubator for 30 min. At this time the tubes were centrifuged for 1.5 min, and 0.8 ml of the supernatant was removed. Approximately 0.5 ml of the supernatant was injected into the detector to determine protein content by adsorption at 280 nm. The remainder of the supernatant was kept for a β-gal activity assay and for a second protein determination by the BCA assay. Next 0.3 ml of tris buffer and 0.5 ml of the β-gal solution were added back to the tubes; these loading steps were repeated until the absorbances at 280 nm showed that the β-gal was not being entirely adsorbed by the support. At this point the tubes were microcentrifuged again, and all the supernatant was removed. In a wash step 1.0 ml of the tris buffer was added, and the tubes were incubated in the shaker for 30 min. The absorbance at 280 nm of the supernatant was read to determine if any protein had leached during washing.

To determine the loading of BSA, 0.5 ml of a 5 mg protein/ml buffer BCA solution was added, and the tubes capped and taped in the shaking incubator for 30 min at 30°C. The tubes were then centrifuged for 1.5 min, and the absorbance read at 280 nm. The β-gal activity assay and the BCA protein assay were also performed. This experiment was repeated three times. Blanks consisted of support which had received incremental addition of buffer in place of the β-gal solution. These blanks did have the BSA addition.

Results and Discussion

Separation of BSA from β-galactosidase is readily achieved using a salt gradient (Figure 1). The elution of β-galactosidase at the latter part of the salt gradient is consistent with a prior result, also for this type of ion exchanger (*3*). Collection of fractions and analysis of the fractions for protein concentration and β-gal activity showed that all of the β-gal activity was in the second peak and there was no β-gal activity associated with the BSA peak (Figure 2). The two peaks eluted at 0.23 and 0.33 M NaCl, respectively. It should be noted that gradient delay volumes are negligible: the sample introduction was begun when the gradient was at the pump inlet, thus eliminating delay in reaching the pump. Micro tubing was used to connect the pump outlet to the column, minimizing dead volume. The shoulders on the β-gal peak may be due to small amounts of other proteins present, since electrophoresis of the starting protein indicated several light bands of lower molecular weight in addition to the main β-gal band.

Another separation examined for this stationary phase was insulin from β-galactosidase (Figure 3) and from insulin chain A and B (Figure 4). Satisfactory resolution is demonstrated, and suggests this anion exchanger to be applicable to smaller polypeptides, as well as large proteins such as β-gal. In these figures, the EDTA peak was identified by comparing its retention time to those of separate injections of only EDTA into the same column.

Equilibrium measurements (batch) were carried out to test BSA loading on DEAE 650M in the presence of increasing amounts of NaCl. The maximum loading observed for an initial concentration of BSA of 5 mg/ml was about 124 mg protein per g stationary phase (dry weight basis) as determined by the equilibrium measurement, (batch) procedure.

The loading of the BSA is sensitive to salt concentration (Figure 5). The minimum loading, at high salt concentration, approaches zero within the experimental error of the microscale adsorption procedure. While desorption in a column chromatography run is achieved at 0.23 M NaCl, the adsorption study shows that loading is already small at 0.1 M NaCl. Hence, the data suggest that there may be a difference between the salt concentration which prevents adsorption, compared to that which gives significant desorption. This phenomena is being investigated further. In the absence of salt, the equilibrium loading of β-galactosidase was found to be about 220 mg/g support (dry weight basis) or approximately 80% higher than the BSA. The β-gal activity assay confirmed the results of the β-gal adsorption. There was

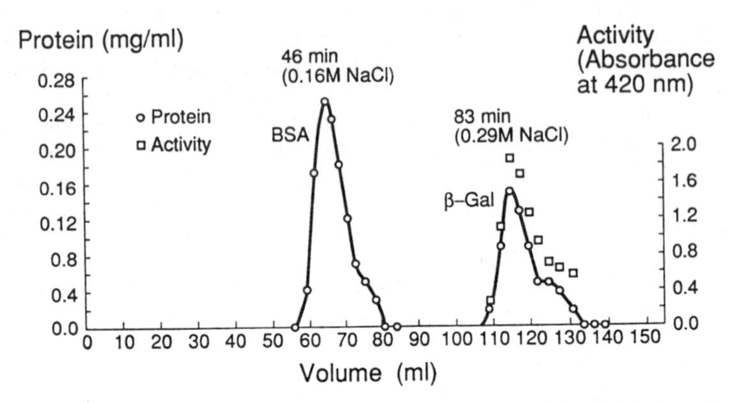

Figure 1. Separation of BSA from β-galactosidase over DEAE 650M. Sample size of 1 ml. Protein concentrations in sample are 2 mg/ml each of BSA and β-gal. Eluent flow rate of 1.38 ml/min. Other conditions as given in text.

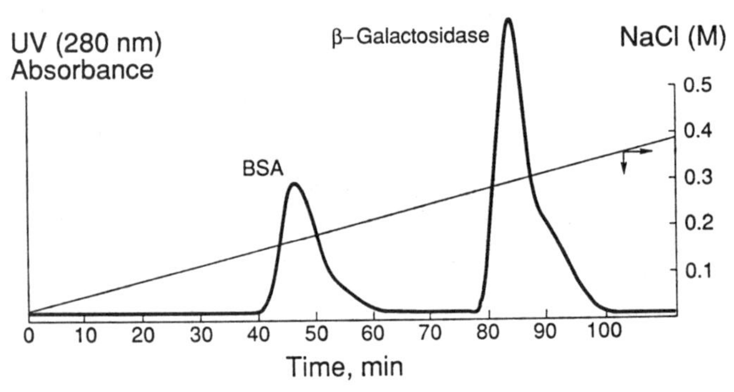

Figure 2. Protein and activity assays show β-gal activity is associated with the second peak.

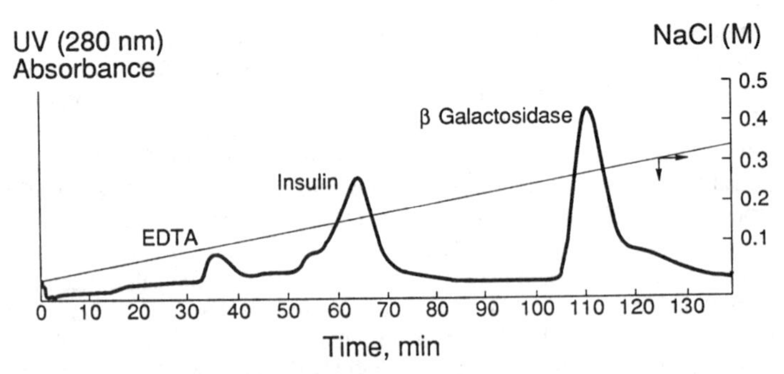

Figure 3. Separation of insulin from β-galactosidase. Injection volume of 1 ml. Protein concentrations in sample of 1 mg/ml each. Eluent, flow rate of 1.4 ml/min. Other conditions given in text.

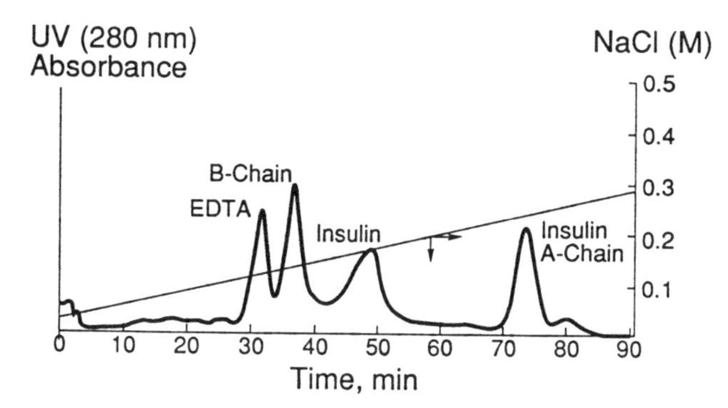

Figure 4. Separation of insulin, and insulin A and B chains. Sample volume of 0.5 ml. Protein concentration in sample of 1 mg/ml of each component. Eluent flow rate of 1.27 ml/min.

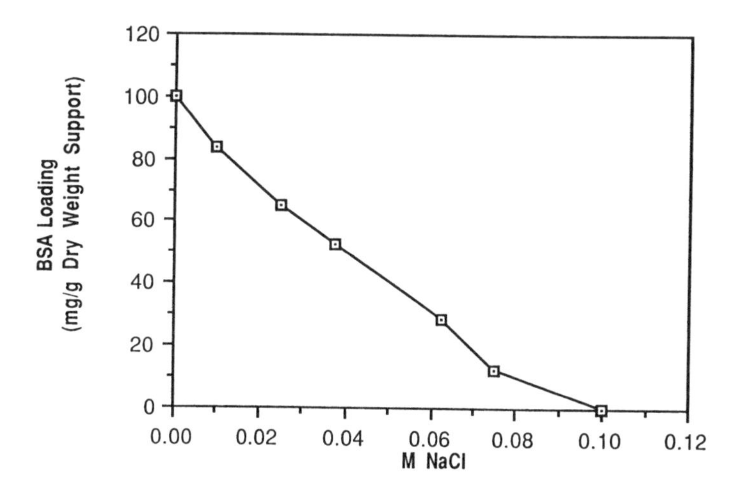

Figure 5. Equilibrium loading of BSA on DEAE 650M at salt concentrations ranging from 0 to 0.1 M NaCl using the micro-scale adsorption technique. Conditions as described in text.

increasing β-gal activity in the supernatant as less of the β-gal loaded onto the support. When loading of the β-gal is attempted in 0.4 M NaCl, the β-gal loading is much lower and is on the order of 10 to 20 mg/g support.

The initial concentration of BSA by adsorption on DEAE 650M- micro scale experiments was 2.6 mg/ml (determined by uv absorption). The maximum loading of BSA at this concentration was 100 to 116 mg/g stationary phase (air dry weight basis) as shown in Table I. The variation in observed loading was due to the small volumes and stationary phase weights used in the experiments, and normal experimental error in measuring the protein concentrations. The loading of the BSA in microscale gave similar results, to the loading of BSA in the larger scale (batch) experiments. As mentioned in the Materials and Methods section, the BSA took about 30 minutes to reach its maximum adsorption in the batch equilibrium measurements. The analogous β-gal measurements required a few hours. While the low solubility of β-gal is a factor (repetitive additions of protein were required to obtain the final loading), the 8-fold higher molecular weight of β-gal should also result in slower diffusion than the BSA and possibly be reflected by a slower attainment of equilibrium.

The results of equilibrium loading of BSA by DEAE 650 M (preloaded with β-gal) is presented in Table I (in the "Sample" columns). After β-gal adsorption, β-gal activity and protein was assayed in the supernatant following the washing step. These analyses indicated that the β-gal was not leaching from the support. In addition, there was no β-gal activity detected in the supernatant following addition of the BSA, indicating that the BSA was not displacing β-gal. Thus, the separation factor for β-gal with respect to BSA under these conditions must be very large (which is also supported by the wide gap between their peaks in Figure 2). The competitive adsorption discussed in the Theory section would be such that (with A as β-gal, and B as BSA) $\partial f/\partial B$ is O under the conditions used here, while $\partial g/\partial A$ is slightly negative so that $K_B < K_B^O$. More experiments are necessary to determine the exact functional forms of f(A,B) and g(A,B).

Table I. Loading of BSA on DEAE 650M (support preloaded with β-gal)

	Trial 1		Trial 2		Trial 3	
	Sample	Blank	Sample	Blank	Sample	Blank
β-gal Loaded (mg/g sorbent)	222	-	221	-	223	
BSA Loaded (mg/g sorbent)	45	116	56	112	29	100

The batch equilibrium studies suggests that adsorption of β-galactosidase could affect subsequent sorption of BSA. We hypothesize that adsorption of β-galactosidase in the chromatography mode of high loadings could affect retention of BSA, if the BSA were present in a mixture with β-gal.

Conclusions

Batch equilibria and chromatography experiments show the anion exchanger Toyo-Pearl DEAE 650M is capable of significant protein loading of large molecules and gives clear separation of several bovine insulin/protein mixtures. In large-scale operation, mass overload phenomena may be encountered. Based on batch studies with BSA/β-gal, the Craig distribution concept indicates altered peak retention could result for BSA, if the concentration of the β-gal were to accumulate to a relatively high concentration on the stationary phase, when the column is run in a typical loading/gradient chromatography mode.

Acknowledgment

This work was supported by a grant from the Rohm and Haas Company, Philadelphia, PA. Wet particle size analysis was made possible by NSF grants CBT-8606522 and CBT-8705121. At Rohm and Haas we thank Dr. Edward Firouztale, Dr. Peter Cartier, and Dr. Andy Kielbania for scholarly discussions and assistance with the research. At Purdue, we particularly thank Dr. Ajoy Velayudhan for scholarly discussions and assistance in careful and thoughtful analysis of the experimental results; and Jill Porter for assistance in the electrophoresis of proteins, as well as suggestions during the preparation of the manuscript.

Literature Cited

1. Barth, H. G.; Barber, W. E.; Lochmuller, C. H.; Majors, R. E.; Regnier, F. E. *Anal. Chem.* 1988, *60*, 387R.
2. Knight, P. *Bio/Technology* 1989, *7*, 777.
3. Yamamoto, S.; Nomura, M.; Sano, Y. *J. Chromatogr.* 1987, *396*, 355.
4. Yamamoto, S.; Nomura, M.; Sano, Y. *AIChE J. 1987, 33*(9), 1426.
5. Elbe, J. E.; Grob, R. L.; Antle, P. E.; Snyder, L. R. *J. Chromatogr. 1987, 384*, 25.
6. Elbe, J. E.; Grob, R. L.; Antle, P. E.; Snyder, L. R. *J. Chromatogr. 1987, 405*, 1.
7. Seshadri, S.; Deming, S. N. *Anal. Chem. 1984, 56*, 1567-1572.
8. Norman, A. W.; Witwack, G. *Hormones*; Academic Press: New York, 1987; p. 264.
9. Dolan-Heitlinger, J. *Recombinant DNA and Biosynthetic Human Insulin, A Source Book*; Eli Lilly and Co.; Indianapolis, IN, 1981.
10. Smith, P. K.; Krohn, R. I.; Hermanson, G. T.; Mallia, A. K.; Gortner, F. H.; Provenzano, M. D.; Fujimoto, E. K.; Goeke, N. M.; Olson, B. J.; Klenk, D. C. *Analytical Biochem.* 1985, *150*, 76.
11. Wallenfels, K.; Weil, R. In *The Enzymes*, Boyer, P. D., Ed., Academic Press; New York and London; 1972; p. 617.
12. Richard M. L.; Gray, J. I.; Stine, C. M. *J. Dairy Sci.* 1981, *6*, 1759.
13. Distler, J. J.; Jourdian, G. W. *Methods Enzymol.* 1978, *50*, 514.
14. Steers, E.; Cuetrecasas, P. *Methods Enzymol.* 1974, *34*, 350.
15. *Instruction Manual for packing columns with TSK-Gel Toyopearl Chromatographic Resins*, TosoHaas, Philadelphia, PA, 1988.

RECEIVED February 6, 1990

Chapter 8

Radial-Flow Affinity Chromatography for Trypsin Purification

Wen-Chien Lee[1], Gow-Jen Tsai, and George T. Tsao

School of Chemical Engineering and Laboratory of Renewable Resources Engineering, Purdue University, West Lafayette, IN 47907

The dynamic performance of a radial-flow affinity chromatographic system for trypsin purification has been studied experimentally and theoretically. A mathematical model, which considers both axial dispersion in the mobile phase and pore diffusion inside the fiber, has been developed for simulation and parameter estimation. Based on this model, a design equation has been derived for scale-up of column operation. The design equation employs the concept of volume equivalent of a theoretical stage (VETS) instead of height equivalent of a theoretical plate (HETP) commonly used in the axial-flow chromatography.

Experiments from the frontal elution approach were designed and tested for verification of this design equation. A methodology of estimating system parameters from breakthrough experiments was also proposed and applied to the model system.

For the purification of biomolecules such as proteins, liquid chromatography (LC), in its various forms, is by far the most efficient method. It was reported (*1*) that bioseparation has been the largest segment of the LC market and that a user-wish list of the chromatographic equipment for bioseparation includes high productivity, rigidity, rapid scale-up, high performance without high pressure, reproducibility, biocompatibility and low non-specific adsorption. To meet some of the demands radial flow chromatography has been considered as an alternative because of its new geometry. Two companies, CUNO (*2,3,4*) and Sepragen (*5,6,7*), have marketed their equipment for radial flow chromatography, which is commonly mentioned as one of the most interesting innovations of the last several years although its advantages have not been fully evaluated.

The radial flow column consists of three main parts, i.e., outer channel, column packing, and inner channel (Figure 1). On operation, the sample is distributed into the outer channel and flows radially inward through the column packing. Then, the elution fluid flows down

[1]Current address: Department of Chemical Engineering, Chung Yuan Christian University, Chung Li, Taiwan 32023, Republic of China

0097–6156/90/0427–0104$06.00/0

(a)

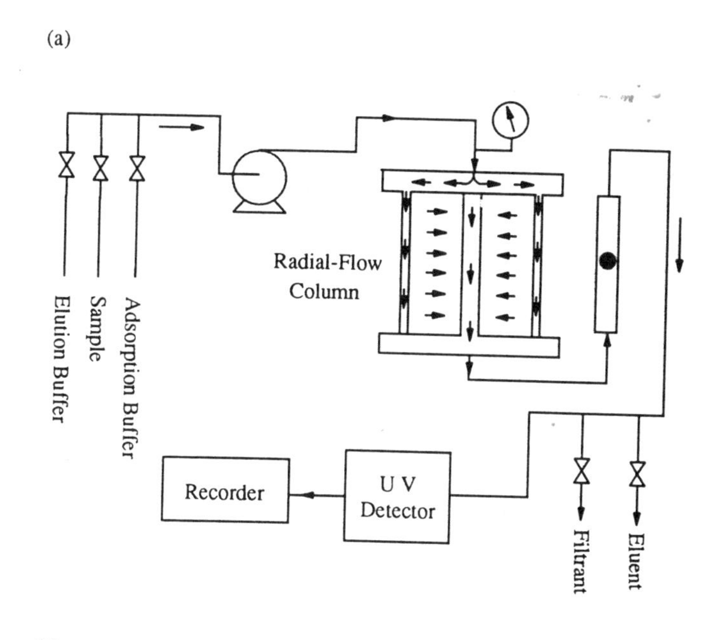

(b)

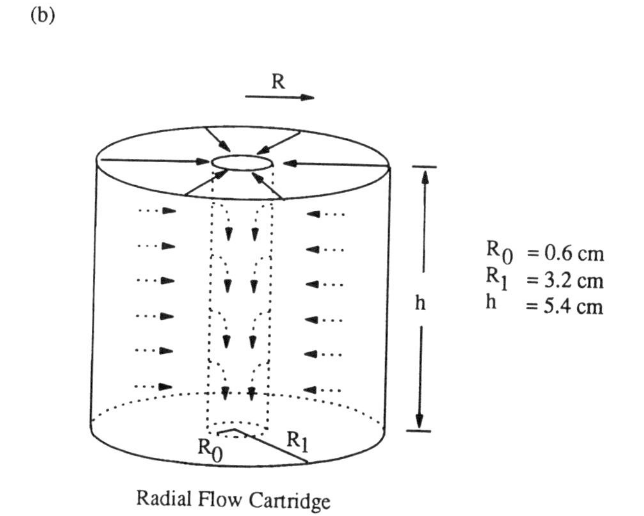

Figure 1. (a)Experimental set-up of radial flow chromatographic system for trypsin purification. (b)Dimensions of the radial flow cartridge.

the inner channel and through a collector to the column outlet. The column packing is where the biospecific adsorption occurs. In the CUNO product, the column packing is a cartridge in which the matrix is fabricated into thin paper sheets and then spirally wound around a center plastic core. Warren (*8*) mentioned that CUNO's radial-flow cartridge is a typical product of the marriage of chromatography and membrane techniques.

Radial flow chromatography is ideal for fast-flow rate systems because of the low pressure drop across the chromatographic packing. The advantages in fast-flow rate and low pressure are especially important in the treatment of large volumes of dilute products from some bioprocesses. Radial flow chromatography becomes more powerful for fast purification when it is combined with the principle of affinity chromatography. The successful scaling-up of affinity chromatography will realize the commercial applications of various proteins and enzymes, which are now too expensive to use in large quantities. Radial-flow affinity chromatography may be one of the best choices for down-stream purification in production scale.

In this work, the dynamic performance of radial-flow chromatography was studied by performing the frontal elution experiments on a trypsin purifying cartridge. Theoretical study of the radial-flow affinity system led to design equations for scaling-up. The results from experiments on cartridge were used to compare those from theoretical predictions. A methodology for parameter estimation was also presented.

Theory

The process by which soluble protein molecules are transported by fluid flow through the void space of porous matrix media and adsorbed on the surfaces of the matrix in radial-flow affinity chromatography can be highly complex. When the method of local volume averaging (*9*) is applied to the column cartridge (Figure 1b), the governing equation of protein (trypsin) concentration in the mobile phase can be simplified as

$$\varepsilon \frac{\partial C}{\partial t} = \frac{Q}{2\pi h R}\frac{\partial C}{\partial R} + \frac{\varepsilon}{R}\frac{\partial}{\partial R}\left(D R \frac{\partial C}{\partial R}\right) + r'' \tag{1}$$

The process is considered to be isothermal and a constant volumetrical flow rate Q is imposed in inward direction. The r'' denotes the net mass transfer rate from the bulk fluid to the adsorbed phase and can be expended as follows.

$$-r'' = \frac{4(1-\varepsilon)}{d_p} D_i \frac{\partial c}{\partial r}\Big|_{r=d_p/2} \tag{2}$$

$$D_i\left(\frac{\partial^2 c}{\partial r^2} + \frac{1}{r}\frac{\partial c}{\partial r}\right) - \varepsilon_p \frac{\partial c}{\partial t} - \rho_p \frac{\partial q}{\partial t} = 0 \tag{3}$$

$$k_f\left(C - c|_{r=d_p/2}\right) = D_i \frac{\partial c}{\partial r}\Big|_{r=d_p/2} \tag{4}$$

$$\frac{\partial c}{\partial r}\Big|_{r=0} = 0 \tag{5}$$

$$\frac{\partial q}{\partial t} = k_a\left(q^* - q\right) \tag{6}$$

In this model, a uniform diameter of d_p is assumed for the cylindrical, fibrous matrix with monodisperse pores, an external film separating the bulk fluid and solid phase is present, and an effective diffusion coefficient D_i based on the entire fiber volume is used to describe the diffusion of solute into the pores. The parameter k_a accounts for the adsorption rate, while the mass transfer rate through the film is accounted for by k_f. The equilibrium isotherm describes the relationship between concentrations c and q. To account for the possibility that all sites can be filled with the adsorbed molecules, many researcher considered the equilibrium relation to be of the Langmuir type,

$$q^* = \frac{q_s\, K_L\, c}{1 + K_L\, c} \tag{7}$$

Equations 1 to 7 are solved numerically by spatially discretizing Equations 1 and 3 into a set of ordinary differential equation (ODE) using finite difference and orthogonal collocation methods, respectively. The ODEs are then solved by the well known Gear's backward difference method *(10)*.

In order to scale-up the radial-flow system, the concept of plate theory from the axial-flow system has been extended *(11)*. The definition of HETP, i.e., H = L/N, is still needed but with small modification. Instead of using the plate height H and column length L, we define a Volume Equivalent Theoretical Stage (VETS) V_s in the following equation to replace the plate height H and use the bed volume V_B to replace the column length L.

$$V_s = \frac{V_B}{N} = \frac{\mu_2'}{\mu_1^2}\, V_B \tag{8}$$

By following the generalized methodology developed in the previous work *(11)*, the design equation for this particular radial-flow affinity chromatographic system can be obtained as

$$V_s = \frac{A}{Q/\varepsilon} + B + C\,\frac{Q}{\varepsilon} \tag{9}$$

where

$$A = 4\,\gamma_1 D_m\, \pi^2\, h^2\, (R_1^2 + R_0^2) \tag{10}$$

$$B = \frac{8}{3}\,\gamma_2 d_p\, \pi h\left(\frac{R_1^3 - R_0^3}{R_1^2 - R_0^2}\right) \tag{11}$$

$$C = -\,\frac{f_2}{f_1^2} \tag{12}$$

with

$$f_1 = 1 + \left(\frac{1-\varepsilon}{\varepsilon}\right)(\varepsilon_p + \rho_p\, q_s\, K_L)$$

and

$$f_2 = \left(\frac{1-\varepsilon}{\varepsilon}\right)\left[\frac{2\,\rho_p q_s^2 K_L^2}{k_a} + \left(\frac{d_p}{2k_f} + \frac{d_p^2}{16 D_i}\right)(\varepsilon_p + \rho_p\, q_s\, K_L)^2\right]$$

In the first step of scaling-up, fixed N means that the chromatographic efficiency is preserved. The volume of a theoretical stage V_s can be evaluated by the design equations once the parameters in these equations are estimated experimentally. Finally, the packed bed volume for a given operating flow rate is obtained by a product of N and V_s. The value of N can be estimated experimentally by injecting a standard sample into the chromatography system.

When the isotherm nonlinearity is considered, the equations derived above are not proper but still useful. Equations 9 to 12 provide the first approximation of the true values of V_s. In order to calculate μ_1 and μ_2' in Equation 8, the following two definitions are needed in frontal elution.

$$\mu_1 = \int_0^\infty \left(1 - \frac{C}{C_0}\right) dt \tag{13}$$

$$\mu_2' = \int_0^\infty \left(1 - \frac{C}{C_0}\right) t\, dt - \mu_1^2 \tag{14}$$

When local equilibrium is reached and a Langmuir isotherm is assumed, the first moment can be expressed as (*12*)

$$\mu_1 = \frac{\pi\, h\, (R_1^2 - R_0^2)}{Q} \left[\varepsilon + (1-\varepsilon)\, \varepsilon_p + \frac{(1-\varepsilon)\, \rho_p q_s K_L}{1 + K_L C_0} \right] \tag{15}$$

Numerical method is used to obtain μ_2' because no analytical solution is available when the isotherm is of the Langmuir-type.

Experimental

Purification System. The experimental set-up is schematically shown in Figure 1a. A para-aminobenzamidine Zetaffinity cartridge manufactured by CUNO, Inc., was used in the study. The dimensions of this cartridge are shown in Figure 1b. The equipment is connected in such a way that influent solution can be delivered by a peristaltic pump into the cartridge. On the upstream side of the cartridge, a pressure gauge is installed to measure the operating pressure of the system. The effluent is constantly monitored by a UV detector at 280 nm and is fraction collected for activity analysis.

The radial-flow cartridge is formed by spirally wrapping the paper, which is fabricated from a composite matrix, around a perforated center core. The fibrous matrix is a composite of cellulose containing grafted vinyl polymers and contains amino derivatives, which are ready for being reacted with a bifunctional reagent glutaraldehyde to yield active aldehyde groups at the end of spacer arms. The p-aminobenzamidine is then coupled to the spacer arms. The pores in the fiber matrix are large enough for large protein molecules to enter and to interact with the ligand. In this work, the cartridge was coupled 1,503.2 mg p-aminobenzamidine according to the producer.

Materials and Buffers. Trypsin Type I, trypsin Type II, and benzoyl-L-arginine-para-nitroanilide hydrochloride (L-BAPNA) were obtained from Sigma Chemical Co., St. Louis, MO. The following three buffer solutions were prepared for the study: (1) Adsorption and wash buffer: 0.05 Tris-HCl, 0.25 M NaCl, 0.002 M $CaCl_2$, pH = 7.6; (2) Elution buffer: (1) 0.2 M glycine-HCl, pH = 2.3; (2) 0.1 M citric acid, pH = 2.1; (3) Assay buffer: 0.05 Tris-HCl, 0.02 M $CaCl_2$, pH = 8.2.

Trypsin Assay. The activity of trypsin was assayed by the method of Geiger and Fritz (*13*) with only small modifications. The substrate solution was prepared by dissolving 43.5 mg of L-BAPNA in 0.5 ml of dimethylsulfoxide, then diluting with assay buffer to 100 ml (1 m mole/l). A spectrophotometer was used to measure the rate of change in absorbance ($\Delta A/\Delta t$) at a wavelength of 405 nm. Trypsin activity was calculated by 1961 $\Delta A/\Delta t$ (U/l) in which $\Delta A/\Delta t$ is absorbance unit per minute (*13*).

The activities of Type I and Type II trypsins were determined by the trypsin assay. Figure 2 gives the relationship between $\Delta A/\Delta t$ and Type I trypsin concentration C. It can be seen that $\Delta A/\Delta t$ and C are linearly dependent with the range of experiments and the relationship does not vary with change of pH in the range of 7.6 to 8.2. The specific activities of Type I and Type II trypsins were found to be 4.02 U/mg solid and 0.55 U/mg solid, respectively.

Void Fraction Measurement. For the determination of cartridge void volume, V_m, a frontal elution experiment using a large non-penetrating macromolecule, Blue Dextran, was conducted. The total penetrable volume of the radial flow cartridge, V_0, was measured by frontal elution of purified insulin. These V_0 and V_m values are approximately equal to the volumes of effluents corresponding to half inject sample concentrations of Blue Dextran and insulin, respectively (*14*). It has been measured that $V_m = 112$ ml and $V_0 = 150$ ml, while the cartridge has a bed volume (V_B) of 167.6 ml and a dead volume of 70ml. During the course of void fraction measurement, some degree of fouling (contamination) on the CUNO's cartridge was observed. The fouling inherited from the characteristics of the membrane process could be contributed by the reversible or irreversible physical adsorption or filling of microvoids. Furthermore, shallow paths or blockades at some positions in the cartridge were also occasionally observed due to its imperfect packing structure. The diameter of the matrix fiber was measured by a microscope to be 0.002 cm. Other geometrical parameters and packing properties were obtained as: $R_0 = 0.6$ cm, $R_1 = 3.2$ cm, $h = 5.4$ cm, $\varepsilon = 0.67$, and $\varepsilon_p = 0.68$.

Frontal Elution (Breakthrough) Experiments. The cartridge was equilibrated with the adsorption buffer. Trypsin solution of a specified concentration was passed through the cartridge at a specified flow rate. The effluent was monitored by a UV detector at 280 nm for reference. The fractions were collected and assayed. Injection of the sample continues until breakthrough was almost finished, i.e., the activity of the effluent was more than 99% of the sample solution's activity. The breakthrough curve was obtained when effluent activity was plotted versus effluent volume. The cartridge was then washed to the baseline with the wash buffer, then eluted with elution buffer (1) until the baseline was reached, and the eluate was collected in one fraction, which was assayed for total activity. The cartridge was then eluted a second time with elution buffer (2) until the baseline was reached again.

The first moment, μ_1, second central moment, μ_2', and recovery rate were calculated from the assay data. In the calculations of the first and the second moments from data, the cubic spline method was applied to smooth the experimental breakthrough curves, and then a numerical integrating procedure was followed to evaluate the integrations in Equations 13 and 14. The retention volume, V_t, is simply a product of the first moment and the flow rate, Q.

Results and Discussion

Equation 15 shows that, in local equilibrium, the retention volume (first moment) is theoretically not a function of dispersion coefficient and mass transfer resistances. The equilibrium condition was approached by reducing the volumetric flow rate to that less than 20 ml/min. The retention volume was then calculated by counting the area behind the breakthrough curve. Table I reports the experimental retention volumes at various sample concentrations. A reciprocal plot of retention volume vs. sample concentration gives a linear relationship, which agrees well with the following theoretical expression

$$V_t = \mu_1 Q = V_0 + V_B \frac{(1-\varepsilon)\,\rho_p q_s K_L}{1+K_L C_0} \tag{16}$$

where $V_0 = [\,\varepsilon + (1-\varepsilon)\,\varepsilon_p]\,V_B$ and $V_B = \pi\, h(R_1^2 - R_0^2)$.

Table I. Dependence of retention volume on Type I trypsin concentration

Concentration (g Type I trypsin/l)	Retention Volume V_t (ml)	$\frac{V_B}{V_t - V_O}$
2	1339	0.141
1.44	1678	0.11
1.37	1919	0.095
1.1	1982	0.091
0.51	2703	0.066

This plot is shown in Figure 3. The binding (association) constant K_L and maximum binding capacity $(1-\varepsilon)\,\rho_p\,q_s$ were calculated directly from the slope and intercept of this plot to be 1.35 l/g and 20.12 g/l, respectively. This K_L value is the inhibition constant of trypsin and its immobilized inhibitor p-aminobenzamidine. The K_L value obtained in this work is very close to that given by Adamski-Medda et al. (*15*). They reported the inhibition constant of trypsin and p-aminobenzamidine linked to dextran as 1.0 l/g.

Type I trypsin was used and results from breakthrough experiments are given in Table II. Three different volumetric flow rates were employed in these experiments. As expected, an increase in the flow rate decreased the retention volume and increased the effects of finite mass transfer rate on the chromatographic performance. The net effect is an increase of μ_2'/μ_1^2, i.e., an increase of VETS. This concludes that the rate of adsorption and elution of trypsin is dominated by the mass transfer in pores and through external film, i.e., the C term in Equation 9 is dominant. Figure 4 shows the breakthrough curves of the frontal elution study with flow rates of 60 and 19.4 ml/min. In this figure, circles represent data points and solid lines are results from curve fitting. It is obvious that reducing the flow rate tended to decay the breakthrough point and to sharpen the breakthrough curve.

The experimental results in Table II and corresponding breakthrough curves have been used to estimate the parameters which characterized the performance of the radial-flow system. These parameters are the dispersion coefficient D, which is a function of flow rate Q, the pore diffusivity D_i, and the external film mass transfer coefficient k_f. Pore diffusivity and film mass transfer coefficient were determined by fitting experimental breakthrough curves with the mathematical model: Equations 1 to 7. Due to model limitation the curve fitting might not be perfect, but acceptable. This probably results from the imperfect flow pattern in the cartridge and the variation of the mass transfer coefficient with the progress of adsorption. Final results are $D_i = 0.63 \times 10^{-8}$ cm^2/sec and $k_f = 1.45 \times 10^{-4}$ cm/sec.

The affinity cartridge is intended to obtain chymotrypsin-free trypsin. In this work, crude trypsin solution was applied to the purification system as input sample. When the Type II trypsin was used, typical results from breakthrough experiments are given by Figure 5. The abscissa represents the throughput volume, and the ordinate is the protein concentration expressed by optical density and enzyme activity. In the loading stage, the components which did not bind on the matrix were eluted out at the first, then followed by the trypsin. The lag of trypsin breakthrough implies that trypsin was favorably retained in the cartridge. After the cartridge was nearly saturated, the adsorption/wash buffer was used to wash the unbound proteins from the cartridge. Then elution buffer (1) was introduced and a sharp peak appeared in this stage, which clearly showed that trypsin was recovered in a much purer form. From Figure 5, the highest purification factor was obtained as 10-fold when the operating flow rate was 67 ml/min. The purification factor was calculated by the ratio of specific activity in purified

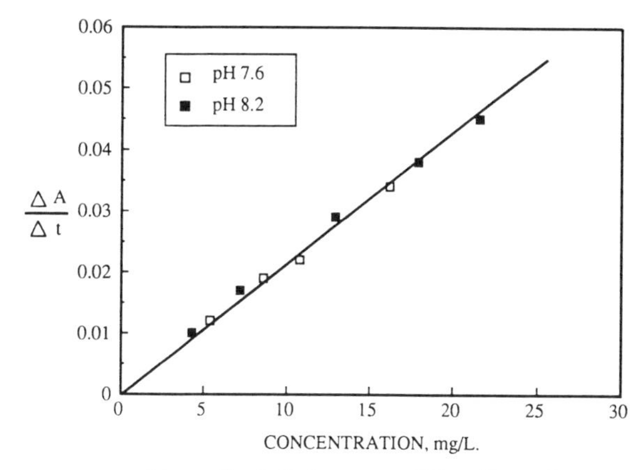

Figure 2. Activity assay of Type I trypsin.

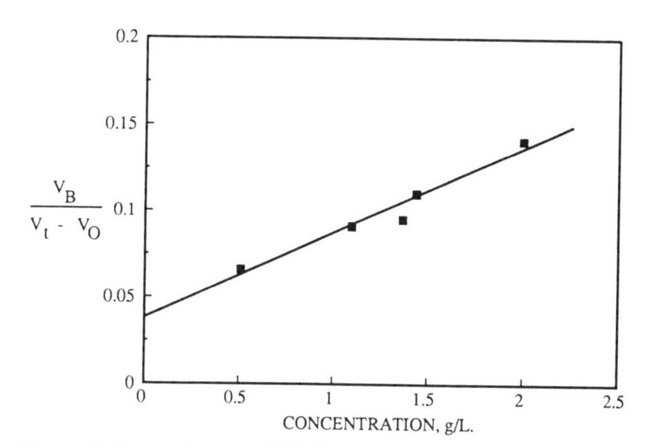

Figure 3. Determination of binding constant and maximum binding capacity.

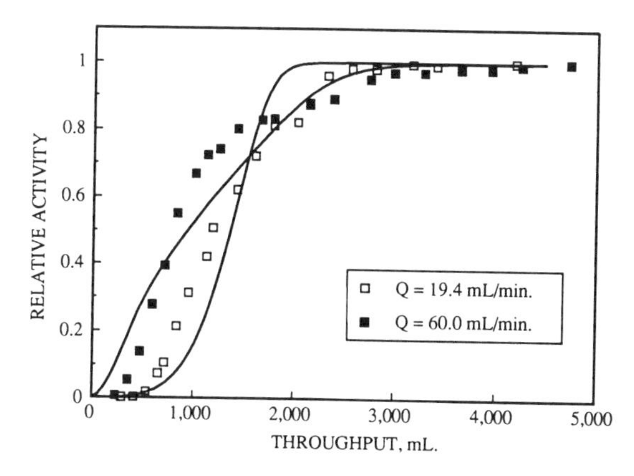

Figure 4. Effect of volumetric flow rate on breakthrough curve with Type I trypsin.

enzyme and that in sample solution, where specific activity is the trypsin activity per unit of protein content. Figure 5 shows that the radial-flow affinity system is successful for trypsin purification. The CUNO cartridges with different packing matrices have also been successfully applied to the purification of $F(ab)_2$ fragments of horse immunoglobulins (*16*) and monoclonal antibody IgG (*17*).

Table III summarizes the results from breakthrough experiments with Type II trypsin as sample. The chromatographic behaviors in these cases are very similar to that in which pure trypsin is applied. From this, one may conclude that the interference of non-specific components on the bio-specific adsorption is insignificant. Figures 6 and 7 report the breakthrough behaviors with Q = 40 and 67 ml/min, respectively. In these two figures, circles are data points, while solid lines represent the predicted values resulting from model simulation with the parameters estimated from the system using Type I trypsin. It is obvious that the model prediction agrees fairly well with the experimental data.

Parameter Estimation. In this work, a methodology of estimating parameters from breakthrough experiments has been developed.

(1). Binding constant K_L and maximum binding capacity $(1-\varepsilon)\,\rho_p\, q_s$ were determined from a reciprocal plot of retention volume vs. sample concentration. In this step Equation 16 has been used.

(2). Dispersion coefficient was estimated from the change of retention volume vs. flow rate. The faster the flow rate, the more significant the dispersion. It is proved that retention volume $(Q\,\mu_1)$ strongly depends on dispersivity, but is nearly independent of finite mass transfer rate. Figure 8 shows the dependency of retention volume on dimensionless reciprocal dispersivity at two different sample concentrations.

(3). Pore diffusivity D_i and film mass transfer coefficient k_f were determined by fitting breakthrough curves with model, i.e. Equations 1 to 7.

In this work, only the results from experiments using Type I trypsin, are used for curve fitting. The adsorption rate constant k_a should be also estimated from this curve fitting. In this affinity system, k_a is very large compared to the mass transfer rate constant due to higher specific interaction. Here we also assume that k_f is not dependent on flow rate.

Notice that some of the system parameters could be estimated from independent experiments other than the breakthrough experiment, depending on which is more convenient. As we have evaluated all parameters, either from literature or from the estimation method mentioned above, we can (1) predict breakthrough behavior and (2) predict VETS. Solid lines in Figures 6 and 7 come from this prediction. Good agreement between predictions and experiments can be observed from these two figures.

Figure 9 shows the experimental and predicted VETS in the system when Type II trypsin is applied. Although they show general agreement, a deviation is observed in the range of small flow rate. The reason probably is model limitation. The model might not exactly simulate the true behaviors in the cartridge, because the made-up packing in the CUNO cartridge is not homogeneous for its special fabrication which makes the flow distribution diversified. It was observed that the occasional blockades contribute imperfection of the flow pattern.

Figure 10 shows the comparison of experimentally obtained VETS with the design equation given by Equation 9. In those experiments, Type I trypsin was used. A close agreement exists because the parameters used in Equation 9 are estimated from the same experiments.

Table II. Results of breakthrough experiments with Type I trypsin

Sample Concentration (g/l)	Activity ($\frac{dA}{dt}$/l)	Flow Rate (ml/min)	Retention Volume ($= \mu_1$ Q, ml)	First Moment μ_1, (min)	Second Central moment μ_2' (min^2)	VETS (ml)	Activity Recovery (%)
2	3.9	19.4	1339	69	938	33	93
2	3.8	37.3	1183	31.7	444	74	94
2	4.1	60	1085	18.1	187	95.7	97

Table III. Results of breakthrough experiments with Type II trypsin

Sample Concentration (g/l)	Activity ($\frac{dA}{dt}$/l)	C_o (g/l) based on Type I trypsin	Flow Rate (ml/min)	Retention Volume ($= \mu_1$ Q, ml)	First Moment μ_1, (min)	Second Central Moment μ_2' (min^2)	VETS (ml)	Activity Recovery (%)
10	2.8	1.39	18.2	2054	112.9	2376	31.2	105
10	2.83	1.41	40	2003	50.1	631	42.1	98
10	2.75	1.37	67	1926	28.7	232	47.2	84
10	4.67	2.32	80	1109	13.9	75.6	65.6	125

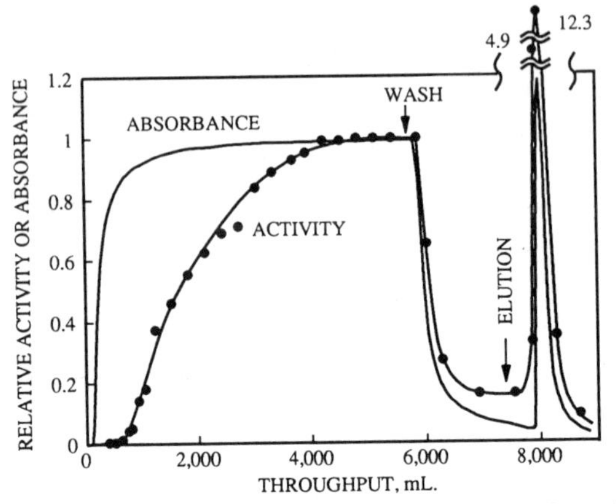

Figure 5. Experimental results of purification of Type I trypsin.

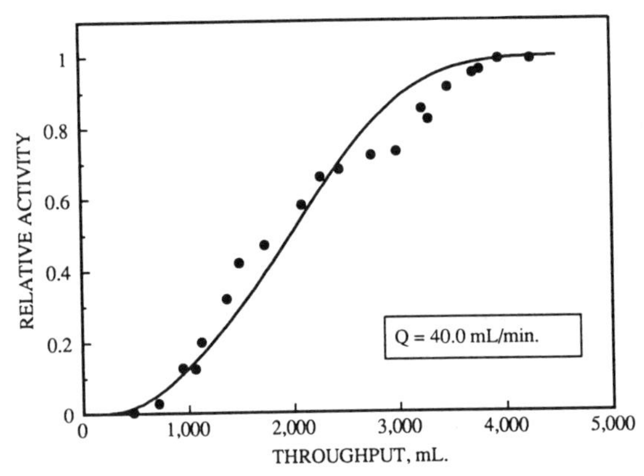

Figure 6. Comparison of experimentally obtained breakthrough curves (circles) with the model prediction (solid line). Flow rate Q = 40 ml/min; sample: Type II trypsin.

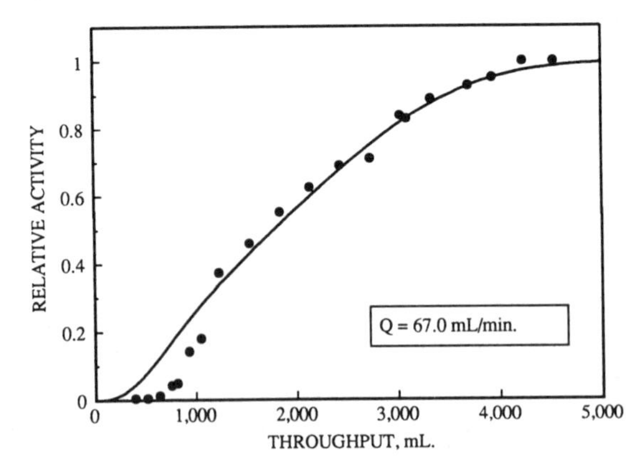

Figure 7. Comparison of experimentally obtained breakthrough curves (circles) with the model prediction (solid line). Flow rate Q = 67 ml/min; sample: Type II trypsin.

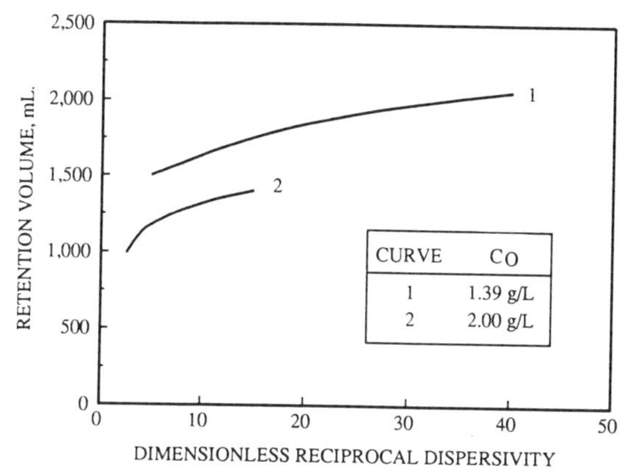

Figure 8. Effect of dimensionless dispersivity on the retention volume.

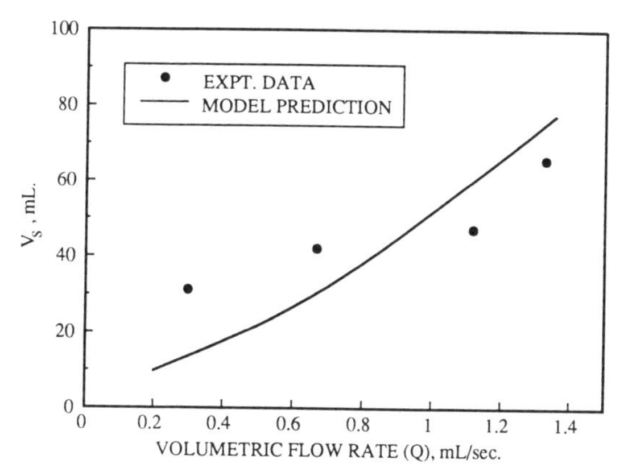

Figure 9. Comparison of experimentally obtained VETS with the model prediction using Equations 1 to 7.

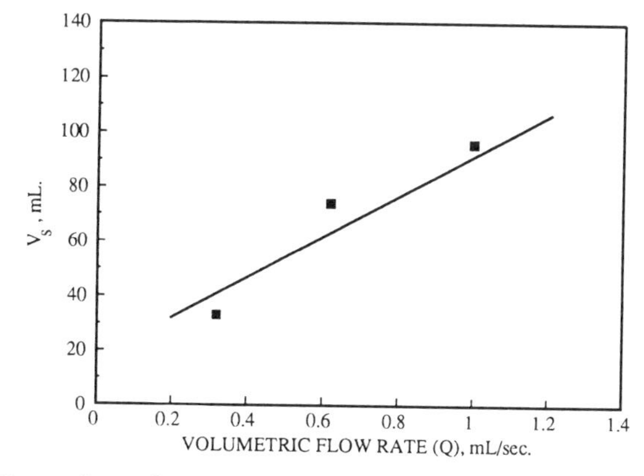

Figure 10. Comparison of experimentally obtained VETS with the design equations given by Equation 9.

Conclusion

Radial-flow affinity Chromatography has been found to be successful for fast purification of trypsin. In frontal elution, as expected, the rate-determining steps of adsorption and elution are the mass transfer in the pores of the fibrous matrix and through the external film. Therefore, the volumetric flow rate Q is the most important factor for production in large scale. The design equations using Q as the design variable have been verified experimentally to be useful and recommended for scaling-up of the radial-flow affinity system. The use of the CUNO cartridge as an example of radial-flow chromatography has exposed a problem of imperfect flow distribution. But this problem becomes minor if we operate the system simply in the adsorption/desorption mode. Despite the flow distribution problem, the experiments on the radial-flow purification system result in useful data. The parameters characterizing the performance of the purification system were evaluated from breakthrough data fitting the proposed model. Once the system parameters are estimated, we can predict breakthrough behavior for any sample concentration and operating condition and predict the values of VETS.

Acknowledgments

This work was supported by the National Science Foundation Grant (EET-8613167A2). We also thank Professor Michael R. Ladisch, Department of Agricultural Engineering, Purdue University, for his helpful comments during preparation of this manuscript, and Dr. Kenneth C. Hou of CUNO, Inc. (Meriden, Connecticut) for the gift of the radial flow cartridge.

Legend of Symbols

C protein concentration in bulk fluid phase
C_0 inlet concentration of protein in frontal elution
c protein concentration in pores
D dispersion coefficient (column dispersivity), $D = \gamma_1 D_m + \gamma_2 d_p v$
D_i effective pore diffusivity of protein in pores
D_m molecular diffusivity of protein
d_p diameter of cylindrical packing fiber
h height of the radial flow chromatography
K_L Langmuir or association (binding) constant
k_a adsorption rate constant
k_f fluid film mass transfer coefficient of soluble protein
Q volumetric flow rate
q^* sorbate concentration, kg/(kg particle)
q_s maximum concentration of sorbate, kg/(kg particle)
R radial coordinate of chromatography
R_0 radius of the inner channel
R_1 radius of the outer channel
r radial distance in fiber
t time
V_0 unretard void volume
V_B volume of packed bed
V_s volume of a theoretical stage
V_t retention (elution) volume
v = linear velocity, $v = \dfrac{Q}{2\pi hR\varepsilon}$

ε void fraction of the packed bed
ε_p porosity of particle
ρ_p fiber density
μ_1 first moment
μ_2' second central moment
ν dimensionless reciprocal dispersivity, defined as $\nu = \frac{Q}{4\pi h \varepsilon D}$
γ_1 constant
γ_2 constant

Literature Cited

1. McCormick, D. *Bio/Technol.* 1988, *6*, 158.
2. Hou, K. C.; Mandara, R. M. *Bio Techniques.* 1986, *4*, 358.
3. Mandara, R. M.; Roy, S.; Hou, K. C. *Bio/Technol.* 1987, *5*, 928.
4. McGregor, W. C.; Szesko, D. P.; Mandara, R. M.; Rai, V. R. *Bio/Technol.* 1985, *4*, 526.
5. Saxena, V. U. S. Patent, 4 676 898, 1987.
6. Saxena, V.; Weil, A. E. *BioChromatogr.* 1987, *2*, 90.
7. Saxena, V.; Subramanian, K.; Saxena, S.; Dunn, M. *BioPharm* 1989, *March*, 46.
8. Warren, D. C. In *Chemical Separation, Vol. II Application*; King; Navratil, Eds., Litarvan Literature; Denver, CO, 1986.
9. Slattery, J. C. *Momentum, Energy and Mass Transfer*; McGraw Hill, New York, NY, 1981.
10. Gear, C. W. *Numerical Initial-Value Problems in Ordinary Differential Equations*, Prentice-Hall, Englewood Cliffs, NJ, 1972.
11. Lee, W.-C. Ph.D. Thesis, Purdue University, West Lafayette, 1989.
12. Geiger, R.; Fritz, H. In *Method in Enzymatic Analysis*; Bergmeyer, Ed., 1984; Vol. V, p 119.
13. Kasai, K.-I.; Oda, Y. *J. Chromatogr.* 1986, *376*, 33.
14. Adamski-Medda, D.; Nguyen, Q. T.; Dellacherie, E. *J. Membrane Sci.* 1981, *9*, 337.
15. Benanchi, P. L.; Gazzei, G.; Giannozzi, A. *J. Chromatogr.* 1988, *450*, 133.
16. Jungbauer, A.; Unterluggauer, F.; Uhl, K.; Buchacher, A.; Steindl, F.; Pettauer, D.; Wenisch, E. *Biotechnol. Bioeng.* 1988, *32*, 326.

RECEIVED December 28, 1989

Chapter 9

Impact of Continuous Affinity–Recycle Extraction (CARE) in Downstream Processing

Neal F. Gordon[1] and Charles L. Cooney

Department of Chemical Engineering and Biotechnology Process Engineering Center, Massachusetts Institute of Technology, Cambridge, MA 02139

As commercialization of products derived from recombinant DNA technology intensifies, the development of manufacturing and large scale processing technology has become a major challenge for the biotechnology industry. Nowhere is the challenge more evident, and yet most lacking, than in the protein recovery and isolation stage. We have attempted to address this challenge, through the development of a continuous, scaleable and integrateable protein purification system.

Rather than packing conventional adsorbent particles in a fixed bed (column), solid/liquid contact is carried out in well-mixed reactors. Continuous operation is achieved by recirculation of the adsorbent particles between two or more contactors. Using a lab scale prototype unit, two continuous protein purification examples were developed; one based on affinity (biospecific) interactions and the second on ion-exchange adsorption.

In an attempt to place CARE in the greater context of downstream processing, two sets of simulations were performed. The first involves early introduction of an adsorptive purification step in a purification train. Pilot plant data for cell debris removal in a continuous centrifuge was contrasted with simulated performance of the CARE system. The second involves a direct comparison of CARE to column chromatography. Adsorptive purification, utilizing the CARE process, can be introduced into a process, at a location where column chromatography is not possible (with solid contaminants and viscous material). This early introduction of a high resolution purification step should positively influence the remaining steps in the process, and lower overall purification costs. For feed streams that do not require clarification or viscosity reduction, the benefits of column operation (high capacity, high purification factor) make it an attractive and in some cases, preferred alternative to CARE.

[1]Current address: PerSeptive Biosystems, 38 Sidney Street, Cambridge, MA 02139

0097–6156/90/0427–0118$06.25/0

MOTIVATION. As the biotechnology industry undergoes a transition from research to product commercialization, cost reductions in process development and large-scale protein purification are emerging as key determinants to commercial success. Techniques used today for purification are mainly chromatographic in nature and employ equipment and material derived directly from laboratory/bench scale experience. With these roots, it is common to find process chromatograms and adsorbents being evaluated on the basis of resolution alone, with little regard to recovery yield or throughput. Process-scale chromatographic purification of proteins requires a different set of design and optimization criteria than those used for laboratory/research work. For example, final purity is a constraint and not an objective. The ultimate objective is minimum cost of a purified product that meets specifications which, in turn, implies maximal recovery and throughput. A different approach to the selection and design of unit operations for manufacturing, is to first consider the entire process at the largest scale, and then scale-down to an intermediate scale which can simulate, with confidence, the larger scale.

Protein purification is most often effected by chromatographic techniques (1). Adsorptive chromatography, which includes ion exchange, affinity, reverse phase and hydrophobic interaction chromatography, accounts for a large portion of preparative chromatography applications. Affinity adsorption, based on molecular recognition, is the most specific of the adsorptive techniques, with large-scale protein affinity purification applications shown in Table I. Traditionally, affinity chromatography is carried out using a fixed bed of adsorbent particles (i.e. column chromatography). While for small molecules, the importance of column length (i.e. number of theoretical plates) on resolution is well characterized, for macromolecules experimental evidence suggests a far lesser need for a large number of plates. Macromolecule separation typically involves strong surface adsorption due to high specificity and/or multiple site interactions. As such, interactions are of the "on/off" type, a separation mechanism that does not require a large number of theoretical stages. Early reports of this observation showed that in surface mediated separations, columns of less than 5 cm long have 80% of the resolving power of 30 cm columns (13). Affinity adsorption is usually governed by "on-off" surface interactions (14-5), and is little more than solid-liquid extraction, a common unit operation in the chemical process industries. As such, a fixed bed is but one of alternative contactors which have been employed. A system, Continuous Affinity-Recycle Extraction (CARE), employing an alternate contacting device (stirred tank), overcoming some the fixed bed's operational limitations, has been developed (16-8) and is described here.

CARE incorporates two features which are not commonly found in protein purification unit operations. These features are continuous, rather than batch, operation and the ability to operate in the presence of suspended solid contaminants. Continuous processes are the norm rather than the exception in the chemical processing industry. In general, they operate with greater throughput, higher purity and lower cost than that possible with an equivalent batch process. Continuous processes are usually more amenable to control and optimization, two important features for large-scale applications. In addition, a continuous purification step is more naturally integrated into a total process that is operating continuously.

For purifications based on adsorption, in order to maximize the extraction efficiency, counter current motion of solid and liquid with minimal axial mixing is desired (14,19,20). In this manner, a component or group of components move preferentially, through the contactor, with the solid adsorbent phase, while the other component or group of components move in the direction of the chromatographic bed. Counter current solid-liquid contact can be achieved in several different manners; one can flow the solid phase, the solid phase can be held in place and the contacting equipment can move, or a fixed bed, which simulates counter current motion by switching valves, can be utilized. All these approaches have been attempted, and selected examples are summarized in Table II.

TABLE I. Examples of Large-Scale Purification by Affinity Adsorption

PROTEIN	LIGAND	REFERENCE
Alcohol Dehydrogenase	Cibacron Blue F3G-A	2
Antibodies (Monoclonal)	Protein A	3
β-Galactosidase	PAPTG (p-aminophenyl-β-D-thiogalactopyranoside)	4
Carboxypeptidase G2	Procion Red H-8BN	5
Factor VIII	anti-VIIIRAg antibody	6
Glucokinase	Procion Brown H-3R	7
Lactate Dehydrogenase	NAD^+	8
Phosphoglycerate Kinase	Cibacron Blue F3G-A	9
Tissue-type Plasminogen Activator	anti-TPA antibody	10
Tissue-type Plasminogen Activator	Zinc Chelate	11
Urokinase	p-Aminobenzamidine and anti-UK antibody	12

While different approaches to protein purification have been taken, several general limitations emerge. Some techniques incorporate continuous processing but cannot operate in the presence of solids, while others can handle solid contaminants but only operate in a batch mode. Many of the reviewed techniques are mechanically complex and do not lend themselves well to scale-up. Some processes that show promise are: two-phase liquid extraction, affinity partition, and affinity precipitation.

DESCRIPTION OF CARE. In Continuous Affinity-Recycle Extraction (CARE), rather than packing affinity support materials in a column, the purification takes place in two (or more) well-mixed contactors (CSTR's). A schematic view of CARE is shown in Figure 1. The process operates as follows: A continuous feed to the adsorption stage contacts the affinity adsorbent; the desired product adsorbs while contaminants are washed out with wash buffer. The adsorbent, with the adsorbed product, is pumped to the desorbing stage where the addition of the desorbing buffer causes detachment of the product from the affinity matrix. The adsorbent, now regenerated, is recycled to the adsorption stage, while the product is removed with the desorbing buffer stream. The system can be operated continuously at steady state. Both vessels are well agitated; the sorbent is retained within the two vessels and the recycle loop by a retaining device, which in this case is a macroporous filter.

Purification performance, while effected by a complicated set of tradeoffs, is conceptually controlled by two key flow rate ratios. A high ratio of adsorption reactor throughput (feed + wash) to adsorbent recycle flow rate, eliminates the bulk of the non-adsorbed contaminants (solids, protein, etc.) with the waste stream and thus results in both purification and clarification. A low eluting buffer to feed flow rate concentrates the proteins that adsorb to the solid phase adsorbent in the first

TABLE II. Adsorptive/Extractive Purification Processes

Approach	Comments	References
Two-Phase Extraction	• Amenable for large-scale • Not very selective • Polymer cost can be prohibitive • Characterized by rapid adsorption equilibria	21,22
Affinity Adsorption with Membrane Filtration	• Only one contact stage • Concern for membrane fouling • Not designed for solid contaminants • Conceptually similar to CARE	23-33
CSTR Adsorption and Filtration	• Almost identical to CARE • Not used for protein purification	34-36
Plug Flow Moving Bed Adsorption	• Cannot process solids • Mechanically complex	37
Fluidized Bed Adsorption	• Single or multi-stage contact • Max. flow rate limited by adsorbent density • Capable of handling solids • Similar to CARE	38-44
Magnetically Stabilized Fluidized Bed Ads.	• True plug flow of liquid and solids • Added mechanical complexity	45,46
Simulated Moving Bed Adsorption	• Not Capable of handling solid • Mechanically Complex	47,48
2D Chromatography	• Not Capable of handling solid • Mechanically Complex • Cyclical operation	49-56
Moving Belt	• Slow adsoprtion kinetics • Poor mixing • Mechanically complex	56
Rotating Column	• Cannot handle solids • Plagued by sealing problems	57,58
Batch Fluidized Bed Adsorption	• Limited by nature of batch operation	59
Two-Phase Liquid Extraction	• Same as for continuous process with lower productivity	60-66
Affinity Partition	• Enhances selectivity • Good scale-up potential	67-77
Affinity Precipitation	• Theoretically sound but not yet widely accepted	78-80

reactor. The desired flow rate ratios are shown qualitatively by the thickness of the arrows in Figure 1. For concentrated feeds, the feed can be diluted with wash buffer in order to maintain the desired throughput to adsorbent recycle flow rate ratio.

Purification schemes incorporating the CARE system provide the opportunity for process integration through the introduction of a highly specific adsorptive purification step early in a purification sequence. It is possible to achieve simultaneously, purification, concentration and clarification (solid/liquid separation), while maintaining high recovery yields (16-8). Consequently, one or more downstream processing steps can be eliminated, potentially resulting in higher overall recovery yields and lower purification cost. In addition, due to the multiple degrees of freedom associated with this unit operation, both the system's design, as well as its operation can be optimized for any one of several performance related objective functions. Finally, as a consequence of operational and design simplicity, predictable system scale-up is anticipated.

OPERATION OF CARE

AFFINITY PURIFICATION EXAMPLE. The CARE process was characterized and developed based on an affinity adsorption example. The enzyme β-galactosidase was continuously recovered from an *E. coli* a homogenate using the affinity adsorbent p-Aminobenzyl-1-Thio-β-D-galactopyranoside-Agarose (Sigma Chemical Co., St. Louis, MO). Adsorption to the affinity support takes place in a pH 7 phosphate buffer with desorption following in a pH 9 borate buffer. Details of this model system can be found elsewhere (18).

A mathematical model describing purification of β-galactosidase in the CARE system was developed and used as a tool to help characterize and experimentally validate this purification approach (17-8). The heart of the model is a description of the adsorption/desorption processes involving β-galactosidase and the affinity adsorbent, PABTG/Agarose. The adsorption step involves the contact of enzyme in solution (C) with adsorbent (A) forming an adsorbed or bound enzyme complex (q), described by equation 1:

$$\mathbf{c} + \mathbf{A} \underset{k_r}{\overset{k_f}{\leftrightarrow}} \mathbf{q} \tag{1}$$

with the rate of adsorption given by equation 2:

$$\frac{dq}{dt} = k_f c(Q_{max} - q) - \left(\frac{k_f}{K}\right) q \tag{2}$$

where Q_{max} represents the saturation capacity of the adsorbent, K is the equilibrium association constant given by $\left(\frac{k_f}{k_r}\right)$, and k_f and k_r are the forward and reverse adsorption rate constants, respectively. The adsorption parameter (K, Q_{max}, k_f) values are obtained from independent batch adsorption experiments (17-8). Of note, the contributions of both the external film and internal pore mass transfer resistances are lumped into the forward rate constant (k_f); hence, the reported value of the forward rate constant does not represent its intrinsic value.

Desorption, in the β-galactosidase affinity purification model system is accomplished by the introduction of borate ions, which are a specific eluent for β-galactosidase. Since the ion concentration (0.1 M) is orders of magnitude greater than the enzyme's, desorption from the ligand is complete and not governed by equilibrium. In addition, external mass transfer resistance has been shown to be negligible. Finally, it has been shown that desorption is a much faster process than adsorption and thus, is considered to occur instantaneously (18).

A model of the CARE process is formulated as a set of material balances involving two well-mixed vessels operating with recycle. The descriptions of both adsorption (equation 2) and desorption are incorporated into the material balances.

Simulations of purification in the CARE system were performed in several ways. For the simple two-stage base case CARE design, a steady-state analytical solution was derived (17-8). The dynamic approach to steady-state, encountered during system start-up, was solved numerically, using the 4th order Runge-Kutta method. These two solutions were developed on a personal computer (PC's Limited 286-8). Simulation of multi-stage operation, as well as optimization was performed using the BioProcess Simulator software (Aspen Technology Inc. Cambridge, MA) on a IBM Mainframe computer.

Extensive simulations of system performance allowed us to evaluate the set of tradeoffs among the various performance measures (purification factor, concentration factor, recovery yield and system throughput) and led to the formulation of several rules of thumb. To increase the purification factor, one must increase the ratio of adsorption reactor throughput relative to adsorbent recirculation rate (e.g. increase the wash or feed flow rates and/or decrease the adsorbent recirculation rate). Concentration of the product can be achieved by decreasing the ratio of desorbing buffer flow rate relative to the feed flow rate, through an increase in feed flow rate and/or a decrease in adsorbent recirculation rate. Finally, recovery is increased most effectively by decreasing the wash flow rate and/or increasing adsorbent recycle. Alternatively, one can decrease the feed flow rate.

An assessment of the mathematical model was undertaken by a series of experiments designed to modify unit performance from a base case run. System design and operating parameters were modified, *a priori*, according to the general rules formulated above, in order to achieve the desired change in performance. Experimental conditions for this set of experiments are shown in Table III. The base case experimental conditions were modified as follows: In order to improve the recovery yield, several actions were taken. The amount of feed to the system was decreased, as indicated by a decrease in the feed flow rate. In addition, the wash flow rate was decreased such that the purification factor and concentration factors would decrease, setting the stage for increased recovery yield.

TABLE III. Experimental Conditions for Model Validation

	Base Case	Increased Recovery	Increased PF & CF
Reactor Volume	100	100	100
Adsorbent Volume Fraction	0.17	0.15	0.20
Flow Rates: (ml/min)			
Feed	0.13	0.018	0.11
Product	1.0	0.90	0.35
Adsorbent Recycle	0.18	0.12	0.29
Wash	6.0	2.9	12

For the second deviation from the base case increasing the amount of adsorbent in the system, while keeping total reactor volume constant, also increases the adsorbent volume fraction. Both of these changes, result in increased performance, across the board. To further increase the purification and concentration factors, the wash flow rate was increased and the product flow rate was decreased, respectively; these improvements are expected to be at the expense of decreasing recovery yield.

The results for this experiment are shown in Table IV. The flow rate ratios, controlling system performance have been normalized to those of the base case. Steady state performance is shown for each case. Overall recovery was increased from 72 % to 81 %. This was achieved at the expense of decreases in both purification and concentration factors. Increased recovery yield results from a decrease in the amount of feed to the system as well as decreases in the (wash+feed)/recycle and feed/product flow rate ratios.

TABLE IV. Experimental Resullts for Model Validation

Experiment	Ratio of Flow Rates			Performance		
	Feed + Wash / Recycle	Feed / Elute	Feed	PF	CF*	Recovery
Base Case	1	1	1	19	0.09	72%
High Recovery	0.70	0.16	0.14	13	0.02	81%
High Purification & High Concentration	1.20	3.10	0.87	40	0.16	58%

* A preconcentrated *E.coli* homogenate feed was utilized for this series of experiments. The concentration factors for this experiment, relative to the original homogenate are 0.9, 0.2, and 1.6, respectively, for the three experimental conditions.

A second example demonstrates the inherent flexibility of the system such that different operating flow rates result in drastically different performance. In the first example only one of the performance parameters was improved. In this second example, both the purification and concentration factors were increased above that of the base case; purification factor was increased from 19 to 40 and concentration factor was increased from 0.09 to 0.16. Improved performance is partially attributable to the increase in the amount of adsorbent as well as the adsorbent volume fraction. In addition, the (wash+feed)/recycle flow rate ratio was increased, resulting in a greater purification factor. In a similar manner the feed/product flow rate ratio was increased, resulting in a greater concentration factor.

ION-EXCHANGE PURIFICATION EXAMPLE. CARE is a generic approach to carrying out continuous protein purification based on adsorption to solid adsorbents. Adsorption mechanisms are not restricted to affinity interactions, although, this type of interaction is well-suited for this type of continuous operation. To demonstrate the generality of this purification approach, the continuous purification of β-galactosidase from *E.coli* using ion-exchange adsorption was examined.

The adsorptive purification of β-galactosidase was established using an FPLC column system (Pharmacia, Piscataway, NJ), with a column packed with DEAE Trisacryl M (IBF Biotechnics, Savage, MD), and sample application in a piperazine-HCl, pH 5.8 buffer followed by desorption in a 0-1 M NaCl salt gradient. Typical chromatograms are shown in Figure 2. Note that β-galactosidase interacts more

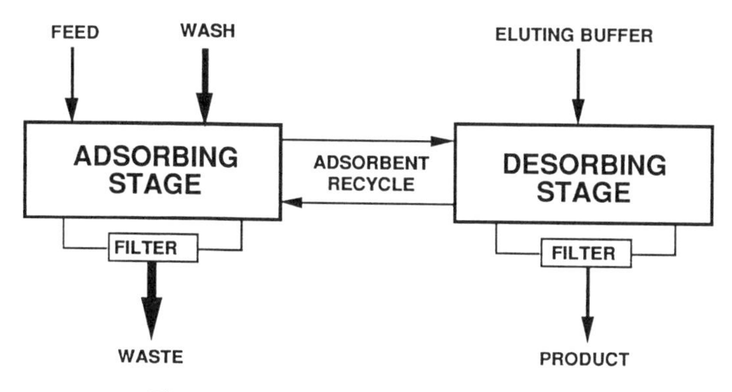

Figure 1. Schematic of the CARE process

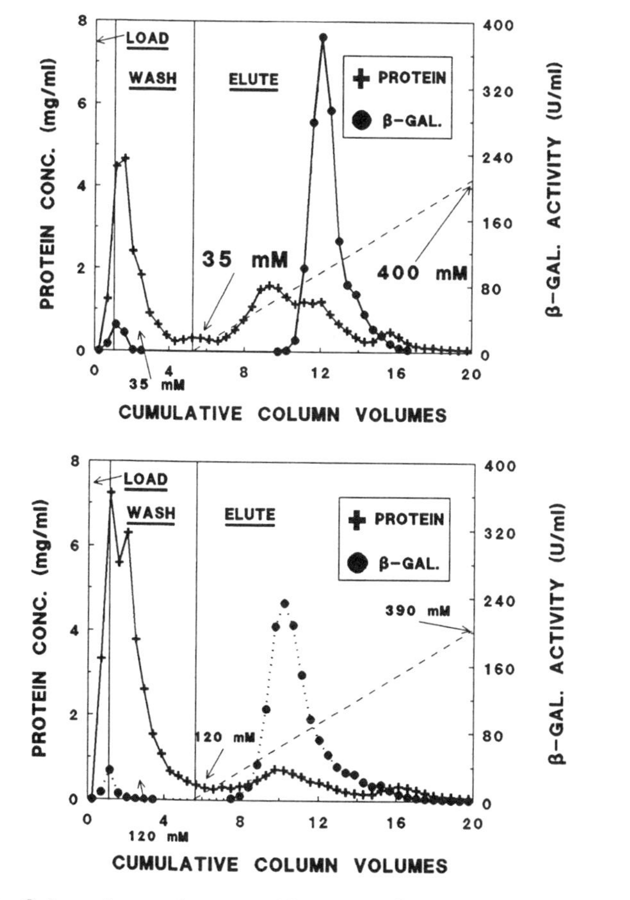

Figure 2. Column ion-exchange purification of β-galactosidase from *E.coli*

strongly with the DEAE surface chemistry and is thus eluted during the latter stages of the chromatogram. Batch adsorption experiments were conducted in order to optimize adsorption conditions. Variables considered were adsorption pH, addition of NaCl and composition of the β-galactosidase preparation.

Batch adsorption experiments for β-galactosidase adsorption to DEAE Trisacryl M were conducted for two preparations of β-galactosidase; a crude, dialyzed *E.coli* homogenate and an affinity purified β-galactosidase preparation. In addition, to the nature of the β-galactosidase preparation, the effect of salt concentration (NaCl) was investigated. The addition of NaCl to the adsorption buffer enhances selectivity for β-galactosidase adsorption by reducing the adsorption of proteins that are eluted from the ion-exchange column at lower NaCl concentration than β-galactosidase (see Figure 2). Increased selectivity is achieved at the expense of the sorbent's total protein binding capacity as the chloride ions compete for adsorption sites. In order to determine the tradeoffs between selectivity and recovery yield, a set of equilibrium adsorption experiments were carried out at increasing NaCl concentration, with results shown in Figure 3. Up to 0.075 M NaCl, β-galactosidase adsorption is unaffected; however, the adsorption of other proteins decreases, and thus, the adsorption selectivity increases. At higher salt concentrations, β-galactosidase recovery is poor.

Since ion-exchange adsorption is not as specific as the affinity adsorption system, additional selectivity, and hence purification, needs to be introduced during the desorption step. Linear elution gradients are typically used to resolve adsorbed proteins. However, since mixed vessels rather than a column is utilized, linear elution gradients cannot be used. Rather, a gradient elution scheme can be substituted with a series of step changes in ionic strength through a series of desorption reactors. A two-step desorption demonstrates this principal of step desorption. Desorption conditions were established by a series of column experiments where the NaCl concentration of the first elution step was varied between 0.1 and 0.3 M, followed by a second elution step operation at 1 M. Experimental results are summarized in Figure 4 and indicate a maximal first step elution concentration of 0.15 M.

A continuous purification experiment was conducted using one adsorption, followed by two desorption, reactors. Reactor NaCl concentrations established above, were relaxed slightly, to ensure high β-galactosidase recovery, with experimental condition shown in Figure 5. The high ionic strength (1 M NaCl) of the second desorption stage, in addition to desorbing the β-galactosidase serves to regenerate the adsorbent before it is returned to the adsorption reactor. Results from this continuous purification experiment are shown in Figure 6 with a recovery yield of sixty percent (60%) and a purification factor of seven (7). In addition, approximately ninety-five percent of incoming solid contaminants were removed from the feed, and carried away in the two waste streams.

Recovery yield was lower than anticipated and likely results from our inability to close the β-galactosidase material balance. Only 82 % of the incoming β-galactosidase in the feed stream could be accounted for in the exiting streams (waste1, waste2, product). It is postulated that enzyme inactivation, probably oxidative in nature due to the vigorous mixing, was occurring and thus decreased the recovery yield of active product. Although the addition of a reducing agent (2-mercaptoethanol) in the affinity purification experiments successfully eliminated

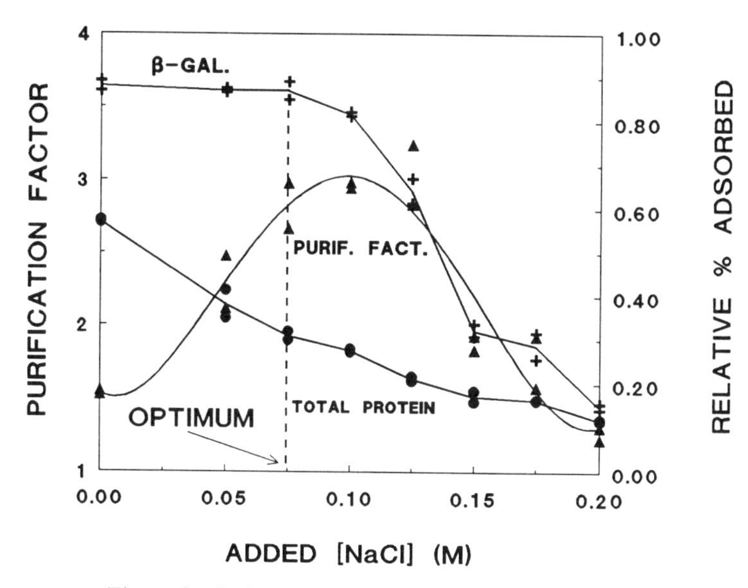

Figure 3. Optimization of batch adsorption conditions

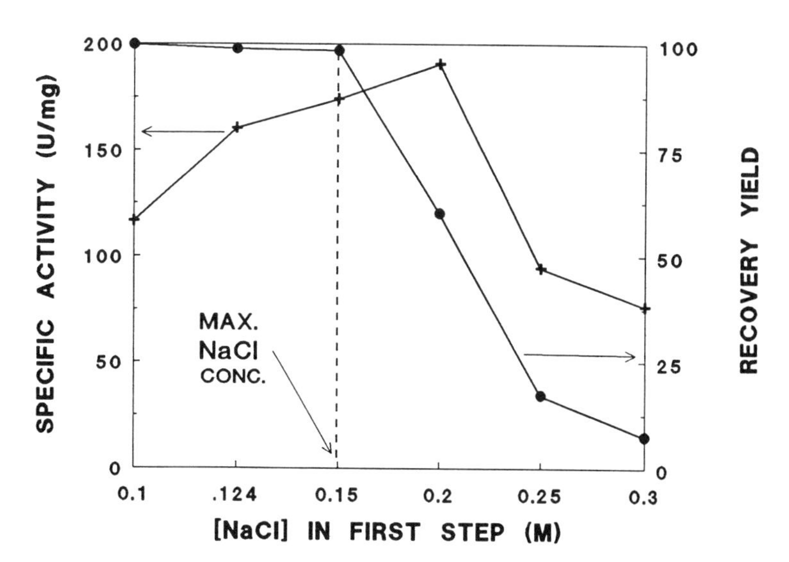

Figure 4. Optimization of two-step desorption process

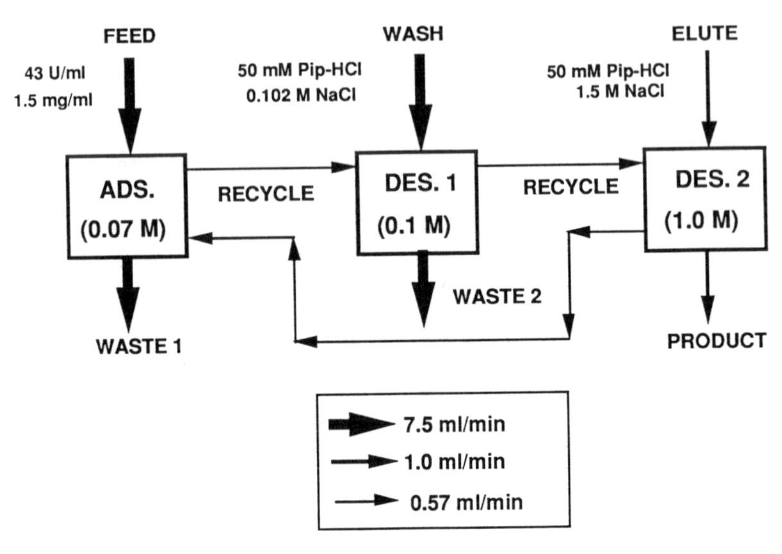

Figure 5. Conditions for continuous ion-exchange purification

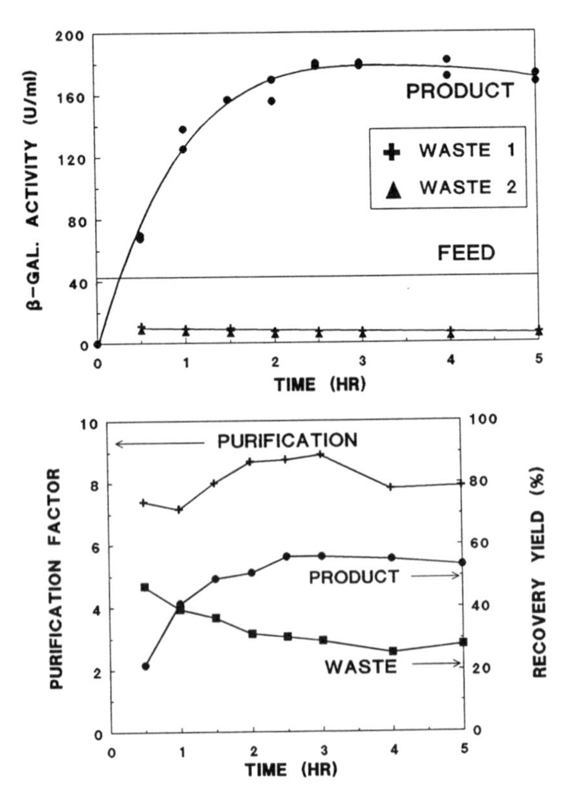

Figure 6. Continuous ion-exchange purification results

enzyme inactivation, this approach was not successful for the ion-exchange case. The addition of reducing agent to the pH 5.8, piperazine-HCl buffer used in the ion-exchange adsorption experiments, had an adverse effect on enzyme stability and resulted in rapid enzyme inactivation, even in the absence of agitation. Hence, experiments were conducted in the absence of reducing agent.

This experiment illustrates the integration of purification, concentration and clarification (removal of solids) in a single operation, eliminating the need for separate clarification and concentration steps. In analogy to the affinity purification example, it is anticipated that this integration will be beneficial to subsequent processing steps and consequently result in higher overall recovery yields.

CARE IN THE CONTEXT OF DOWNSTREAM PROCESSING

PLACEMENT of CARE in a DOWNSTREAM PROCESSING SEQUENCE. CARE can be utilized at several points in the sequence; however, the benefits of the CARE system are best exploited by its early introduction in a purification train. Through early introduction, several objectives can be integrated into this first purification step, increasing the efficiency of subsequent processing steps. This point is demonstrated by the following example.

The performance of CARE is contrasted with that of centrifugation for the clarification step following cell disruption (removal of cell debris) in the recovery band isolation process of β-galactosidase produced in *E.coli*. Pilot plant data for cell debris removal in a continuous centrifuge (81) was contrasted with simulated performance of the CARE system (details of the simulation procedure are provided in reference (18). Two CARE configurations were simulated; the base two stage design and a three stage process incorporating a wash stage between the adsorption and desorption reactors. The second CARE design operates with higher purification; this includes higher clarification as well as protein enrichment.

Simulation results are shown in Figures 7. The CARE system can remove an equivalent amount of cell debris while accomplishing higher recovery yield and concentration. More importantly, through the introduction of a highly specific adsorptive step, significant purification is achieved simultaneously

The impact of early introduction of CARE in the purification sequence has many implications. We believe there is significant opportunity to cut out one or several processing steps resulting in higher overall recovery yield. In addition, early elimination of contaminating proteins, such as proteases as well as introduction of a buffered environment can enhance product stability, and hence improve recovery of active product. Finally, reduction of contaminants can enhance the performance of subsequent purification steps in the downstream processing operation, further increasing the overall recovery yield. The ultimate measure for this unit operation is its ability to reduce the overall cost of purification. This measure, however, is case dependent and requires extensive optimization and experimental validation of several alternate purification sequences, a task which is beyond the scope of this paper. Based on our evaluation of the benefits and tradeoffs inherent to the CARE system, we have reason to speculate that implementation of CARE in a downstream processing sequence will reduce the overall cost of purification.

RELATIVE PERFORMANCE of the CARE SYSTEM. For difficult separation problems, where selectivity is low, or two components of similar interaction strength with the adsorbent need to be resolved, the necessity for a large number of contacting stages is well established. Such purifications are typically conducted in fixed bed contactors (column), where the liquid feed travels down the bed in plug flow (assuming negligible axial dispersion and bulk mixing), creating as many as several

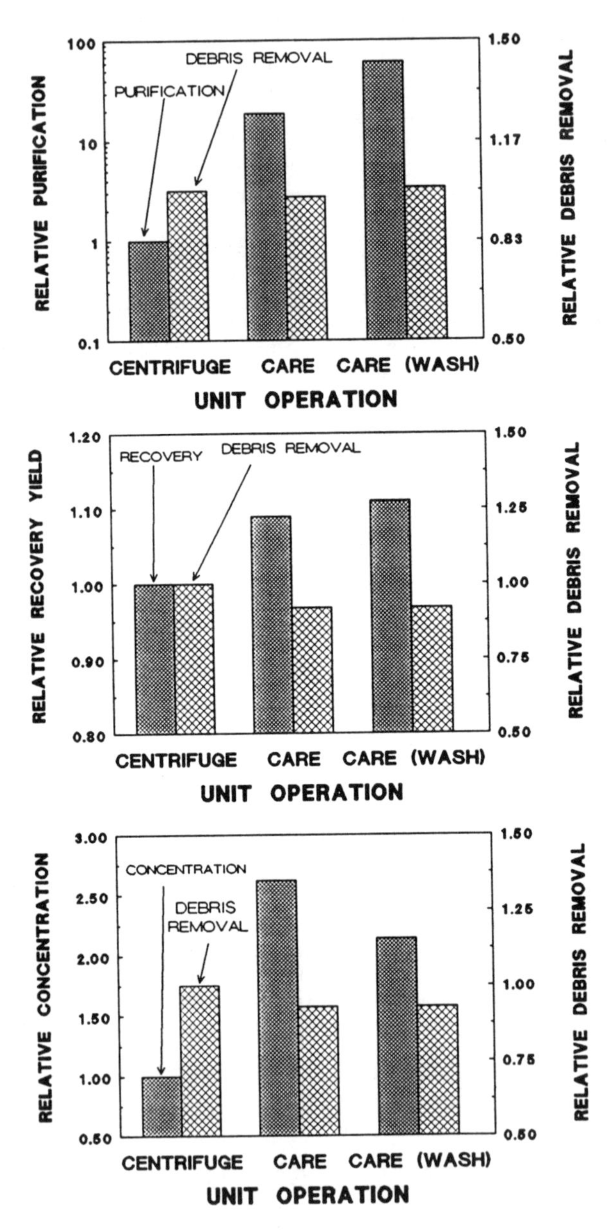

Figure 7. Comparison of CARE and centrifugation

thousand theoretical adsorption plates or stages. In contrast, CSTR adsorption (CARE) provides fewer contacting stages (one for each reactor in the process), and thus, can result in a significant drop in adsorptive capacity relative to the fixed bed mode of operation. The relative performance of these two contactors depends on the adsorption kinetics of the system, the selectivity of the adsorption process, and flow rate through the contactors. Affinity adsorption of proteins is typically characterized by favorable equilibria, high association constants and high selectivity. Consequently, the adsorptive capacity provided by a column's multiple contacting stages might be approached in only a few CSTR contacting stages.

When solute is continually fed to a constant volume adsorber, solute initially adsorbs to the adsorbent. As more solute is introduced the adsorbent loading increases and solute begins to emerge in the adsorber effluent. The plot of effluent concentration versus time (processed volume) is known as a breakthrough curve. When the adsorbed enzyme (sorbate) concentration attains equilibrium with the solute concentration in the feed, no more product is adsorbed and the effluent concentration attains that of the feed. The maximum (equilibrium) capacity of the adsorber is equal to the area above the breakthrough curve. CSTR adsorption breakthrough curves were simulated for systems of increasing number of contact stages, with simulation results shown in Figure 8 (simulation conditions are shown in Table V). Mass transfer resistances were combined into the forward adsorption rate constant (k_f), and axial dispersion was assumed to be negligible. These systems, all operate with fixed total volumetric residence time, feed concentration and amount of sorbent material, but a varying number of contacting stages. For example, a single CSTR containing 50 ml of sorbent material can be compared to 5 CSTR's operating in series, with 10 ml of sorbent per CSTR, where the residence time per stage is five fold lower in the latter case. CSTR adsorption behavior is contrasted to that for a column (simulation conditions are shown in Table VI), operating with the same amount of adsorbent.

TABLE V. Conditions for Simulation of CSTR Adsorption

Assumptions	Conditions
• Simplified adsorption model applies	• 60 ml adsorbent
• Reactor is well-mixed	• V_{fg}: 0.5
	• Q_{max}: 19,500 U/ml
	• K: 0.085 ml/U
	• k_f: 0.00035 ml/U-s
	• C_0: 100 U/ml
	• Flow: 34 ml/min
	• τ*: 3.5 min

* Residence time calculated as the ratio of contactor volume to flow rate

For single stage CSTR contact, breakthrough occurs early and would result in poor recovery yields if adsorption were carried out under these conditions. However, splitting up the CSTR into 2 stages dramatically effects the shape of the breakthrough curve, resulting in higher adsorbent saturation as a function of time, and thus, less solute loss due to premature breakthrough. As the number of CSTR's is increased, breakthrough behavior becomes sharper and approaches the plug flow (column) limit. The incremental benefit of each additional adsorption stage decreases rapidly, indicating column adsorption efficiency can be approached with as few as five adsorption stages.

TABLE VI. Conditions for Simulation of Column Adsorption

Assumptions	Conditions
• Simplified adsorption model applies • Plug flow • No axial dispersion • Column lenght/diameter = 5	• 60 ml adsorbent • ε: 0.5 • Q_{max}: 19,500 U/ml • K: 0.085 ml/U • k_f: 0.00035 ml/U-s • C_o: 100 U/ml • Flow: 34 ml/min • τ*: 2.9 min

* Residence time calculated as the ratio of contactor volume to flow rate

Once adsorption or "loading" is completed, non-adsorbed components, such as proteins, nucleic acids,etc. are removed from the adsorber, prior to elution and recovery of the protein product. The wash step continues until the outlet concentration, from the adsorber, decreases to a specified level. Contaminant removal, in analogy to the adsorption step, can be characterized by a breakthrough curve, where contaminant concentration in the adsorber effluent decreases from an initial value to zero as shown in Figure 9. The area above the breakthrough curve gives the total mass of contaminant removed from the adsorber, and the area under the curve indicates the amount of contaminant still remaining in the adsorber. As was the case for adsorption capacity, contaminant removal is more efficient (requires less wash buffer) for plug flow relative to CSTR contactors.

Having assessed the comparative adsorptive capacity and contaminant removal performance of CSTR and fixed bed (column contactors), we are in a position to draw the following conclusions regarding the relative performance of CARE: Adsorptive purification, utilizing the CARE process, can be introduced into a process, at a location where column chromatography is not possible (with solid contaminants and viscous material). This early introduction of a high resolution purification step should positively influence the remaining steps in the process, and lower overall purification costs. If the reduction in total adsorptive capacity, associated with single stage, CSTR contact (base case CARE design) is important (i.e. if the cost of adsorbent dominates the purification costs), the adsorbent can be split up into several CSTR's, increasing the adsorptive capacity. Simulations have shown that the adsorptive capacity of a five-stage CSTR contacting device approaches the maximal adsorptive capacity for the adsorbent (obtained with an infinite number of contacting stages).

In contrast, column contactors, operating in plug flow are an effective device for obtaining multiple contacting stages, maximizing adsorptive capacity and hence, recovery yield. Contaminant wash-out is accomplished within one to two column wash volumes, for nonadsorbing contaminants, resulting in large purification factors. For feed streams that do not require clarification or viscosity reduction, the benefits of column operation (high capacity, high purification factor) make it an attractive and in some cases, preferred alternative to CARE.

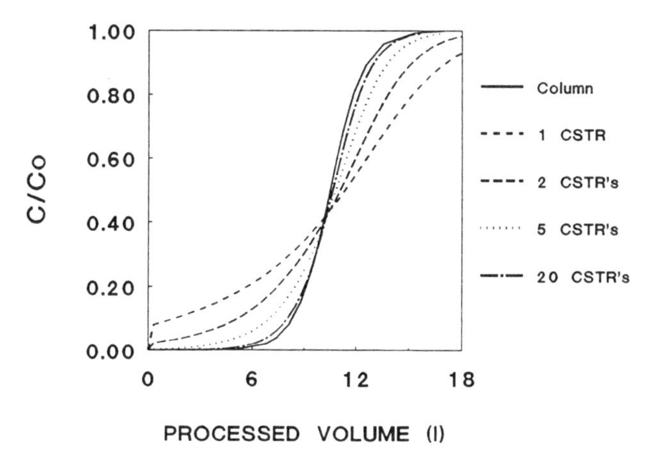

Figure 8. Simulated solute breakthrough curves

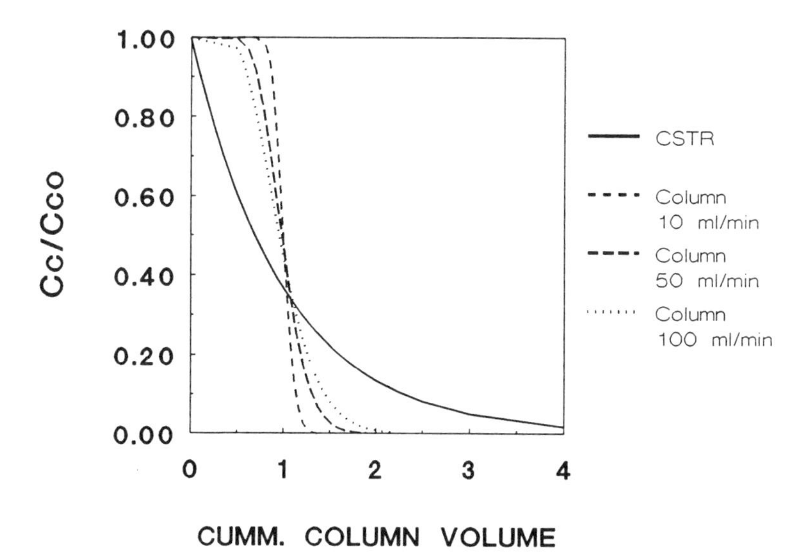

Figure 9. Simulated contaminant wash-out profiles

ACKNOWLEDGMENTS

The authors would like to acknowledge Aspen Technology Inc. of Cambridge, MA for use of the BPS simulation software as well as their computing facilities. In addition, the contributions of Jeff Kolodney, Rolf Jansen and Philip Gomez III during the experimental portion of this work, as well as Hideo Tsujimura for conducting mathematical simulations, are gratefully acknowledged. Project funding was obtained from two sources; the National Science Foundation under the Engineering Research Center Initiative to the Biotechnology Process Engineering Center (Cooperative Agreement CDR -88-0314) and Alfa Laval. In addition, Neal Gordon was supported in part by the National Science and Engineering Research Council of Canada.

LITERATURE CITED

1. Pfund, N.E. (1987). *The Wheat From The Chaff: The Separations Industry Comes of Age"*, Hambrecht & Quist Inc., San Francisco.
2. Roy, S.K. and A.H. Nishikawa (1979). Large-Scale Isolation of Equine Liver Alcohol Dehydrogenase on Blue-Agarose Gel. *Biotechnol. Bioeng.* 21, 775-785.
3. Ostlund, C. (1986). Large-Scale Purification of Monoclonal Antibodies. *Trends Biotechnol.* November, 288-293.
4. Robinson, P.J., M.A. Wheatley, J.C. Janson, P. Dunnill and M.D. Lilly (1974). Pilot Scale Affinity Chromatography: Purification of b-Galactosidase. *Biotechnol. Bioeng.* 16, 1103-1112.
5. Sherwood, F., G. Melton, S.M. Alwin and P. Hughes (1985). Purification and Properties of Carboxypeptidase G_2 From *Pseudomonas sp.* Strain RS-16. Use of a Novel Triazine Dye Affinity Method. *Eur. J. Biochem.* 148, 447-453.
6. Fulcher, C.A. and T.S. Zimmerman (1982). Characterization of the Human Factor VIII Procoagulant Protein With a Heterologous Precipitating Antibody. *Proc. Natl. Acad. Sci USA* 79, 1648-1652.
7. Goward, R., R. Hartwell, T. Atkinson and M.D. Scawen (1986). The Purification and Characterization of Glucokinase From the Thermophile *Bacillus stearothermophilus*. *Biochem. J.* 237, 415-420.
8. Wikstrom, P. and P.-O. Larsson (1987). Affinity Fibre - A New Support for Rapid Enzyme Purification by High-Performance Liquid Affinity Chromatography. *J. Chromatogr.* 388, 123-134.
9. Kulbe, K.D. and R. Schuer (1979). Large-Scale Preparation of Phosphoglycerate Kinase from *Saccharomyces cervisiae* Using Cibacron Blue-Sepharose 4B Pseudoaffinity Chromatography. *Anal. Biochem.* 93, 46-51.
10. Einarsson, M., J. Brandt and L. Kaplan (1985). Large-Scale Purification of Human Tissue-Type Plasminogen Activator Using Monoclonal Antibodies. *Biochim. Biophys. Acta* 830, 1-10.
11. Dodd, I., S. Jalalpour, W. Southwick, P. Newsome, M.J. Browne and J.H. Robinson (1986). Large Scale, Rapid Purification of Recombinant Tissue-Type Plasminogen Activator. *FEBS Letters*209(1), 13-17.
12. Stump, D.C., M. Thienpont and D. Collen (1986). Urokinase-Related Proteins in Human Urine. *J. Biol. Chem.* 261(3), 1267-1273.
13. Pearson, J.D. (1986). High-Performance Liquid Chromatography Column Length Designed For Submicrogram Scale Protein Isolation. *Anal. Biochem.* 152, 189-198.
14. Wankat, P.C. (1986). *Large-Scale Adsorption and Chromatography*, Vol. 2, CRC Press, Inc., Boca Raton.

15. Brewer, S.J. and B.R. Larsen (1987). Isolation and Purification of Proteins Using Preparative HPLC. In *"Separations for Biotechnology"*, M.S. Verrall and M.J. Hudson, eds., Ellis Horwood Ltd., Chichester, pp. 113-126.
16. Pungor, E., Jr., N.B. Afeyan, N.F. Gordon and C.L. Cooney. (1987). Continuous Affinity-Recycle Extraction: A Novel Protein Separation Technique. *Bio/Technol.* 5(6), 604-608.
17. Afeyan, N.B., N.F. Gordon and C.L. Cooney (1989). Mathematical Modelling of the Continuous Affinity-Recycle Extraction Purification Technique. *J. Chromatogr.* 478, 1-19.
18. Gordon, N.F. (1989). Characterization and Development of Continuous Affinity-Recycle Extraction (CARE). PhD Thesis, The Massachusetts Institute of Technology, Cambridge, MA
19. Lerch, R.G. and D.A. Ratkowsky (1967). Optimum Allocation of Adsorbent inStagewise Adsorption Operations. *I&EC Fundam.* 6(2), 308-310.
20. Lucas, J.P. and D.A. Ratkowsky (1969). Optimization in Various Multistage Adsorption Operations. *I&EC Fundam.* 8(3), 576-581.
21. Hustedt, H., K.H. Kroner and M.-R. Kula (1984). Extractive Purification of Enzymes. In *"Enzyme Technology"*, R.M. Lafferty, ed., Springer-Verlag, Berlin, pp. 135-145.
22. Hustedt, H., K.H. Kroner, H. Schutte and M.-R. Kula (1983). Continuous Enzyme Purification by Crosscurrent Exctraction. In *"3rd European Congress on Biotechnology"*, Verlag Chemie, Aweinheim, pp. I-597 - I-605.
23. Mattiasson, B. and M.J. Ramstorp (1984). Ultrafiltration Affinity Purification - Isolation of Concanavalin A From Seeds of Canavalia Ensiformis. *J. Chromatogr.* 283, 323-330.
24. Mattiasson, B. and M. Ramstorp (1983). Ultrafiltration Affinity Purification. *Ann. N.Y. Acad. Sci.* 413, 307-309.
25. Mattiasson, B., T.G.I. Ling and J.L. Nilsson (1983). Ultrafiltration Affinity Purification. In *"Affinity Chromatography and Biological Recognition"*, I.M. Chaiken, M. Wilchek and I. Parikh, eds., Academic Press, Inc., New York, pp. 223-228.
26. Adamski-Medda, D., Q.T. Nguyen. and E. Dellacherie (1981). Biospecific Ultrafiltration: A Promising Purification Technique for Proteins?. *J. Membrane Sci.* 9, 337-342.
27. Geahel, I. and M.-R. Kula (1984). Integration of Ion Exchange and Ultrafiltration Steps Studied During Purification of Formate Dehydrogenase Using DEAE Dextran. *Biotech. Letters*6(8), 481-486.
28. Luong, J.H.T., A.L. Nguyen and K.B. Male (1987a). Affinity Cross-Flow Filtration for Purifying Biomolecules. *Bio/Technol.* 5, 564-566.
29. Luong, J.H.T., A.L. Nguyen and K.B. Male (1987b). Recent Developments in Downstream Processing Based On Affinity Interactions. *Trends Biotechnol.* 5, 281-286.
30. Luong, J.H.T., K.B. Male and A.L. Nguyen (1988a). A Continuous AffinityUltrafiltration Process for Trypsin Purification. *Biotechnol. Bioeng.* 31, 516-520.
31. Luong, J.H.T., K.B. Male and A.L. Nguyen (1988b). Synthesis and Characterization of a Water-Soluble Affinity Polymer for Trypsin Purification. *Biotechnol. Bioeng.* 31, 439-446.
32. Luong, J.H.T., K.B. Male, A.L. Nguyen and A. Mulchandani (1988c). Mathematical Modeling of Affinity Ultrafiltration. *Process. Biotechnol.* Bioeng. 32, 451-459.
33. Herak, D.C. and E.W. Merrill (1989). Affinity Cross-Flow Filtration: Experimental and Modeling Work Using the System of HSA and Cibacron Blue-Agarose. *Biotechnol. Prog.* 5(1), 9-17.

34. Prout, W.E. and L.P. Fernandez (1961). Performance of Anion Resins in Agitated Beds. *Ind. Eng. Chem.* 53(6), 449-452.
35. Hutchins, R.A. (1979) Activated-Carbon Systems for Separation of Liquids. In *"Handbook of Separation Techniques for Chemical Engineers"*, P.A. Schweitzer, ed., McGraw-Hill Book Co., New York, pp. 1-415-1-447.
36. Eisele, J.A., A.F. Columbo and G.E. McClelland (1983). Recovery of Gold and Silver From Ores by Hydrometallurgical Processing. *Sep. Sci. Technol.* 18(12&13), 1081-1094.
37. Arehart, T.A., J.C. Bressee, C.W. Hancher and S.H. Jury (1956). Countercurrent Ion Exchange. *Chem. Eng. Progr.* 52(9), 353-359.
38. Bartels, C.R., G. Kleiman, J.N. Korzun and D.B. Irish (1958). A Novel Ion-Exchange Method for the Isolation of Steptomycin. *Chem. Eng. Progr.* 54(8), 49-51.
39. Belter, P.A., F.L. Cunningham and J.W. Chen (1973). Development of a Recovery Process for Novobiocin. *Biotechnol. Bioeng.* 15, 533-549.
40. Berg, C. (1951). Hypersorption design. Modern Advancements. *Chem. Eng. Prog.* 47, 585-591.
41. Cloete, F.L.D. and M. Streat (1963). A New Continuous Solid-Fluid Contacting Technique. *Nature* 200, 1199-1200.
42. Dodds, R., P.I. Hudson, L. Kershenbaum and M. Streat (1973). The Operation and Modelling of a Periodic, Countercurrent, Solid-Liquid Reactor. *Chem. Eng. Sci.* 28, 1233-1248.
43. Selke, W.A. and H. Bliss (1951). Continuous Countercurrent Ion Exchange. *Chem. Eng. Progr.* 47(10), 529-533.
44. van der Wiel, J.P. and J.A. Wesselingh (1989). Continuous Adsorption in Biotechnology. To be Published In *"NATO ASI Adsorption: Science and Technology"*, A.E. Rodriques et al., eds., Kluwer Academic Publishers, Dodrecht.
45. Burns, M.A. and D.J. Graves (1985). Continuous Affinity Chromatography Using a Magnetically Stabilized Fluidized Bed. *Biotechnol. Prog.* 1(2), 95-103.
46. Siegell, J.H., G.D. Dupre and J.C. Prikle, Jr. (1986). Chromatographic Separations in a Crossflow Magnetically Stabilized Bed. *AIChE Symp. Ser.* 82(250), 128-134.
47. Broughton, D.B. (1984-85). Production-Scale Adsorptive Separations of Liquid Mixtures by Simulated Moving-Bed Technology. *Sep. Sci. Technol.* 19(11&12), 723-736.
48. Broughton, D.B. (1968). Molex: Case History of a Process. *Chem. Eng. Progr.* 64(8), 60-65.
49. Martin, A.J.R. (1949). Summarizing Paper. *Discuss. Faraday Soc.* 7, 332-336.
50. Sussman, M.V. (1976). Continuous Chromatography. *Chemtech* April, 260-264.
51. Fox, Jr., J.B. (1969). Continuous Chromatography Apparatus - II. Operation. *J. Chromatogr.* 43, 55-60.
52. Fox, Jr., J.B., R.C. Calhoun and W.J. Eglinton (1969). Continuous Chromatography Apparatus - I. Construction. *J. Chromatogr.* 43, 48-54.
53. Nicholas, R.A. and J.B. Fox, Jr. (1969). Continuous Chromatography Apparatus - III. Application. *J. Chromatogr.* 43, 61-65.
54. Scott, C.D., R.D. Spence and W.G. Sisson (1976). Pressurized Annular Chromatograph for Continuoius Separations. *J. Chromatogr.* 126, 381-400.
55. Canon, R.M., J.M. Begovich and W.G. Sisson (1980). Pressurized Continuous Chromatography. *Sep. Sci. Technol.* 15(3), 655-678.56.
56. Hughes, J.J. and S.E. Charm (1979). Method for Continuous Purification of Biological Material Using Immunosorbent. *Biotechnol. Bioeng.* 21, 1439-1455.

57. Frost, F. and D. Glaser (1970). An Apparatus for Fractional Ion-Exchnge Separation. In *"Ion Exchange in the Process Industries"*, Society of Chemical Industry, London, p.189
58. Barker, P.E, S.A. Barker, B.W. Hatt and P.J. Somers (1971). Separation by Continuous Chromatography. *Chem. Process. Eng.* 52(1), 64-66.
59. Buijs, A. and J.A. Wesselingh (1980). Batch Fluidized Ion-Exchange Column for Streams Containing Suspended Solids. *J. Chromatogr.* 201, 319-327.
60. Datar, R. and C.-G. Rosen (1986). Studies on the Removal of *Escheridhia coli* Cell Debris by Aqueous Two-Phase Polymer Extraction. *J. Biotechnol.* 3, 207-219.
61. Kroner, K.H. and M.-R. Kula (1978). Extraction of Enzymes in Aqueous Two-Phase Systems. *Process Biochem.* April, 7-10.
62. Kroner, K.H., H. Hustedt and M.-R. Kula (1984). Extractive Enzyme Recovery: Economic Considerations. *Process Biochem.* Oct., 170-179.
63. Kroner, K.H., H. Hustedt and M.-R. Kula (1982a). Evaluation of Crude Dextran as Phase-Forming Polymer for the Extraction of Enzymes in Aqueous Two-Phase Systems in Large Scale. *Biotechnol. Bioeng.* 24, 1015-1045.
64. Kula, M.-R. (1985). Liquid-Liquid Extraction of Biopolymers. In *"Comprehensive Biotechnology: The Principles, Applications and Regulations of Biotechnologyu in Industry, Agriculture and Medicine"*, M.M. Young, ed. in chief, Vol 2, Pergamon Press, New York, pp. 451-471.
65. Kula, M.-R., K.H. Kroner and H. Husted (1982). Purification of Enzymes by Liquid-Liquid Extraction. *Adv. Biochem. Eng.* 24, 73-118.
66. Mattiasson, B. (1983). Applications of Aqueous Two-Phase Systems in Biotechnology. *Trends Biotechnol.* 1(1), 16-20.
67. Cordes, A. and M.-R. Kula (1986). Process Design for Large-Scale Purification of Formate Dehydrogenase From *Candida Boidinii* by Affinity Partition. *J. Chromatogr.* 376, 375-384.
68. Flanagan, S.D., S.H. Barondes and P. Taylor (1976). Affinity Partitioning of Membranes. *J. Biol. Chem.* 251(3), 858-865.
69. Hedman, P.O. and J.-G. Custafsson (1984). Protein Adsorbents Intended for Use in Aqueous Two-Phase Systems. *Anal. Biochem.* 138, 411-415.
70. Johansson, G., M. Joelsson (1985). Partial Purification of D-Glucose-6-Phosphate Dehydrogenase From Bakers' Yeast by Affinity Partitioning Using Polymer-Bound Triazine Dyes. *Enzyme Microb. Technol.* 7, 629-634.
71. Johansson, G., M. Joelsson and H.-E. Åkerlund (1985a). An Affinity-Ligand Gradient Technique for Purification of Enzymes by Counter-Current Distribution. *J. Biotechnol.* 2, 225-237.
72. Johansson, G., M. Joelsson and B. Olde (1985b). Affinity Partitioning of Biopolymers and Membranes in Ficoll-Dextran Aqueous Two-Phase Systems. *J. Chromatogr.* 331, 11-21.
73. Kopperschlager, G. and G. Birkenmeier (1986). Affinity Partitioning: A New Approach for Studying Dye-Protein Interactions. *J. Chromatogr.* 376, 141-148.
74. Kopperschlager, G. and G. Johansson (1982). Affinity Partitioning with Polymer-Bound Cibacron Blue F3G-A for Rapid, Large-Scale Purification of Phosphofructokinase from Baker's Yeast. *Anal. Biochem.* 124, 117-124.
75. Kroner, K.H., H. Schutte, W. Stach and M.-R. Kula (1982b). Scale-Up of Formate Dehydrogenase by Partition. *J. Chem. Tech. Biotechnol.* 32, 130-137.
76. Mattiasson, B. and T.G.I. Ling (1986). Efforts to Integrate Affinity Interactions With Conventional Separation Technologies: Affinity Partition Using

Biospecific Chromatographic Particles in Aqueous Two-Phase Systems. *J. Chromatogr.* 376, 235-243.
77. Tjerneld, F., G. Johansson and M. Joelsson (1987). Affinity Liquid-Liquid Extraction of Lactate Dehydrogenase on a Large Scale. *Biotechnol. Bioeng.* 30, 809-816.
78. Larsson,P.-O. and K. Mosbach (1979). Affinity Precipitation of Enzymes. *FEBS Letters* 98(2), 333-338.
79. Mosbach, K., M. Glad, P.-O. Larsson and S. Ohlson (1982). Affinity Precipitation (B) and High Performance Liquid Affinity Chromatography (C). In *"Affinity Chromatography and Related Techniques"*, T.C.J. Gribnau, J.Visser and R.J.F. Nivard, eds., Elsevier Scientific Publishing Co., Amsterdam, pp. 201-206.
80. Schneider, M., C. Guillot and B. Lamy (1981). The Affinity Precipitation Technique. Application to the Isolation and Purification of Trypsin from Bovine Pancreas. *Annals N.Y. Acad. Sci.*, 369, 257-263.
81. Datar, R. (1985). A Comparitive Study of Primary Separation Steps in Fermentation. PhD Thesis, The Royal Institute of Technology, Sweden

RECEIVED December 28, 1989

Chapter 10

Novel Metal-Affinity Protein Separations

S.-S. Suh, M. E. Van Dam, G. E. Wuenschell, S. Plunkett, and F. H. Arnold

Division of Chemistry and Chemical Engineering 210–41, California Institute of Technology, Pasadena, CA 91125

The affinity exhibited by proteins for chelated metals has been exploited in two new protein purification techniques: metal affinity aqueous two-phase extraction and metal affinity precipitation. Metal-chelating derivatives of polyethylene glycol have been used to selectively enhance the partition coefficients of proteins in aqueous two-phase separations. When the metal chelate is attached to both ends of the water-soluble polymer, the resulting bis-chelate is an effective precipitating agent for proteins that contain multiple surface-accessible histidines. Parameters important in the design and application of these affinity separations include pH, the number of accessible histidines on the proteins, and the design of the metal affinity ligand. These techniques offer the potential for large-scale application and integration of an affinity separation early in the purification process.

Interactions between particular surface amino acids and immobilized metal ions provide the basis for metal affinity protein separations. The first purification technique which exploited the affinity exhibited by proteins for metal ions was introduced by Porath (1), who used the divalent cations Zn and Cu chelated to a chromatographic matrix to fractionate serum proteins. Immobilized metal affinity chromatography (IMAC) is an attractive alternative to affinity chromatography with biospecific ligands. The small, stable metal chelate ligand is amenable to modification to prepare high-capacity, easily-regenerated affinity supports. The principles and applications of IMAC have been reviewed by Lönnerdal et al. (2), Sulkowski (3), and Porath (4).

Two new types of metal affinity separations have recently been introduced: metal affinity aqueous two-phase extraction (Wuenschell, G. E.; Naranjo, E.; Arnold, F. H. *Bioprocess Engineering*, in press), and metal affinity precipitation

0097–6156/90/0427–0139$06.00/0

(Van Dam, M. E.; Wuenschell, G. E.; Arnold, F. H. *Biotechnol. Appl. Biochem.*, in press). These techniques retain the high selectivity of an affinity separation, but may be more easily applied to large-scale protein purification and integrated into the earlier stages of the purification process.

METAL AFFINITY AQUEOUS TWO-PHASE EXTRACTION

Basic Principles and Methods

The binding of proteins to transition metals occurs via the electron-donating side chains of residues such as histidine and cysteine, which substitute water molecules coordinated to the metal. Figure 1 illustrates a complex formed between a copper atom bound by polyethylene glycol-iminodiacetic acid (PEG-IDA) and the ϵ-nitrogen from the imidazole ring of a histidine on the protein. The ability of this complex to form and its binding constant are affected by the environment of the participating histidine. At a given pH, the apparent binding affinity to a histidine accessible on the protein surface depends on the pK_a of the individual histidine.

Typical two-phase systems for protein separations consist of aqueous mixtures of PEG and dextran or PEG and a salt such as potassium phosphate or sodium sulfate (5). When a PEG-bound metal is added to the two-phase system containing the protein mixture, proteins which interact with the metal accumulate preferentially in the PEG-rich phase. Since the metal is attached to a soluble polymer, all binding sites exposed on the protein surface are accessible and contribute to partitioning.

Modeling and Binding Studies

The processes involved in metal affinity partitioning of histidine-containing proteins are represented below. Protonation of the histidine side chain competes with metal binding; the PEG-bound metal binds only to the free base form of the imidazole. The association constants for these competing reactions are K_H and K_a, where $'$ and $''$ represent the top and bottom phases, respectively. Since the phase environments can differ substantially, binding constants in the two phases are not necessarily the same. The individual species are in equilibrium between the two phases.

$$\begin{array}{c} P_i + H^+ \overset{K'_H}{\rightleftharpoons} P_iH^+ \\ P_j + M \overset{K'_a}{\rightleftharpoons} P_jM \\ \hline P_i + H^+ \overset{K''_H}{\rightleftharpoons} P_iH^+ \\ P_j + M \overset{K''_a}{\rightleftharpoons} P_jM \end{array} \qquad (1)$$

The partition coefficient in the presence of the metal affinity ligand is given by the ratio of the total protein concentration in the top and bottom phases, where n is the number of metal binding sites.

$$K = \frac{[P_{tot}]_{top\ phase}}{[P_{tot}]_{bottom\ phase}} = \frac{\sum_{i=0}^{n} \sum_{j=0}^{n-i} [PM_i H_j]_{top}}{\sum_{i=0}^{n} \sum_{j=0}^{n-i} [PM_i H_j]_{bottom}} \tag{2}$$

The following expression for the partition coefficient has been derived for the case of n identical binding sites and $K'_H = K''_H = K_H$ (Suh, S.-S. and Arnold, F. H., *Biotechnol. Bioeng.*, in press).

$$\ln\left(\frac{K}{K_o}\right) = n \ln \left[\frac{1 + K'_a \frac{(R+1)}{(R+\frac{1}{K_m})} [M_{tot}] + K_H [H^+]}{1 + K''_a \frac{(R+1)}{(K_m R+1)} [M_{tot}] + K_H [H^+]}\right] \tag{3}$$

K_o is the partition coefficient in the absence of the metal chelate, R is the phase volume ratio, K_m is the partition coefficient of the polymer-metal chelate, and $[M_{tot}]$ is the total metal concentration in the two-phase system. Equation 3 predicts a linear dependence of $\ln(K/K_o)$ on the number of metal binding sites, when other system parameters are fixed.

Metal affinity partitioning data for a series of heme-containing proteins are shown in Table I. The system composition is 7% PEG 8000, 4.4% dextran T500, 0.1 M NaCl, 0.01 M sodium phosphate, 1.0 mg protein. The partition coefficients in the absence of copper, K_o, do not vary greatly for the different proteins at the three pH values. In contrast, partition coefficients in the presence of a small amount of affinity ligand (1.6×10^{-4} M Cu(II)IDA-PEG) are strongly dependent on the number of surface histidines and the system pH. Exposed histidines are those which have more than 1Å of imidazole surface accessible to a 3Å radius probe, determined from protein crystal structures.

$\ln(K/K_o)$ is plotted versus the number of accessible histidines in Figure 2. Over a very wide range, the increase in partitioning due to the presence of the polymer-bound metal is proportional to the number of binding sites, as predicted by Equation 3. This observation supports the assumptions that (1) histidine is the only contributor to metal binding (there are no free cysteines in these proteins) and (2) all of the exposed histidines are accessible for interaction with PEG-bound copper under these conditions. The strong dependence of $\ln(K/K_o)$ on pH reflects the protonation behavior of histidine. As the pH is reduced to 5.5, nearly all the metal-binding sites become occupied by H^+ and thereby unavailable for coordination to the metal.

We have used Equation 3 and metal affinity partitioning experiments to measure binding constants for protein-metal interactions (Suh, S.-S.; Arnold, F.

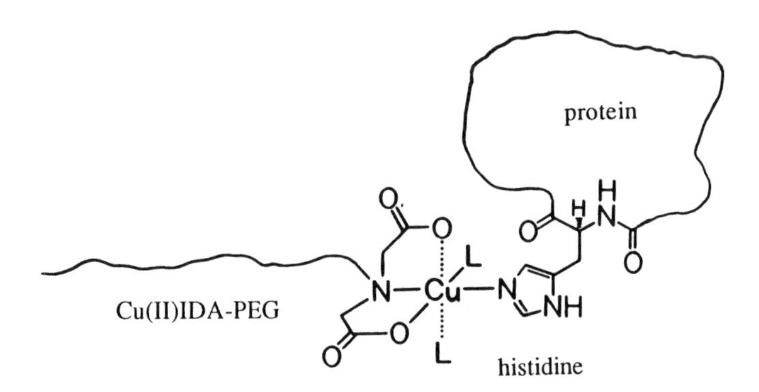

Figure 1. Complex formed by Cu(II)IDA-PEG and exposed histidine on protein. Histidine displaces metal-bound water.

Table I. Partition coefficients for heme-containing proteins in PEG/dextran (K_o) and Cu(II)PEG-IDA/dextran (K) two-phase systems.

	pH 8.0		pH 7.0		pH 5.5		Exposed histidine side chains
Proteins	K_o	K	K_o	K	K_o	K	
cytochromes c							
tuna	0.48	0.46	0.56	0.53	0.60	0.58	0
horse	0.41	0.39	0.48	0.46	0.51	0.49	1
c. krusei	0.56	0.75	0.59	0.72	0.68	0.69	2
myoglobins							
horse	0.36	0.64	0.37	0.57	0.38	0.40	4
whale	0.42	0.85	0.40	0.73	0.45	0.50	5
hemoglobins							
cow	0.25	6.0	0.25	1.8	0.34	0.45	20
horse	0.25	7.2	0.26	2.4	0.36	0.50	~ 24
human	0.38	14	0.35	3.4	0.46	0.63	~ 24

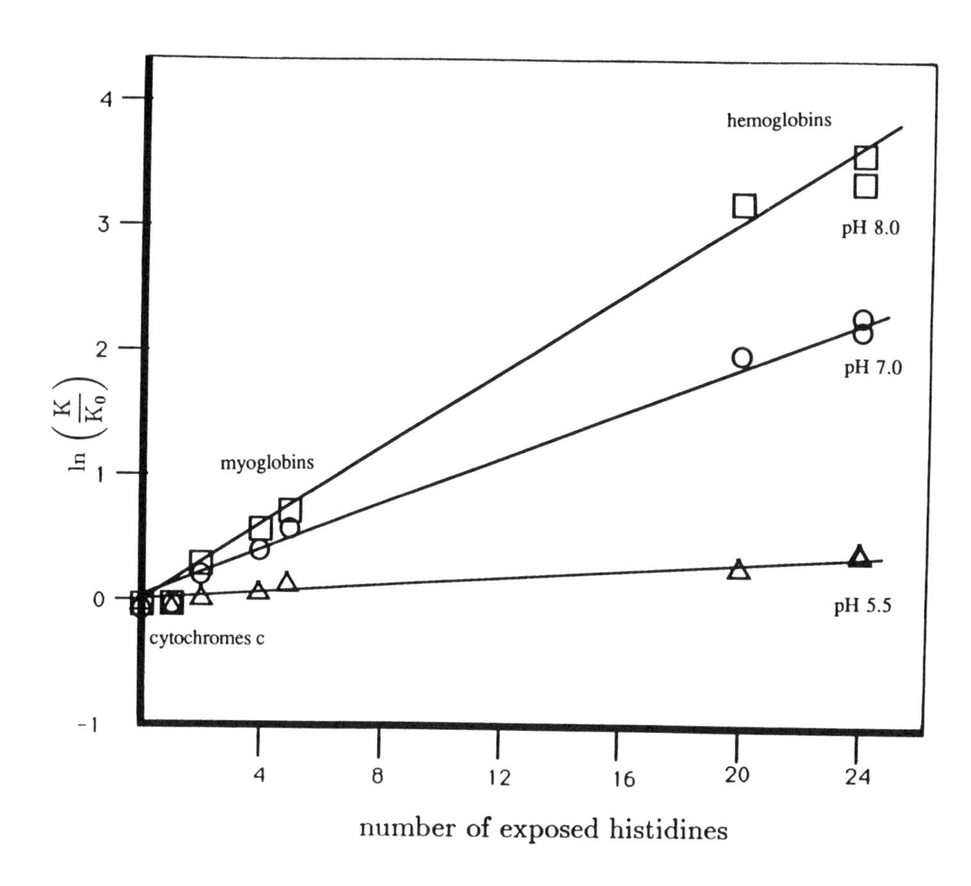

Figure 2. Plot of $\ln(K/K_o)$ versus number of accessible histidine residues at pH 8.0 (□), pH 7.0 (o) and pH 5.5 (△). Data from Table I. (Reproduced with permission from Wuenschell, G. E.; Naranjo, E.; Arnold, F. H. *Bioprocess Engineering*, in press. Copyright 1989 Springer International.)

H. *Biotechnol. Bioeng.*, in press). For the set of proteins tuna heart cytochrome c, *candida krusei* cytochrome c, horse myoglobin, and sperm whale myoglobin, an average value of 6.5 was found for the pK_a of the surface histidines ($K_H = 3.1 \times 10^6$). This value is well within the range of values commonly found for surface histidine pKa's (6.4-6.7) in titration studies. The association constant for the histidine-Cu(II)IDA-PEG complex was found to be 4500 in the dextran-rich phase (logK = 3.65). This value is very close to association constants measured by conventional potentiometric methods for analogous complexes between N-methyl-IDA-Cu(II) and various histidine derivatives. Aqueous two-phase protein partitioning appears to provide a convenient method for determining binding constants between proteins and immobilized metals. These macromolecular binding constants are difficult to measure by conventional methods due to the small concentrations.

PEG/Salt Systems

The relative high expense of the dextran component has promoted the use of two-phase systems formed from PEG and inexpensive inorganic salts. Unlike many other affinity interactions which are disrupted in high concentrations of salts, coordination to metals in aqueous solution is promoted by phase-forming salts. As a result, PEG/salt systems appear to provide an excellent medium for metal affinity extraction.

Hemoglobin partitioning in two-phase systems of PEG and Na_2SO_4 is shown in Table II (Plunkett, S.; Arnold, F. H., unpublished results). In the absence of copper, the partition coefficients are extremely small. When 1% of the PEG is substituted with Cu(II)-IDA, the partition coefficients increase dramatically, with the vast majority of the protein partitioning to the PEG phase at pH 8.0.

In order to test the ability of metal affinity aqueous two-phase extraction to separate proteins which have different affinities for copper, a single stage extraction was carried out on a mixture of human serum albumin and human hemoglobin, with the phase composition shown in Table II. The results of extractions with and without Cu(II)IDA-PEG are shown in the electrophoresis gel in Figure 3. Lanes 5 and 6 represent samples taken from the top and bottom phases of a PEG/Na_2SO_4 two-phase system which contained no metal affinity ligand. Both proteins partition almost entirely to the bottom phase, and no separation can be achieved. In contrast, extraction in the presence of Cu(II)IDA-PEG (lanes 7 and 8) causes the hemoglobin to partition strongly to the top phase, while the serum albumin remains in the bottom phase. The separation between the two proteins effected in a *single stage* is nearly quantitative.

Table II. Hemoglobin partitioning in 14% MPEG 500/8% Na_2SO_4, pH 8.0

Protein	K_o [no Cu(II)]	K [1% Cu(II)IDA-PEG]
horse hemoglobin	0.10	33
human hemoglobin	0.09	95

Future Perspectives

The structure and chemical properties of the metal chelate ligand can affect protein partitioning in two-phase systems. If one can design metal chelates for binding to histidines in particular environments, it will be possible to alter the selectivity of the extraction. Factors to consider in the design of the metal chelate for protein recognition include net charge, size, coordination number, hydrophobicity and chirality of the complex. Significant alterations in selectivity have been observed using different metal chelates (Wen, E.; Wuenschell, G. E.; Arnold, F. H., unpublished results). These differences in selectivity are manifested in deviations from the ideal linear relationship between $\ln(K/K_o)$ and the number of exposed histidines. For example, partitioning with Cu(II)dien-PEG shows considerable preference for sheep myoglobin over whale myoglobin, although the two proteins contain the same number of accessible histidines (and partition similarly in Cu(II)IDA-PEG).

Metal affinity aqueous two-phase extraction is potentially attractive for the purification of proteins with multiple surface histidines. For proteins which contain few accessible histidines, the partitioning enhancement brought on by the metal ligand is so small that multi-stage processes would be required to attain a reasonable degree of separation. Histidine is a relatively rare amino acid, and many natural proteins contain no surface histidines at all. However, progress in genetic engineering has made it possible to alter the amino acid composition of proteins, and this can be used to great advantage in the purification of recombinant proteins by metal affinity techniques. A high-affinity binding site for metals has been engineered into the surface of *S. cerevisiae* cytochrome c without disrupting its biological activity (Todd, R.; Van Dam, M.; Casimiro, D.; Haymore, B.; Arnold, F. H., submitted for publication). This protein and similar metal-binding variants of bovine growth hormones containing two histidines on adjacent turns of an α-helix exhibit partition coefficients in metal affinity extraction.that are equivalent to hypothetical proteins with 10-15 surface histidines (Suh, S.-S.; Haymore, B.; Arnold, F. H., in preparation). The partition coefficients for these engineered proteins are sufficiently high to allow substantial purification and recovery in a single extraction stage.

METAL AFFINITY PRECIPITATION

Basic Principles and Methods

Bis-metal chelates have been shown to be effective precipitating agents for multiple-histidine proteins (Van Dam, M. E.; Wuenschell, G. E.; Arnold, F. H., *Biotechnol. Appl. Biochem.*, in press). The bis-chelate contains two copper atoms at the ends of the polymer chain which serve to bind and crosslink proteins. This leads to the formation of a macromolecular network and, eventually, to precipitation. This precipitation mechanism is similar to the immunoprecipitation of multivalent antigens with bivalent antibodies.

Factors Affecting Metal Affinity Precipitation

The number of binding sites on the protein (histidines), the chelate-polymer structure, and pH influence metal affinity protein precipitation. In Figure 4 the percentage protein precipitated is plotted versus the added copper, calculated as the ratio of copper to accessible histidines (histidines/protein x protein concentration). Human hemoglobin is 100% precipitated at Cu/His = 1, due to the large number of accessible crosslinking sites on hemoglobin. Sperm whale myoglobin contains many fewer exposed histidines than hemoglobin (5-6 versus 24-26) and does not precipitate until the concentration of added copper significantly exceeds the concentration of histidines. Horse heart cytochrome c, with only one surface histidine, does not precipitate even when large quantities of the bis-chelate are added.

The precipitation curves for different bis-chelates are compared in Figure 4, where one sees that the high molecular weight $Cu(II)_2PEG$ 20000-$(IDA)_2$ is more effective than the low molecular weight $Cu(II)_2EGTA$, on a molar basis. The long PEG ligand carrier likely allows greatly flexibility for attachment to binding sites on the protein surface. The additional polymer may also decrease the solubility of the complex. As with extraction using monofunctional PEG-metal chelates, the effect of decreasing pH on protein precipitation arises from histidine protonation and fewer available binding sites.

The amount of copper in the precipitate, reported in Table III, illustrates the specific nature of the resulting protein-bis-chelate complex. When excess copper is added to the protein-containing solution, the precipitates contain approximately 26 coppers per protein for hemoglobin and 6 for whale myoglobin. These numbers are essentially equal to the number of accessible histidines, supporting the argument that precipitation is a result of metal binding to surface histidines.

Future Perspectives

The choice of ligand carrier can have a large influence on the performance of an affinity precipitation. Instead of relying on a mechanism of network formation and precipitation of the macromolecular complex, one may wish to manipulate the solubility of the ligand carrier to effect precipitation. This approach has been used effectively by Mattiasson in the precipitation of trypsin with chitosan-STI (soy bean trypsin inhibitor) (6). In principle, the metal affinity interaction

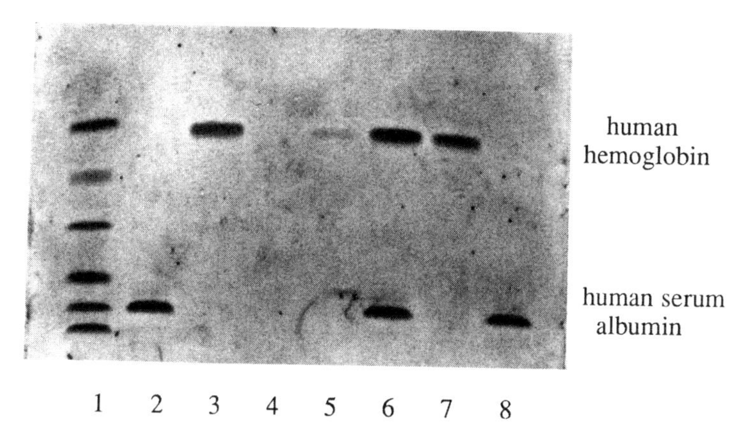

Figure 3. Separation of human hemoglobin and human serum albumin by polyacrylamide gel electrophoresis. Lane 1: molecular weight marker proteins, Lane 2: human serum albumin, Lane 3: human hemoglobin, Lane 4: empty, Lane 5: top phase (no Cu(II)IDA-PEG), Lane 6: bottom phase (no Cu(II)IDA-PEG), Lane 7: top phase (1% Cu(II)IDA-PEG), and Lane 8: bottom phase (1% Cu(II)IDA-PEG).

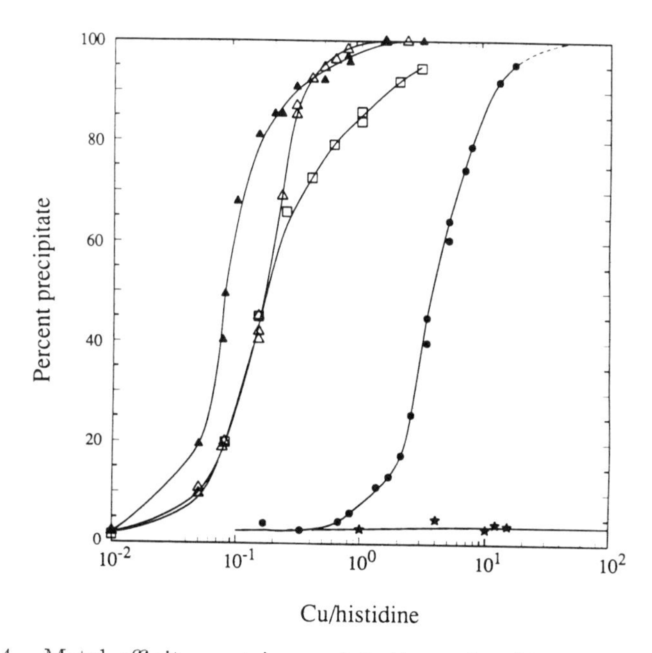

Figure 4. Metal affinity protein precipitation using bis-copper chelates. (▲) human hemoglobin, $Cu(II)_2PEG20000\text{-}(IDA)_2$, pH 8; (△) human hemoglobin, $Cu(II)_2EGTA$, pH 8; (□) human hemoglobin, $Cu(II)_2EGTA$, pH 5.5; (•) sperm whale myoglobin, $Cu(II)_2EGTA$, pH 8; (⋆) horse heart cytochrome c, $Cu(II)_2EGTA$, pH 8.

that leads to precipitation is reversible, as for the extraction process. However, the recovery of precipitated proteins has not been investigated in detail.

Table III. Copper Contents of Protein Precipitates with $Cu(II)_2EGTA$

Protein	Precipitating Agent	Cu/His[a]	Cu/protein[b] in precipitate
human hemoglobin	$Cu(II)_2EGTA$	0.77	12[c]
			13
		2.31	23
			26
		4.60	26
			26
whale myoglobin	$Cu(II)_2EGTA$	8.3	6.7
			6.0
		15	5.7
			5.3

[a] amount of precipitating agent added, calculated as ratio of copper added to number of surface-accessible histidines.

[b] determined by atomic adsorption, Cu/Fe.

[c] results of duplicate experiments.

CONCLUSIONS

A significant advance in protein purification in the last decade has been the widespread application of immobilized metal affinity chromatography, which achieves reasonable selectivities with inexpensive substitutes for natural biological affinity ligands. The two new approaches to making use of metal chelates reviewed in this paper offer the potential for high-capacity affinity separations. One of the main disadvantages of these methods is that they are limited to applications in which the proteins (or major contaminants) exhibit multiple accessible metal-binding amino acids (histidines). However, high-affinity sites engineered into the protein by site-directed mutagenesis or via polypeptide fusions (7) have been utilized effectively in metal affinity separations of recombinant proteins.

ACKNOWLEDGMENTS

This research has been supported by the National Science Foundation Grant No. EET-8807351. F. H. A. is the recipient of an NSF Presidential Young Investigator Award.

LITERATURE CITED

1. Porath, J.; Carlsson, J.; Olsson, I.; Belfrage, G. *Nature* 1975, **258**, 598.
2. Lönnerdal, B.; Keen, C. L. *J. Appl. Biochemistry* 1982, **4**, 203.
3. Sulkowski, E. *Trends Biotechnol.* 1985, **3**, 1.
4. Porath, J. *Trends Anal. Chem.* 1988, **7**, 254.
5. Albertsson, P. A. *Partitioning of Cell Particles and Macromolecules*; Wiley Interscience: New York, 1986.
6. Senstad, C.; Mattiasson, B. *Biotechnol. Bioeng.* 1989, **33**, 216.
7. Smith, M. C.; Furman, T. C.; Ingolia, T. D.; Pidgeon, C. *J. Biol. Chem.* 1988, **263**, 7211. Hochuli, E., Bannwarth, W.; Dobeli, H.; Gentz, R.; Stuber, D. *Bio/Technology* 1988, **6**, 1321.

RECEIVED January 18, 1990

Chapter 11

Recovery of Recombinant Proteins by Immunoaffinity Chromatography

Pascal Bailon and Swapan K. Roy

Department of Protein Biochemistry, Roche Research Center, Hoffman–La Roche, Inc., Nutley, NJ 07110

An overview of the use of immunoaffinity chromatography as a purification tool in downstream processing of therapeutically useful recombinant proteins is presented. Using the recovery process of recombinant interferon-alpha 2a as the model system, a systematic approach to the development and optimization of immunoaffinity purification systems is discussed. Advantages and problem areas of immunoaffinity chromatography are pointed out. Overall, the immunoaffinity purification method is shown to be a viable, scalable separation method for the purification of recombinant proteins.

The explosion of biotechnology research and its evolution into pioneering therapeutic and diagnostic products have made great demands on the fledgling bioprocess technology. The clinical use of recombinant proteins requires the capabilities of producing biologically and biochemically homogeneous materials, reliably and economically. Immunoaffinity chromatography, based upon the specificity and reversibility of the antigen-antibody interactions, is an ideal method for the selective purification of recombinant proteins.

Immunoaffinity chromatography is in fact the predecessor of affinity chromatography which became very popular in the 1970's. Most of the earlier work in this field involved the use of solid-phase antigen derivatives for the purification of homologous antibodies from immune sera (1). The first well-characterized immunoadsorbent was prepared by Campbell et al., (2) in which the antigen ovalbumin was chemically bonded to an insoluble polymeric carrier and used for the isolation of antibodies to ovalbumin. In 1964, Gurvich and Drizlikh (3) used an immobilized antibody for the detection of radiolabeled antigens.

Since the discovery of monoclonal antibody-producing hybrid cell lines by Kohler and Milstein in 1975 (4), a large number of murine monoclonal antibodies have been prepared. It is now common

0097–6156/90/0427–0150$06.00/0

practice to develop monoclonal antibodies which bind proteins at different antigenic sites and possess different binding affinities. Consequently, the use of monoclonal antibodies in biomedical industries has been increasing steadily. One such application is the use of immobilized monoclonal antibodies as immunoadsorbents for the purification of biomolecules such as interferons of various origins (5,6,7), interleukin-2 (8), blood clotting factors (9), membrane antigens (10) and interleukin-2 receptor (11) among others. The advent of fast, efficient and cost-effective technologies for the large-scale production (12), purification (13) and immobilization (14, 15) of monoclonal antibodies have made immunoaffinity chromatography a full-fledged technology for the industrial scale purification of recombinant protein therapeutics (16).

In this paper we present an overview of the use of immunoaffinity chromatography as a purification tool in the processing of recombinant DNA proteins from E.coli and mammalian cell culture supernatants. A systematic approach to the development and optimization of immunoaffinity purification systems is presented. Most of the methodologies developed for the industrial production of recombinant proteins are proprietary. Consequently, we will rely on a model system developed in our laboratory, in order to explain the systematic approach needed for purifying a recombinant protein to homogeneity and rendering it suitable for therapeutic purposes.

Immunoaffinity Concept

Molecular Recognition Property of Monoclonal Antibodies. Molecular recognition between a monoclonal antibody and its antigen occurs through the formation of a non-covalently bonded immunological complex. The above mentioned antigen-antibody interaction is the basis of immunoaffinity chromatography. Since most antibodies are bivalent, theoretically two antigen molecules can bind to one molecule of monoclonal antibody. However, this theoretical binding capacity is seldom achieved by immunoadsorbents for a variety of reasons.

The inherent properties of antigen-antibody recognition impart the following characteristics to immunoaffinity chromatography:

a. Rapid formation of a stable but reversible complex
b. High selectivity which permits the isolation of a single molecule from a complex mixture of components
c. Equivalent applicability for large or small molecules
d. Usually high recovery of biological activity

Schematic Outline of Immunoaffinity Chromatography. An overall scheme of immunoaffinity chromatography is given in Figure 1.

Monoclonal antibodies chemically bonded to an inert polymer support are used as immunoadsorbents. After proper equilibration, the crude antigen is passed through the immunosorbent column and the unadsorbed materials are washed away. The specifically bound antigen is then eluted with mild desorbing agents.

Operational Parameters in Immunoaffinity Chromatography. In order to take full advantage of the molecular recognition characteristics of immunoaffinity chromatography, proper design of the system is of utmost importance. Using immunoadsorbents made for the purification of recombinant interferon as a model system, we determined the general operational parameters in the immunoaffinity purification of recombinant proteins.

Selection of Suitable Monoclonal Antibodies. The monoclonal antibodies chosen for immobilization should possess the following characteristics: (1) they should form a reversible immunocomplex with the desired antigen; and (2) they should have high enough affinity to bind the antigen from even dilute solutions, yet low enough affinity to allow dissociation under relatively mild conditions, thus permitting the release of bound antigen.

Determination of Relative Affinities. Often antibodies which show high affinity in solid-phase immunoassays exhibit little or no affinity in immunoaffinity chromatography or vice versa (7,17). Listed in Table I are the results of some of the monoclonal antibodies screened for immobilization.

The above list does not take into account the large number of clones screened after cell fusion to identify the ones which produce antibodies specific for the antigenic site.

Competition ELISA has been used to determine the relative affinities of monoclonal antibodies (17). This is helpful in narrowing down the number of monoclonal antibodies screened for immobilization. However, this would not indicate affinities retained after immobilization. The most practical approach is the empirical method, i.e., immobilization of the monoclonal antibody on a small scale followed by the determination of the immunosorbent's antigen binding capacity and selectivity, as well as its reversibility with mild desorbing agents.

Polymer Supports for Immobilization

Physical, Chemical and Mechanical Properties. Insoluble carriers chosen for the preparation of immunosorbents should be: (1) inert, hydrophilic, chemically stable and should contain an optimum number of functional groups (e.g., hydroxyl, carboxyl, aliphatic or aromatic primary amines, hydrazide, etc.) which can be easily activated for efficient antibody coupling (2) rigid beads with high porosity, thereby allowing the rapid passage of potentially viscous fluids at moderate pressure; and (3) capable of providing a microenvironment in terms of biocompatibility and hydrophilicity, which favors optimal antigen-antibody interactions.

Flow Properties of Various Support Media. The flow properties of various column support media were determined (data not shown). After careful evaluation we chose two NuGels (Separation Industries, Metuchen, NJ) and Sepharose CL-6B (Pharmacia LKB Biotechnology, Inc., Piscataway, NJ) for the immobilization of monoclonal antibodies. The NuGel support has the added advantage of being a durable bed support and allowing a 4-fold flux over agarose supports.

Immobilization Methods

Common Strategy. The common strategy for immobilization involves treatment of the polymer support with a reagent or sequence of reagents to convert it into a chemically reactive form. The protein is then allowed to react with the support so that stable covalent bonds are formed. Numerous immobilization methods have been reported in the literature (14,15).

Activation Procedures

We chose the following four activation procedures for the preparation of immunosorbents.

N-Hydroxysuccinimide Ester (NHS) Derivatives. An N-hydroxysuccinimide ester derivative of crosslinked agarose having the formula

$$\text{Agarose-OCH}_2\text{CH(OH)CH}_2\text{NHCOCH}_2\text{CH}_2\text{CONHS}$$

has been prepared according to the published procedures (18,19,20,).

The commercially available NHS ester derivative of NuGel (NuGel P-AF Poly-N-Hydroxysuccinimide, Separation Industries, Metuchen, NJ) has the following structure.

$$\text{NuGel-(CH}_2)_3\ \text{OCH}_2\text{CH}_2\text{NHCH}_2\text{CONHS}$$

Carbonyl-imidazole (CI) Derivatives. The imidazolyl-carbamate derivatives of cross-linked agarose and NuGel were prepared according to the procedure of Bethel et al. (21).

The Aldehyde (CHO) Derivatives. Crosslinked agarose and NuGel are subjected to periodate oxidation to convert them into their respective aldehyde derivatives (22,23).

These derivatives readily react with amine nucleophiles to yield covalently coupled ligand in aqueous solutions.

Hydrazide Derivative. Periodate oxidized monoclonal antibody is coupled to the commercially available adipic dihydrazide derivative of NuGel (Separation Industry, Metuchen, NJ) (24). Theoretically, oriented coupling can be achieved by coupling through the sugar moieties of the Fc region of the IgG molecule. This, in turn, should result in higher antigen binding capacity than that achieved for immunosorbents prepared by coupling via the primary amino groups (see Table IV).

Immobilization Procedure. The activated gel is quickly washed with three volumes of ice-cold water in a coarse sintered glass funnel. The gel is then mixed with an equal volume of protein solution (monoclonal antibody of known concentration) made up in the coupling buffer (0.1M potassium phosphate containing 0.1M NaCl, pH, 7.0) in an appropriate vessel and shaken for 4-16 hours at 4°C. The unbound protein is collected by filtering the reaction mixture. The gel is washed with two volumes of phosphate buffered saline (PBS), pH, 7.4. The filtrate and washes are combined and a small aliquot is dialyzed against PBS. The gel is then treated with ethanolamine-HCl, pH, 7.0 to neutralize any remaining activated groups. It is washed with PBS and stored as a suspension in PBS in the presence of 0.1% sodium azide, at 4°C. The volume of the unbound protein solution is recorded and the protein concentration is determined from the dialyzed aliquot by the method of Lowry et al (25). From these two values the total unbound protein is calculated. The difference between the starting amount and uncoupled antibody in the pooled filtrate and washes, divided by the gel volume gives the antibody coupling density (mg/ml gel).

Factors Affecting Coupling Efficiency and Residual Immunoreactivity. Factors affecting the covalent immobilization of polyclonal antibodies have been studied extensively by Comoglio et al. (26). Coupling conditions such as pH, activated group density on the matrix and the amount of antibody coupled per unit volume of gel (coupling density) influence the loss of immunoreactivity usually accompanied by immobilization.

Residual Immunoreactivity. The amount of antigen binding capacity retained by the monoclonal antibody after immobilization is defined as the residual immunoreactivity of the immunoadsorbent. This can be determined experimentally. A small immunosorbent column of known volume (0.5-1.0 ml) is saturated with an excess of antigen (purified or crude) , the unadsorbed materials are washed away and the specifically bound antigen is eluted with mild desorbing agents. From the protein content of the bound antigen and its M_r, the number of nanomoles of the antigen bound is calculated. The number of nanomoles of antigen bound per unit volume of gel is taken as the antigen binding capacity of the immunosorbent.

Effect of coupling pH. At a constant ratio of antibody to activated matrix (15 mg/ml gel) the coupling efficiencies and residual immunoreactivities of the immunosorbents derived from NHS-ester and aldehyde derivatives were studied. Results are illustrated in Figure 2.

At lower pHs the coupling is essentially through the low pKa alpha-amino groups, whereas at higher pHs the high pKa epsilon-amino groups of the lysines of the IgG molecule are also available for coupling, which accounts for the increased coupling efficiencies. The residual immunoreactivity is optimum at pH, 6-7. At this pH fewer amino groups are non-protonated and consequently, fewer multipoint attachments occur. This should result in less steric hindrance to the antibody, which in turn should allow greater accessibility of the antigen to the antibody binding sites. When the activated groups are stable, as in the CHO-derivatives, the pH of the coupling reaction is not as important as in the labile NHS-ester derivatives.

Effect of Activated Group Density. The effect of functional group density on the matrix is summarized in Table II.

At very low functional group density (2-5 μ moles/ml gel) coupling efficiency is very poor and the antigen binding capacity

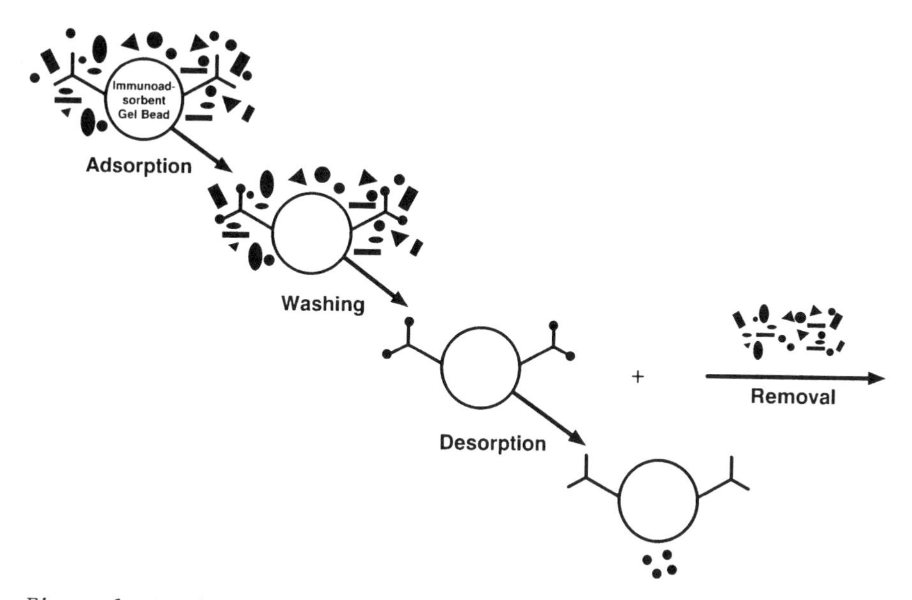

Figure 1. A Schematic Illustration of Immunoaffinity Chromatography.

Table I. Results of Monoclonal Antibodies Screened for Immobilization

Antigen	No. Antibodies Screened Which Bind The Antigen	No. Antibodies Suitable For Immunoaffinity Chromatography
rIFN-alpha A (Roche)	13	4
rIFN-beta (7)	5	2
rIL-2 (Roche)	6	1
Urokinase (Purification Eng.)	2	1

Table II. Effect of Functional Group Density

Activated Groups µmoles/ml gel	Coupling Efficiency	Immunoreactivity
2-5	Poor	High
20-40	Optimum	Optimum
100-200	High	Poor

is high. At the medium density (20-40 μ moles/ml gel) both the coupling efficiency and the residual immunoreactivity are optimal. At relatively high functional group density (100-200 μ moles/ml gel) coupling efficiency is the highest but there is substantial loss in immunoreactivity, possibly due to multipoint attachment. With the imidazolyl-carbamate-derivatives of agarose, 40-50 μ moles of activated groups/ml gel were needed for efficient coupling (data not shown). This may be due to the lack of a spacer arm between the matrix and the IgG molecule.

Effect of Antibody Coupling Density. When the activated matrix is labile as in the NHS-derivative, contact ratios of protein (mg) to support (ml gel) determine the coupling efficiency. When the antibody concentrations ranged from 1-20 mg/ml gel, >90% coupling efficiency was achieved. The residual immunoreactivities at various antibody loadings on the matrix are summarized in Table III.

In accordance with the observation of others, (26) high antibody loadings resulted in lower residual immunoreactivities, possibly due to steric hindrance.

Residual Immunoreactivities of Various Immunosorbents

The immunoreactivities retained by the immunosorbents depend not only upon the factors discussed previously but also upon the chemical nature of the activated matrices used. Table IV lists the various immunosorbents with respect to their residual immunoreactivities. For practical reasons, the antibody loading was chosen at 8-15 mg/ml gel. At this antibody density approximately 1 mg recombinant protein (interferon-alpha 2a) can be purified per ml gel. Neutral pH was chosen for antibody coupling.

Among the immunosorbents studied, the adipic dihydrazide derivative had the best antigen binding capacity due to the oriented coupling through the carbohydrate moieties of the Fc domain of the monoclonal antibody. Since the NHS-derivatives of agarose and NuGel were readily available, we used them for all of our studies. The NuGel immunosorbent has the added advantages of being a durable bed support and allowing a four-fold increase in flux (23) as compared to the cross-linked agarose. These are important factors to consider when scaling up the production of recombinant proteins.

Detection and Prevention of Antibody Leaching from Immunosorbents

The covalent bond formed between the antibody and the matrix during immobilization may not be completely stable. Trace amounts of immobilized antibody may leach from the column during the immunoaffinity purification process, thereby contaminating the final bulk product.

Detection Method. A sensitive, non-competitive, sandwich ELISA is used to detect antibody leaching from the immunosorbent during column operations (data not shown). The lower limit of the assay's sensitivity is 0.1 ng/ml.

Stabilization of Immobilized Monoclonal Antibodies. The use of glutaraldehyde crosslinking to prevent immobilized protein leakage has been reported previously (27). We have successfully used this

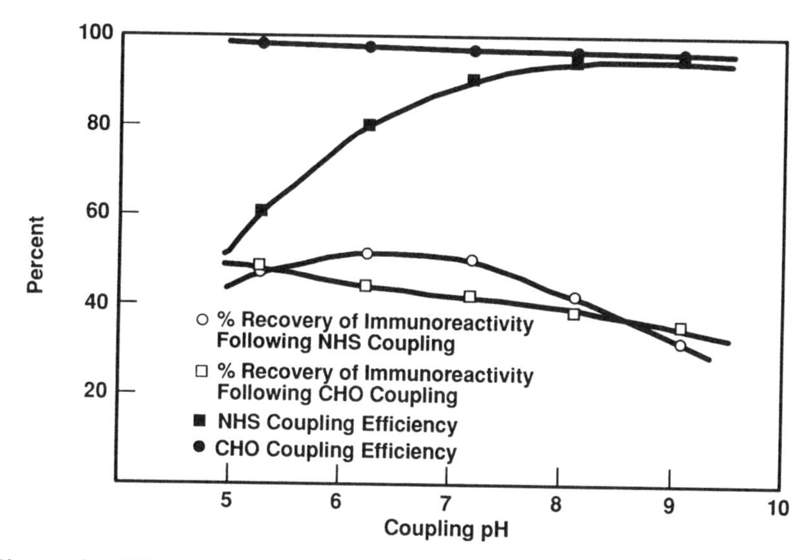

Figure 2. Effect of pH on Coupling Efficiency and Immunoreactivity.

Table III. Effect of Antibody Concentration

Ab Coupled (mg/ml)	Binding Capacity (mg/ml)	Residual Immunoreactivity Expected	Observed (nmoles/ml)	Recovery (%)
0.6	0.13	8	7	89
3.1	0.45	39	23	60
6.2	0.76	78	40	51
11.7	1.43	148	74	50
17.7	1.86	224	97	43
24.0	1.76	304	92	30

Monoclonal antibody to interferon-alpha A at various concentrations (1-25 mg/ml) was immobilized on NuGel-NHS ester derivative according to the procedure described in the text.

Immunoreactivities are calculated taking into account the two binding sites of IgG per molecule and the molecular weight of 158 Kd. Molecular weight of rIFN-alpha 2a is taken as 19.2 Kd for the calculation of observed immunoreactivities.

technique to stabilize the covalently bonded antibody and the results are summarized in Table V.

At low concentrations of glutaraldehyde and under controlled contact time the antibody leaching from immunosorbents was reduced to non-detectable levels, without significant loss in immunoreactivities.

Fast Assays for Monitoring Downstream Purification Steps

The success of large-scale purification processes depends upon the availability of rapid and reliable assays for monitoring the downstream purification steps. Bioassays and ELISAs are time consuming and often give ambiguous results, especially in the early stages of the purification scheme. Roy et al (28) reported the use of an automated high-performance immunosorbent assay for monitoring recombinant leukocyte A interferon purification steps. Similar immunoadsorbent assays are found to be a fast and simple way of monitoring the downstream purification of other recombinant proteins.

Immunoaffinity Purification Procedures

In this section, we focus our attention on the procedures involved in the immunoaffinity purification of clinical grade recombinant proteins from *E. coli* and cultured cell supernatants. Theoretical aspects of the factors involved in the large-scale immunoaffinity purification process have been reviewed by Chase (29). In his review article, Sharma (30) has presented an excellent overview of the recovery of recombinant proteins from *E.coli*. Use of immunoaffinity chromatography in the purification of biomolecules from cultured cells is described by Boschetti et al (31).

General purification schemes for the production of pharmaceutical grade recombinant proteins from microbial and mammalian sources are given below in Figure 3.

Solubilization and Renaturation of Recombinant Proteins

Extraction of the desired protein from *E. coli* in its native form poses unique problems, especially when the protein is expressed in high concentrations in an insoluble form within the inclusion bodies. The cells are disrupted by mechanical, enzymatic or chemical means. For example, interferon is extracted by simply stirring the frozen and thawed cells in buffers containing denaturants like guanidine hydrochloride (GuHCl) and non-ionic detergents such as Triton X-100, Tween-20, etc. Recombinant interleukin-1 is solubilized by simple extraction after homogenization. Since recombinant interleukin-2 (rIL-2) is expressed in *E. coli* in high concentrations within the inclusion bodies, special treatments are required. These include homogenization, isolation of inclusion bodies, washing the inclusion bodies to remove unwanted cellular proteins, solubilizing the rIL-2 with strong denaturants such as 7M GuHCl and finally, diluting the extract and giving the protein enough time to refold. Studies conducted by Light (32) have shown that optimal refolding occurs when the protein concentration is at or below the micromolar range. Consequently, relatively large dilutions of the denatured extracts, followed by aging for various

Table IV. Residual Immunoreactivities of Various Immunosorbents

Immunosorbents	Ab Coupled (mg/ml)	Binding Capac. (mg/ml)	Residual Immunoreactivity Expected (nmoles/ml)	Observed (nmoles/ml)	Recovery (%)
Agarose-NHS	11.7	1.43	148	74	50
NuGel-NHS	11.1	1.34	140	70	50
Agarose-CI	12.4	1.03	157	54	34
NuGel-CI	12.5	0.77	158	40	25
Agarose-CHO	14.7	1.48	186	77	41
NuGel-CHO	12.4	1.45	157	76	48
NuGel-Hydrazide	8.4	1.47	106	77	73

Immunoreactivities are calculated as in Table III.

Table V. Detection and Prevention of Antibody Leaching

Glutaraldehyde % (v/v)	Agarose-NHS ng/ml*	Agarose-NHS %**	Agarose-CHO ng/ml	Agarose-CHO %**
0.0	8.2	100	10.1	100
0.1	2.2	97	2.7	99
0.5	0.5	96	0.0	96
1.0	0.1	96	0.0	95

* Antibody leaching
** Residual Immunoreactivities

Immunoadsorbents were treated with bifunctional glutaraldehyde followed by $NaBH_4$ reduction.

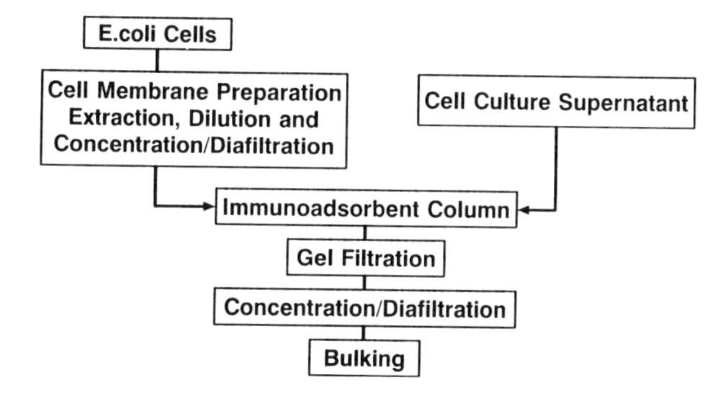

Figure 3. Immunoaffinity Purification Schemes for Recombinant Proteins

periods of time are required. The results of such an aging or refolding experiment conducted with 7M GuHCl extracted rIL-2 after 40-fold dilution with PBS is given in Table VI.

In some instances, the inclusion of reducing agents like DTT in the extraction buffer is helpful in solubilizing the recombinant proteins. When reducing agents are used in the extraction buffer, care should be taken to lower the concentration of the reducing agents so that they will not adversely affect the affinity adsorbents which contain disulfide bonds; for example antibody or receptor adsorbent columns. In general, each recombinant protein may require customized extraction procedures. Usually no special treatments are needed for the cell culture supernatants and they can be applied to the adsorbent column after a simple filtration step.

Adsorption

Adsorption is one of the most critical aspects of immunoaffinity chromatography. During adsorption, the crude material is kept in a buffer which allows maximum adsorption. In order to ensure that no product is wasted during the adsorption phase, sufficient contact time between the soluble antigen and the immunosorbent is maintained by carefully choosing the flowrate.

Washing

The purpose of washing the immunosorbent immediately after adsorption is two-fold: (1) to remove the crude materials from within or surrounding the immunosorbent beads and (2) to remove materials non-specifically bound either to the support or to the immobilized antibody. Non-specific binding to the support can be minimized, but is rarely eliminated completely. Electrostatic as well as hydrophobic interactions between the IgG molecule and extraneous materials in the crude extract are another source of non-specific binding. These non-specifically bound contaminants can usually be reduced to low levels by washing extensively with buffers containing salts at neutral or slightly alkaline pH or by inclusion of low concentrations of non-ionic detergents in the starting materials and in all buffers used for washing.

Elution

The elution of adsorbed antigen from the immunosorbent is achieved by causing the dissociation of the antigen-antibody complex. Non-specific elution methods are commonly used for the desorption of antigens. These eluents involve low or high pH buffers, protein denaturants such as urea or GuHCl and chaotropic agents like potassium thiocyanate. If the antigen involved is stable at acidic pH and is readily eluted from the immunosorbent under these conditions, an eluent of choice is a low pH (<3) buffer.

Size-exclusion Chromatography

Gel filtration as a final step in the purification process is a convenient way of preparing recombinant proteins free of trace high molecular weight contaminants (e.g., leached antibody, pyrogens, oligomers, contaminant proteins of microbial or mammalian origin, etc.) as well as low molecular weight contaminants such as salts, detergents and denaturants. Quite often, gel filtration is a convenient method for exchanging the protein into the final formulation buffer. This step is carried out under aseptic conditions.

Concentration/Diafiltration and Bulking

The gel filtered recombinant protein is concentrated by diafiltration using appropriate M_r cut-off membranes under sterile conditions. This step can also be used to exchange the final product into the final formulation buffer before bulking.

Scale-up and Automation of Immunoaffinity Chromatography

The design of large-scale immunoaffinity purification systems is based on the performance data obtained from small scale operations. In a true linear scale-up as the volume of crude material to be processed increases, the size and flowrates of the column employed are increased proportionally. This is seldom achieved in conventional column mode operations. In order to process large volumes at increased flowrates, increasing the cross-sectional area is more desirable than increasing the column height. When the column height of the soft gel column is increased, a pressure drop occurs across the column bed, leading to flow problems. The non-compressible rigid bead supports like NuGel-PAF (Separation Industries, Metuchen, NJ) and the Superflow™ column (Sepragen Corporation, Hercules, CA) make linear scale-up a distinct possibility.

Advances made in the hybridoma technology as well as improved purification procedures have provided sufficient quantities of monoclonal antibodies which are necessary for the construction of industrial scale immunoadsorbent columns.

Automation

The continuous use of the immunoaffinity purification process can be managed, often unattended, by a microcomputer based control system. The basic outline of a microcomputer based immunoaffinity purification system is shown in Figure 4. It basically consists of a collection of valves which control the input of fluids to the column bed as well as the output from the column. Included in the system are a pump to control fluid flow, a spectrophotometer to monitor the protein and a programmable fraction collector for the collection of eluted proteins. A column regeneration step after each cycle is also included in the system. A more sophisticated microprocessor based automated immunoaffinity purification system has been described by Eveleigh (33) and Chase (29).

Longevity and Stability of Immunosorbents During Long-term Use

The immunosorbents are usually subjected to a stability and longevity study involving at least 100 cycles of operation (automated) to determine their efficacy for long-term use. The qualitative and

Table VI. rIL-2 Refolding Studies

Age of Extract (Days)	Amount of rIL-2 Recovered (mg/g Cells)
0	1.536
1	1.918
2	2.300
3	2.480
4	2.604
5	2.572

The diluted extract was aged for varying periods of time. See text for details.

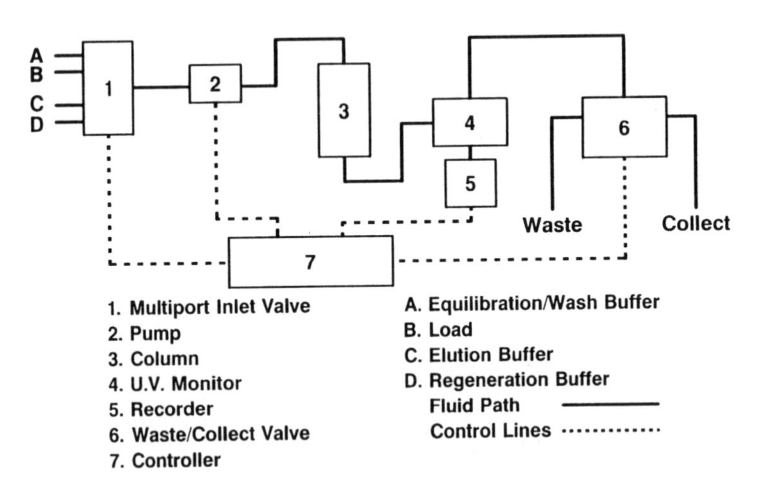

Figure 4. Automated Immunoaffinity Chromatography System

quantitative recoveries of the recombinant proteins are determined by the criteria presented in the following section. All the immunosorbents listed in Table IV performed satisfactorily with no significant impairment in the functionality during long-term use.

Quality Control Aspects of Pharmaceutical Grade Recombinant Proteins

FDA's Center for Drugs and Biologics has published several "Points to Consider" bulletins (34,35) regarding the quality control requirements for producing recombinant proteins for medicinal use in humans. For further insight into this subject the reader is referred to a publication by Bogdansky (36).

Biological Potency. Biological potency data expressed as activity units/mg protein (specific activity) is a good indicator of the nativeness of the purified proteins. Unnatural disturbances in the secondary or tertiary structure of the intact molecule can render the molecule partially or completely inactive.

Protein Structure. The identity, quality and biological potency of a protein are dependent upon its primary, secondary and tertiary structures. Partial amino acid sequence analysis is useful in determining the primary structure, whereas secondary structure such as the location of the intramolecular disulfide bonds can be determined by tryptic mapping under reducing and non-reducing conditions. Full biological activity usually confirms that at least the tertiary structure of the biologically relevant (interacting) domains is correct.

Determination of Purity. SDS-PAGE under reducing and non-reducing conditions is usually used as a primary test for purity (see Figure 5). Fragments, oligomers and foreign proteins in trace amounts are often detected and separated by a variety of methods. Isoelectric focusing, two-dimensional gel electrophoresis and Western blotting are good techniques for detecting heterogeneity in the recombinant protein preparation.

Analysis of Trace Contaminants in the Final Product. Recombinant proteins used for therapeutic purposes should be analyzed for trace levels of contaminants such as foreign proteins of bacterial origin, oligomers, antibiotics, etc. Sensitive customized assays are needed to detect these trace contaminants in the parts per million range. Other types of potentially antigenic contaminants are alternate and derivatized N-terminal amino acids (e.g., N-terminal methionine or derivatized N-terminal methionine) which can be detected by sequence analysis. Nucleic acids, which are potential sources of oncogenes, are another potential contaminant in the recombinant proteins. Nucleic acids can be detected at the picogram level by hybridization analysis. Many of the trace contaminants and pyrogens in the final product can also be removed by the inclusion of ion-exchange chromatography steps followed by size-exclsion liquid chromatography.

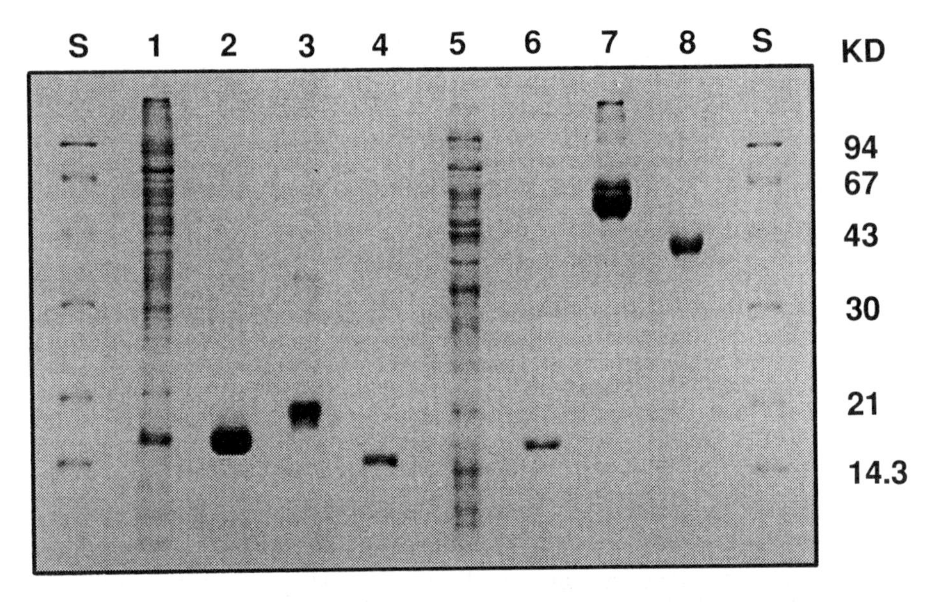

Figure 5. SDS-PAGE Analysis of Immunoaffinity Purified Recombinant Proteins. Lane S: M_r Marker Proteins; Lanes 1: Crude rIFN-alpha 2a, 2: rIFN-alpha 2a; 3: rIFN-gamma; 4: rIL-2; 5: Crude rIL-1; 6: rIL-1; 7: Crude IL-2 Receptor (conditioned media); 8: Soluble IL-2 Receptor

Problem Areas of Immunoaffinity Chromatography

Monoclonal Antibodies. Monoclonal antibodies are produced by a cell line derived from a cancerous cell which is a major concern when the antibodies are to be used in the purification of products for medicinal purposes. Because the final product could be contaminated with potential oncogenic materials such as nucleic acids and viruses, the monoclonal antibodies used for immobilization must be of high purity.

Immobilization Methods. To date, no immobilization method has been reported which retains the theoretical binding capacity of two antigen molecules per one IgG molecule. The reasons for the loss of binding capacity during immobilization are described elsewhere in this paper. We assume that in most instances immobilized monoclonal antibodies are randomly oriented, resulting in loss of antigen binding capacity due to a variety of reasons. The oriented coupling achieved by immobilizing through the sugar moieties of the Fc region of the antibody molecule may alleviate this problem.

Immunoaffinity Purification Procedure. Considerable amounts of time and effort are spent before the crude microbial extract is ready for application onto the immunosorbent column. Centrifugation, addition of strong denaturants for extraction, removal or dilution of denaturants, allowance of adequate time for refolding, and reduction of large volumes are often unavoidable steps due to the nature of the products origin (i.e. micro-organism). At the present time, no general methods are available for the washing and elution stages of the process. Consequently, they are developed on an individual application basis. It is anticipated that the second generation recombinant proteins will involve mammalian cells and that the product will be secreted into the medium. This should allow the product to be processed directly without any pretreatments.

Cost-Effectiveness. The initial cost of installation of immunoaffinity purification systems is still high. The extent to which this technique will be used for the industrial scale production of biomolecules depends upon considerably reducing the production costs of monoclonal antibodies suitable for immobilization.

FDA Regulations. At the present time, the stringent regulations set forth by the FDA require the monoclonal antibody preparations be tested for (1) polynucleotides (2) mycoplasma and xenotropic retroviruses and (3) ecotropic murine leukemia virus. The tests for such contaminants include (1) hybridization analysis, (2) S+L-focus assay and (3) XC plaque assay, respectively. The final product should also be tested for murine antibody production.

Concluding Remarks

Immunoaffinity chromatography, utilizing immobilized monoclonal antibodies, has proven to be a viable and scalable separation technique for the purification of therapeutically useful recombinant DNA proteins, for the following reasons: (1) The method is well-suited for scaled-up continuous operations. (2) High-flow, low pressure, industrial scale columns and sophisticated automatic control systems are readily available commercially. (3) Non-compressible, large pore, rigid bead supports (e.g. silica) which allow high flowrates are available for the preparation of immunoadsorbents. (4) Efficient immobilization methods have been developed. (5) The cost of monoclonal antibody production has been reduced to affordable levels due to recent advances made in hybridoma technology. (6) Product recoveries are usually high.

It should be mentioned that the successful implementation of immunoaffinity purification has involved a close working relationship between biochemists, immunologists and biochemical engineers.

Acknowledgments

We thank Mrs. Alisa Rao and Ms. Nellie Dlugazima for technical assistance and Ms. Michele Nachman and Mrs. Lisa Nieves for their invaluable help in preparing the manuscript. Review of this paper by Drs. John E. Smart and Nancy Tobkes is greatly appreciated.

Literature Cited

1. Silman, I. H.; Katchalski, E. Ann. Rev. Biochem. 1966, 35, 873-908.
2. Campbell, D. H.; Luesher, E.; Lerman, L. S. Proc. Natl. Acad. Sci. USA 1951, 37, 575-8.
3. Gurvich, A. E.; Drizlikh, G. J. Nature (London) 1964, 203, 648-9.
4. Kohler, G.; Milstein, C. Nature 1975, 256, 495-7.
5. Tarnowski, S. J.; Roy, S. K.; Liptak, R. A.; Lee, D. K.; Ning, R. Y. Methods in Enzymology, Interferons, Part C; Pestka, S., Ed.; Academic Press: Orlando, FL, Vol. 119, 153-65.
6. Secher, D. S.; Burke, D. C. Nature (London) 1980, 285, 446-50.
7. Novick, D.; Eshhar, Z.; Gig, O.; Marks, Z.; Revel, M.; Rubinstein, M. J. Gen. Virol. 1983, 64, 905-10.
8. Robb, R. J.; Kutny, R. M.; Chowdhry, V. Proc. Natl. Acad. Sci. USA 1983, 80, 5990-4
9. Katzman, J. A.; Nesheim, M. E.; Hibbard, L. S.; Maan, K. G. Proc. Natl. Acad. Sci. USA 1981, 78, 162-6.
10. Hermann, S. H.; Mescher, M. F. J. Biol. Chem. 1979, 254, 8713-16.
11. Leonard, W. J.; Depper, J. M.; Crabtree, G. R.; Rudikoff, S.; Pumphrey, J.; Robb, R. J.; Kronke, M.; Svetlik, P. B.; Peffer, N. J.; Waldmann, T. A.; Greene, W. C. Nature 1984, 311, 626-31.
12. Posillico, E. G. Bio/Technology 1986, 4, 114-17.
13. Bailon, P.; Dlugazima, N.; Nishikawa, A. H. "Large-scale purification of monoclonal antibodies from intracapsular supernatant and ascites fluids", Conference on Bio/Technology Looks to the Next Decade, 1986, New Orleans, LA, paper no. 210.
14. Affinity Chromatography: a practical approach; Dean, P.D.G.; Johnson, W. S.; Middle, F. A., Eds.; IRL Press Ltd.: Oxford, England, 1985; 16-148.
15. Methods in Enzymology; Mosbach, K., Ed.; Academic: New York, 1987; Vol. 135, Part B, p 3.

16. Physicians' Desk Reference; Barnhart, E. R., Publisher; Medical Economics Company, Inc.: Oradell, NJ, 1987; 41M, 1676-7.
17. Vetterlein, D.; Calton, G. Affinity Chromatography and Biological Recognition; Chaiken, I. M.; Wilcheck, M.; Parikh, I., Eds.; Academic Press Inc.: Orlando, FL, 1984; 393-4.
18. Nishikawa, A. H.; Bailon, P. J. Solid-Phase Biochem. 1976, 1, 33-49.
19. Cuatrecasas, P.; Parikh, I. Biochem. 1972, 11, 2291-9.
20. Cuatrecasas, P. J. Biol. Chem. 1970, 245, 3059-65.
21. Bethell, G. C.; Ayers, J. S.; Hearn, M. T. W.; Hancock, W. S. J. Chrom. 1981, 219, 361-72.
22. Fischer, E. A. Affinity Chromatography and Biological Recognition; Chaiken, I. M.; Wilcheck, M.; Parikh, I., Eds., Academic Press Inc.: Orlando, FL, 1983; 399-400.
23. Roy, S. K.; Weber, D. V.; McGregor, W. C. J. Chrom. 1984, 303, 225-8.
24. Bailon, P., unpublished data.
25. Lowry, O. H.; Rosebrough, N. J.; Farr, A. L.; Randall, R. J. J. Biol. Chem. 1951, 193, 265-75.
26. Comoglio, S.; Massaglia, A.; Rolleri, E.; Rosa, U. Biochem. Biophys. Acta 1976, 420, 246-57.
27. Kowal, R.; Parsons, R. G. Anal. Biochem. 1980, 102, 72-6.
28. Roy, S. K.; McGregor, W. C.; Orichowskyj, S. T. J. Chrom. 1985, 327, 190-2.
29. Chase, H. A. Chem. Eng. Sci. 1984, 39, 1099-1125.
30. Sharma, S. K. Separation Science and Technology 1986, 21, 701-26.
31. Boschetti, E.; Egly, J. M.; Monsigny, M. Trends in Anal. Chem. 1986, 5, 4-10.
32. Light, A. Bio/Techniques 1985, 3, 298-306.
33. Eveleigh, J. W. Anal. Chem. Symp. Ser. 1982, 9, 293-303.
34. FDA, "Points to consider in the production and testing of new drugs and biologicals produced by recombinant DNA technology," Office of Biologics Research and Review Center for Drugs and Biologics, April 10, 1985.
35. FDA, "Points to consider in the characterization of cell lines used to produce biological products," Office of Biologics Research and Review, Center for Drugs and Biologics, June 1, 1984.
36. Bogdansky, F. M. Pharmaceutical Technology 1987, 11, 72-4.

RECEIVED January 30, 1990

Chapter 12

Chelating Peptide-Immobilized Metal-Ion Affinity Chromatography

Michele C. Smith[1], James A. Cook[1], Thomas C. Furman[1], Paul D. Gesellchen[1], Dennis P. Smith[2], and Hansen Hsiung[2]

[1]Biochemistry Research and [2]Molecular Biology Research, Lilly Research Laboratories, Eli Lilly and Company, Lilly Corporate Center, Indianapolis, IN 46285

A new method for purifying recombinant proteins called, Chelating Peptide-immobilized Metal Ion Affinity Chromatography, or CP-IMAC is described. This method is based on the hypothesis that a specific chelating peptide (CP) with a high affinity for immobilized metal ions can be used to purify a recombinant protein by extending the gene sequence of the recombinant protein to code for the extra amino acids in the chelating peptide sequence. The resulting fusion protein can then be purified using immobilized metal ion affinity chromatography (IMAC). A four step strategy was followed to develop CP-IMAC: identifying chelating peptides, attaching a CP to a polypeptide to test the feasibility of this concept, purifying a recombinant CP-protein, and lastly purifying a CP-X-protein where X is an enzymatic cleavage site which allows the CP to be removed so as to generate the desired protein.

Recombinant DNA methods have provided the means for obtaining large quantities of interesting proteins whose supply has been limited or nonexistent. The prospects of such a supply and the potential applications of such proteins, have spawned a large biotechnology industry. Recombinant protein purification is a time consuming process critical for obtaining large amounts of protein. In addition, protein purification is the one step in the process that must be continually repeated, compared to the construction of the expression vector, transformation of the host organism, and optimization of fermentation conditions.

The need for more efficient purification methods is clear and presents some unique opportunities for proteins synthesized by recombinant DNA methods. Access to the gene which codes for the protein affords one the opportunity to extend the gene to code for additional amino acids and create a fusion protein which exaggerates a given property that can be exploited for purification purposes. These additional amino acids which provide for a more efficient purification step, can then be removed either chemically or enzymatically after the purification, to generate the desired protein sequence.

The first examples of recombinant proteins with purification handles were reported in 1984 (1-2). Sassenfeld and Brewer (1) efficiently purified a urogastrone

0097–6156/90/0427–0168$06.00/0

fusion protein with a C-terminal polyarginine sequence over an anion exchange column. The C-terminal position of the polyarginine sequence allowed it to be removed enzymatically with carboxypeptidase B. Bastia and co-workers (2) fused ß-galactosidase to collagen followed by the protein of interest, the R6K replication initiator. The ß-galactosidase portion allows the fusion protein to be purified over an affinity column, while the collagen sequence which is sensitive to collagenase, provides the cleavage site. More recent examples of recombinant proteins with purification handles, include staphylococcal protein A fusion proteins with specific chemical cleavage sites (3), and chimeric proteins with an eight residue peptide which is recognized by a specific antibody for affinity purification and the enzyme enterokinase for subsequent cleavage (4).

The property we set out to transfer to a recombinant protein to facilitate the purification process is the ability to bind immobilized transition metal ions. We hypothesized that a few extra amino acids which bind metal ions, could be cloned onto N-terminus of the recombinant protein, and allow the expressed protein to be purified using immobilized metal ion affinity chromatography (IMAC). The same amino acid sequence, or chelating peptide (CP) could be cloned on the end of new recombinant proteins and the same purification scheme applied, thus eliminating extensive methods development for each new recombinant protein. This approach, called CP-IMAC, would take advantage of the untapped power of IMAC by making it predictable and therefore a more useful purification method. The relative cost and availability of simple metal salts used to prepare IMAC columns is a significant advantage over the cost of antibodies and other reagents used to prepare affinity columns.

IMAC was introduced in 1975 by Porath who showed that serum proteins could be fractionated over columns containing immobilized metal ions, such as copper, nickel, and zinc (5). A number of proteins have been purified using IMAC since its introduction (6). Figure 1 is an illustration of the immobilized iminodiacetic acid complex formed with metal ions, which leaves up to three open coordination sites for binding to a protein with accessible donor atoms or ligands. The open coordination sites are occupied by rapidly exchanging water molecules which can be displaced by the incoming protein which has a higher affinity for the metal ion than water molecules. The bound protein can be eluted by either lowering the pH of the buffer or by introducing a displacing ligand which competes with the protein for the metal coordination sites and displaces the protein.

Porath attributes protein binding to immobilized metal ions to the presence of cysteine, histidine, and tryptophan residues (5). There is ample precedence for cysteine and histidine binding metals in well characterized metalloproteins, metal peptide complexes, and metal amino acid complexes (7-8). There are no known examples of tryptophan coordinating metal ions through the indole group, either as amino acid complexes, peptide complexes, or in metalloproteins.

Studies aimed at understanding the mechanism of protein binding to immobilized metals have lead to the hypothesis that the number of exposed histidine and/or cysteine residues determines the strength of the interaction (9). Frontal analysis of model proteins on IMAC columns has been carried out recently, which provides quantitative estimates of the affinity constants of these proteins for immobilized metal ions and binding capacities (10-11). This approach provides a basis for comparing the chromatographic behavior of proteins on IMAC columns and establishes a database to better understand how proteins interact with immobilized metal ions. Spectroscopic structural studies of the complexes formed on the resin, would provide unambiguous answers to questions raised about the binding mechanism of proteins to immobilized metal ions. If the crystal structure of a protein has been determined then the surface exposure of histidine residues can be established and used to design a separation using IMAC. For proteins whose three dimensional structure is unknown, which includes the vast majority of proteins, trial and error is the best that can be done. Most proteins synthesized by recombinant DNA methods are those that are not readily available and have been studied the least, compared to such classics as myoglobin, cytochrome c, lysozyme, and ribonuclease whose crystal structures have been solved. IMAC would probably be used more often if it were possible to predict which proteins bind immobilized metal ions.

The first step in engineering a recombinant protein to contain an accessible metal binding site, was to identify a chelating peptide sequence (12). With a suitable sequence in hand, its ability to bind immobilized metal ions when attached to a larger polypeptide could be tested (13). The third step was to splice the chelating peptide gene sequence on the front end of a recombinant protein gene sequence and determine whether the metal binding properties were conserved in the expressed CP-protein (13). The last step was to demonstrate that the chelating peptide could be removed chemically or enzymatically when an appropriate cleavage site was included in the expression product. CP-IMAC must be applied to many different systems to become a general method for purifying recombinant proteins. Recently, Hochuli and co-workers have prepared a chelating gel with nitrilotriacetic acid as a tetradentate ligand for immobilizing nickel and have used polyhistidine chelating peptides cloned on the end of dihydrofolate reductase to purify the fusion protein over a nitrilotriacetato Ni(II) column (14).

Experimental Section

The preparation of IMAC columns and measurement of peptide elution pH values has been described in detail (12). The construction of the expression plasmid for His-Trp-proinsulin and the chromatography of His-Trp-proinsulin on IMAC columns has been described (13).

Peptide Synthesis and Purification. Met-His-Gly-His was prepared by solid phase peptide synthesis using a Beckman 990A Automated Peptide Synthesizer and following the general outline of a previously published protocol (15). The two tripeptide amides, Met-His-Trp-NH_2 and Met-His-Tyr-NH_2, were prepared and purified using previously published methodology(16-17). The purified peptides gave satisfactory amino acid, mass spectral, and thin layer chromatographic analyses.

Plasmid Construction and Expression of CP-X-protein. The expression plasmid for Met-His-Trp-His-ompA signal peptide-IGF-II, pHS235 was derived from pPRO-IGF-II (Hsiung, H., Eli Lilly & Company, personal communication, 1989.) and contained a lambda PL promoter, an *Escherichia coli* lpp ribosome binding site sequence, and the cI857 repressor gene which was used to regulate the activity of the PL promoter. A synthetic DNA linker shown below was synthesized on an Applied Biosystems model 380B DNA synthesizer and used to modify the 5' end of the IGF-II coding region

*Xba*I

5' CTAGAGGGTATCAT ATG CAT TGG CAT AAA AAG ACA GCT ATC GCT ATT GCC GTG

3'TCCCATAGTA TAC GTA ACC GTA TTT TTC TGT CGA TAG CGA TAA CGG CAC

*Taq*I

GCG CTA GCT GGT TTC GCT ACT GTA GCT CAG GCC GCT TAT 3'

CGC GAT CGA CCA AAC CGA TGA CAT CGA GTC CGG CGA ATA GC 5'

The plasmid was used to transform *E. coli* RV308 cells (18) which were grown overnight at 32°C in 2xTY medium, then diluted into fresh broth (20ml overnight culture/1 L medium) and grown until the O.D. was 0.3. The culture was then switched to a 41°C shaker bath for three additional hours of incubation. The cell pellet was harvested for protein isolation.

CP-IMAC Purification of Met-His-Trp-His-ompA signal peptide-IGF-II. Granules containing the expression product from pHS235 were collected and the protein sulfitolysed as described previously for IGF-II (19). The sulfitolysis reaction mixture was applied to a 1 X 8 cm Cu(II) IMAC column equilibrated in 0.1M Na_2HPO_4, 0.1M NaCl, 7M urea, pH 7.7 buffer. The bound material was eluted from the column with a pH gradient generated with 0.1M acetic acid, 0.1M NaCl, 7M urea, pH 3.6 buffer on an FPLC as described previously (9). The bound pool of protein was desalted into 0.1M NH_4HCO_3 pH 7.9 buffer on a Sephadex G-25 column and then lyophilized. The cleavage reaction with signal peptidase using lyophilized CP-X-protein was carried out as previously described (20). After 1.3 hours the reaction mixture was applied directly to the Cu(II) IMAC column as described above. Samples of the crude sulfitolysis reaction mixture and pools from the IMAC steps were analyzed by SDS-PAGE. The gel consisted of 15% acrylamide and 18% glycerol in 0.1M Tris-phosphate buffer, pH 6.8. Electrophoresis was carried out at room temperature with a constant current of 35 mA. The proteins in the gel were then fixed with TCA and stained with Coomasie.

Results and Discussion

Identify a Chelating Peptide

A chelating peptide intended to serve as a purification handle for recombinant proteins, must have certain thermodynamic and kinetic properties. It must have a high affinity for immobilized metal ions and the kinetics of complex formation must be fast. A slow rate of complex formation would defeat the purpose of using a chelating peptide as a purification handle, since the complex must form during the time of the chromatographic separation. First row transition metals, Fe(II), Co(II), Ni(II), Cu(II), and Zn(II) were chosen as candidates for immobilized metal ions, since they have fast water exchange rates and ligand exchange rates in general (7). The affinity of a peptide for a given metal will be largely determined by the amino acid composition of the peptide. Certain amino acids bind metals in metalloproteins, form complexes with metals either as amino acids or as part of a peptide, and include, histidine, cysteine, aspartate, glutamate, methionine, lysine, and tyrosine (7). Structural studies have been carried out on a limited number or metal peptide complexes, which show that a dipeptide contains a sufficient number of donor atoms to occupy three coordination sites on a metal (21). Since only three coordination sites are available for binding immobilized metal ions, a dipeptide should be sufficient.

Potential chelating peptides were obtained commercially and included a series of di- and tripeptides which contained either histidine, aspartic acid or lysine in at least one position (12). A convenient method for measuring the relative affinity of these peptides for immobilized first row transition metals was to carry out an IMAC separation and elute the peptides from the column by lowering the pH of the buffer. The elution pH, or pH required for elution, is a relative measure of affinity for the immobilized metal ion on a particular column. In other words, the lower the elution pH, the higher the affinity for the metal ion, since more protons are required to compete with the metal ion to protonate the donor atoms. A survey of approximately fifty peptides on Fe(II), Co(II), Ni(II), Cu(II), and Zn(II) IMAC columns revealed that histidine containing peptides had the highest affinity for these immobilized metal ions (12). None of the peptides bound the Fe(II) or Zn(II) columns in phosphate buffer. The aspartic acid and lysine containing peptides were retained only on the Ni(II) and Cu(II) columns but eluted at pH 7.5 with various retention times. Of the twenty one histidine containing peptides studied, a little over half bound Cu(II) and Ni(II) with various affinities, but three had unusually high affinities for Co(II), Ni(II), and Cu(II). Figure 2 shows the separation that can be achieved between histidine containing peptides on a Ni(II) IMAC column and one of these peptides, His-Trp. These three peptides, His-Trp, His-Tyr-NH_2, and His-Gly-His, are therefore suitable candidates for chelating peptide purification handles.

Attach a CP to a Polypeptide and Purify using IMAC

Luteinizing Hormone Releasing Hormone (LHRH) analogs were used to test the ability of a CP, His-Trp, to purify a large polypeptide through binding immobilized metal ions (13).The sequences of the LHRH analogs used is shown below:

2-10 LHRH	His-Trp-Ser-Tyr-Gly-Leu Arg-Pro-Gly-NH_2
3-10 LHRH	Trp-Ser-Tyr-Gly-Leu Arg-Pro-Gly-NH_2
4-10 LHRH	Ser-Tyr-Gly-Leu Arg-Pro-Gly-NH_2

The 2-10 LHRH analog contains an N-terminal dipeptide which corresponds to the chelating peptide sequence, His-Trp. 4-10 LHRH has the same sequence as 2-10 LHRH but lacks the N-terminal CP. A mixture of these LHRH analogs was applied to a Ni(II) IMAC column and the analogs without the CP eluted in the wash at pH 7.5, while the CP containing 2-10 LHRH analog required a pH of 4.4 for elution, as shown in Figure 3. This experiment demonstrated that the ability to bind immobilized Ni(II) could be transferred to a polypeptide which otherwise has no affinity for Ni(II) by simply extending the sequence to include a CP, His-Trp.

Purify a Recombinant CP-Protein

His-Trp-proinsulin was used as a model system for a recombinant protein to determine whether such a small chelating peptide could be used to immobilize and purify a larger recombinant protein on an IMAC column (13). The S-sulfonates of His-Trp-proinsulin and proinsulin were isolated from *Escherichia coli* engineered to overproduce these proteins as trpLE' fusion proteins. A mixture of these two proteins was applied to a Ni(II) IMAC column and the proteins were eluted with a pH step gradient. Figure 4 shows the separation of His-Trp-proinsulin from proinsulin which can be achieved by virtue of the CP, His-Trp, on the N-terminus of His-Trp-proinsulin. Proinsulin was a good test case, since it has an endogenous transition metal binding site and binds IMAC columns. The affinity of His-Trp-proinsulin for immobilized Ni(II) is greatly increased over that for proinsulin, as a much lower pH is required for elution. This increase in binding affinity can only be attributed to the presence of the CP, His-Trp, which therefore has a higher affinity for Ni(II) than the B10 histidine in proinsulin.

The examples presented have been used to prove the feasibility of CP-IMAC as a purification method for recombinant proteins. In an actual practice, most foreign proteins expressed in host organisms such as *E. coli* contain an N-terminal methionine whose presence might prevent an N-terminal chelating peptide from binding immobilized metal ions. In fact the histidine containing peptides with the highest affinity for immobilized metal ions, all contain an N-terminal histidine residues (12). Those with histidine in the second position only bound immobilized

Figure 1. Schematic representation of an immobilized iminodiacetic acid (IDA) complex with a six coordinate metal ion. Three open coordination sites are occupied by water and are displaced by the incoming protein, represented as H_3L_3, where L is the ligand or donor atom on the protein. Bound protein can be eluted by lowering the pH of the buffer, thereby protonating the donor atoms so they can no longer coordinate the metal ion.

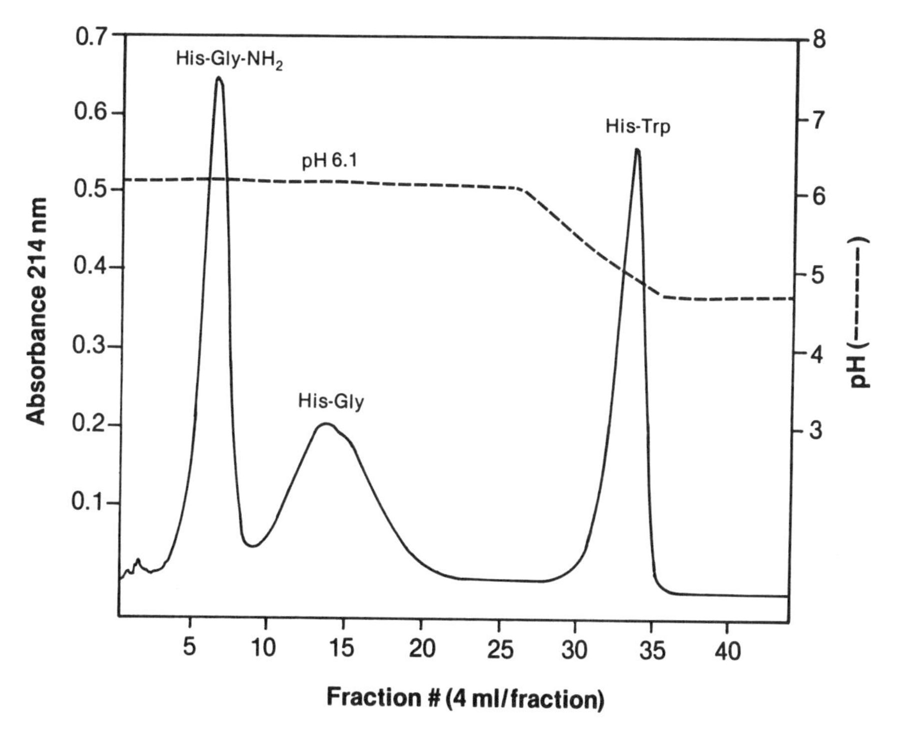

Figure 2. Separation of histidine containing peptides on a Ni(II) IDA Sephadex G-25 column. Peptides were eluted with the pH gradient shown. (Reproduced with permission from Ref. 12. Copyright 1987 ACS)

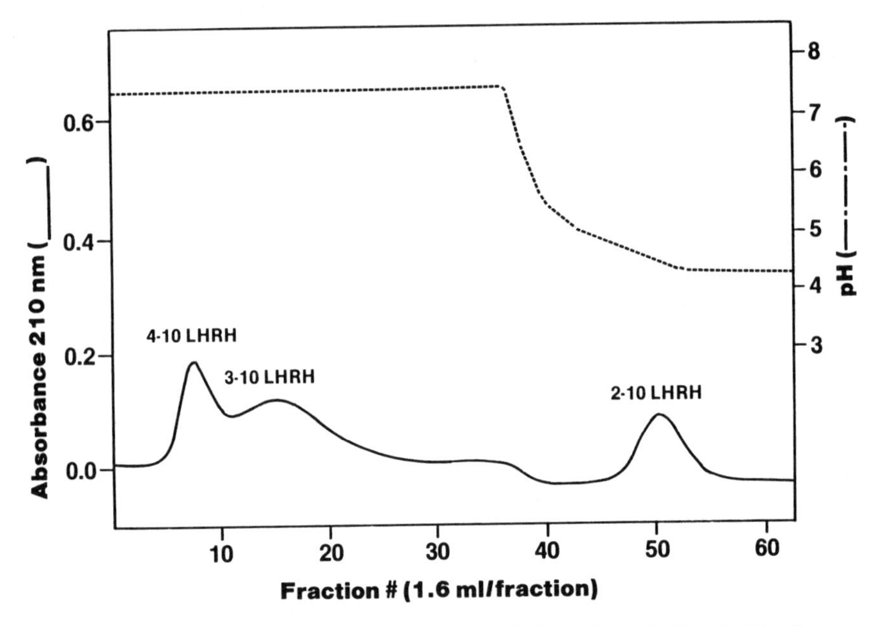

Figure 3. Separation of LHRH analogs on a Ni(II) IDA Sephadex G-25 column. The analogs were eluted with the pH gradient shown. (Reproduced with permission from Ref. 13. Copyright 1988 The American Society for Biochemistry and Molecular Biology)

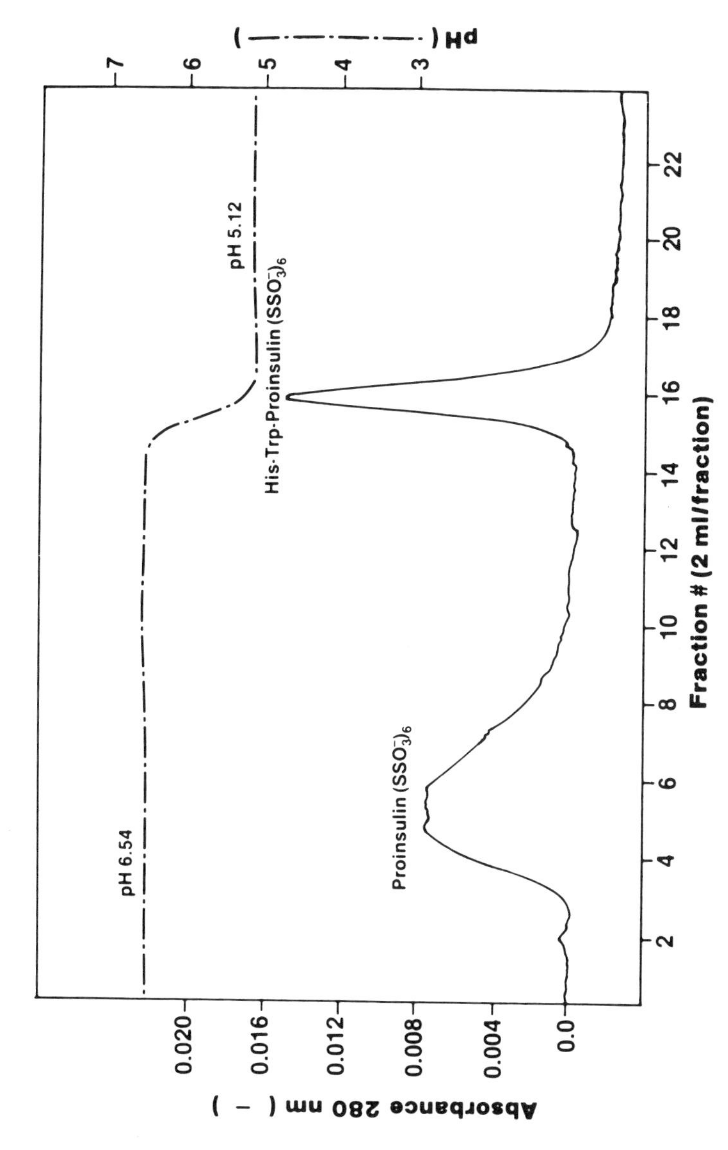

Figure 4. Separation of the S-sulfonates of His-Trp-proinsulin and proinsulin on a Ni(II) IDA 4% agarose column. The proteins were eluted with the pH gradient shown. (Reproduced with permission from Ref. 13. Copyright 1988 The American Society for Biochemistry and Molecular Biology)

Ni(II) and less strongly than peptides with N-terminal histidines. Methionyl analogs of the three chelating peptides identified were therefore synthesized for further study.

The chromatographic behavior of these peptides on Co(II), Ni(II), and Cu(II) IMAC columns was studied as described previously for the histidine containing peptides (12). A summary of the elution pH values for these chelating peptides is given in Table I.

Table I. Comparison of Chelating Peptides to Methionyl Chelating Peptides

		Elution pH	
Peptide	Co(II)	Ni(II)	Cu(II)
His-Gly-His	5.9	4.8	4.4
Met-His-Gly-His	7.5	5.3	4.7
His-Tyr-NH_2	6.4	4.7	4.4
Met-His-Tyr-NH_2	7.5	4.9	7.5
His-Trp	6.4	4.9	4.4
Met-His-Trp-NH_2	7.5	4.8	7.5

The presence of an N-terminal methionine abolished the ability of Co(II) to bind any of the chelating peptides. All three methionyl chelating peptides retained their affinity for immobilized Ni(II) although the relative affinities are somewhat lower as seen by the slightly higher elution pH values. Only one methionyl chelating peptide, Met-His-Gly-His, had a high affinity for immobilized Cu(II), whereas the others eluted from the column during the wash step. The requirement for an N-terminal histidine in single histidine containing peptides is apparently quite strict for immobilized Cu(II). These results indicate that methionine in the N-terminal position of a recombinant protein should not affect the ability of these peptides to bind immobilized Ni(II) and that the presence of more than one histidine will allow immobilized Cu(II) to be used as well. The generality of this method, CP-IMAC, is therefore extended by these findings, since the majority of recombinant proteins contain an N-terminal methionine.

Purify a Recombinant CP-X-Protein and Remove the CP to Generate the Desired Protein

The last step in developing CP-IMAC as a general technique for purifying recombinant proteins involves removing the CP to generate the desired protein sequence. Cleavage of the CP sequence can either be accomplished chemically or enzymatically. Chemical cleavage reactions are usually dependent on the sequence of the protein or amino acid composition. For example, cyanogen bromide has been used to remove leader sequences from fusion proteins, but requires that the protein of interest contain no internal methionine residues. Tryptophan oxidative cleavage reactions could be carried out to remove His-Trp, provided the protein of interest contains no internal tryptophan residues (22). Enzymatic cleavage sites have been used to produce human growth hormone (23), lymphokines (4), urogastrone (1), R6K replication initiator (2), as well as others from fusion proteins (3).

A general enzymatic cleavage site, the ompA signal peptide which is specifically recognized by the enzyme signal peptidase (20), was cloned into the expression product following a CP to show that a CP can be removed to generate the desired protein. The protein in this case was human insulin-like growth factor-II or IGF-II and the CP was Met-His-Trp-His. The N-terminal sequence of the expression product is shown below using the one letter abbreviations for amino acids.

CP ompA signal peptide IGF-II

MHWH-KKTAIAIAVALAGFATVQA-AYRPSE....

The CP-IMAC purification scheme for the model CP-X-protein, involved passing a crude sulfitolysed lysate of *E. coli* over an IMAC column to bind the CP-X-protein and provide purified protein, cleaving the CP and ompA signal peptide with signal peptidase, and then passing the cleavage reaction mixture over the IMAC column to retain contaminating *E.coli* proteins which bound the immobilized metal ion in the first step.

Figure 5 shows the SDS-PAGE analysis of the steps in the purification scheme for Met-His-Trp-His-ompA signal peptide-IGF-II from *E. coli* using CP-IMAC. A band at approximately 11 kDa in the crude sulfitolysis mixture of granules (lane 2) is the expression product and the one below 9.4 kDa is IGF-II present in the granule preparation. Both bands cross reacted with antibodies against IGF-II in a Western blot (data not shown) and the band below 9.4 kDa also migrated in the same position as an authentic sample of IGF-II (lane 1). The presence of IGF-II in the crude lysate, indicated that some cleavage of the expression product had occurred. Preliminary Western blot analysis of the periplasmic fraction indicated that the IGF-II in the granule preparation was not a contaminant of the periplasmic fraction. This result was somewhat surprizing, as cleavage of the ompA signal peptide fusion proteins is reported to occur only in the membrane when the protein is translocated into the periplasmic space (24). The fact that the majority of the expression product was not processed nor secreted into the periplasm may be due to the presence of the chelating peptide sequence, since other recombinant fusion proteins lacking a CP are secreted with the signal peptide removed (24).

Lanes 3 and 4 in Figure 5 show the results of the first Cu(II) IMAC step on crude Met-His-Trp-His-ompA signal peptide IGF-II, from which two pools were made. The proteins with no affinity for immobilized Cu(II) represent the unbound pool (lane 3) and included the processed IGF-II. Lane 4 shows the pool of bound material which was eluted by lowering the pH of the buffer. The expression product, Met-His-Trp-His-ompA signal peptide IGF-II, was the major component while some higher molecular weight *E. coli* proteins were also found in this pool. The pool of bound material was then reacted with signal peptidase. The cleavage reaction products shown in lane 6 illustrate that the expression product has been completely consumed to generate IGF-II. The reaction mixture was then passed over the Cu(II) IMAC column again so as to retain the *E. coli* proteins present in the sample and collect the processed IGF-II. The last two lanes in Figure 5 are fractions from the column flow through which contained the processed and purified IGF-II. The single band in the gel indicated that the protein had been purified to a homogeneous state.

The generality of this approach has been demonstrated with the expression of a removable chelating peptide which was used to purify IGF-II. This same sequence, Met-His-Trp-His-omp A signal peptide, can now be cloned onto the end of

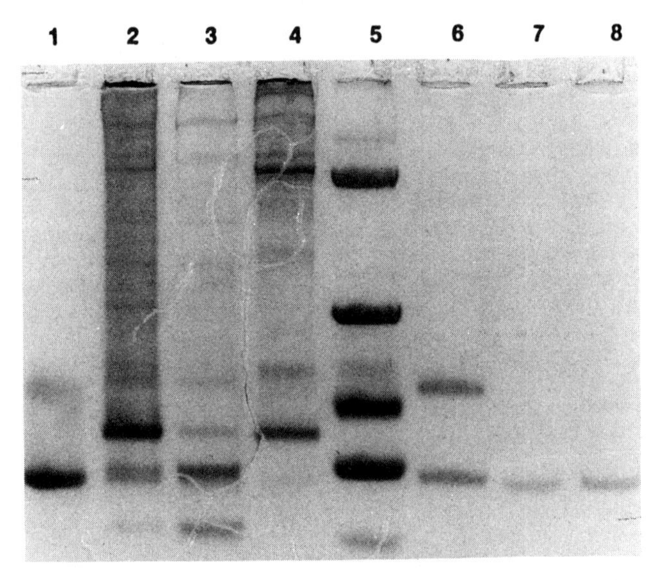

Figure 5. SDS-PAGE analysis of the CP-IMAC purification of Met-His-Trp-His-ompA signal peptide-IGF-II S-sulfonate as described in the text. Lane1, sample of IGF-II S-sulfonate; lane 2, sulfitolysed granules; lane 3, pool of unbound protein from first Cu(II) IMAC step; lane 4, pool of bound protein from first Cu(II) IMAC step; lane 5, molecular weight markers of 43 kDA, 22 kDa, 12.5 kDa, 9.4 kDa, and 3.5 kDa; lane 6, products of signal peptidase cleavage of material in lane 4; lane 7 and 8, fractions of unbound material from the second Cu(II) IMAC step.

recombinant proteins, and the same purification scheme followed to yield pure protein. The capacity of IMAC columns for proteins is very large (10), and should therefore allow large amounts of recombinant proteins to be purified using CP-IMAC.

The design of optimal chelating peptide sequences will depend on our understanding of the mechanism of protein or peptide binding to immobilized metal ions. The chelating peptides we have used were discovered through screening numerous peptides for their ability to bind immobilized transition metal ions. For metals with coordination numbers of six, three open sites are left after the metal has been immobilized with iminodiacetic acid. A peptide with three strong donor atoms in the correct orientation may be the ideal chelating peptide. Whether a simple trihistidine sequence with three imidazole groups, could fulfill the orientation requirement for strong binding is not clear. The presence of more than three histidine residues in a chelating peptide may provide greater opportunity for the metal to encounter a complete set of donor atoms with the correct orientation, and therefore tighter binding. Conversely, a peptide containing three histidine residues separated by other amino acids, which has the proper orientation of donor atoms, may have as high an affinity for immobilized metal ions as peptides containing an excess of histidine residues.

Immobilization of metal ions with tetradentate or pentadentate ligands reduces the number of open coordination sites on the metal ion available for protein or peptide binding. The effect of reducing the number of open coordination sites is to lower the affinity of proteins for that immobilized metal ion (25). Hochuli and co-workers used a tetradentate ligand, nitrilotriacetic acid, to immobilize Ni(II) and polyhistidine peptides with two to six histidine residues as purification handles(14) using the CP-IMAC concept. The apparent affinity of Ni(II) for fusion proteins containing these peptides increases with the number of histidine residues. Stability constants for the complexes formed were not measured but the amount of fusion protein retained on the columns seemed to increase with each successive histidine residue. This may be due to increasing the probability of the immobilized metal ion seeing two imidazole groups with the proper spatial arrangement necessary for optimal binding.

Metal ions with a preferred coordination number of four, such as Cu(II) or Zn(II) have even fewer open coordination sites available after immobilization. The requirements for spatial orientation of the donor atoms in a chelating peptide may be different for metals with square planar or tetrahedral geometries compared to metals which form octahedral complexes. The increase in crystal field stabilization energy for Cu(II) complexes compared to analogous Ni(II) complexes often results in higher affinities, and may therefore be more desirable (7).

The studies described here have demonstrated the feasibility of CP-IMAC and its utility in purifying recombinant proteins. The readily available reagents for performing CP-IMAC on a recombinant CP-protein, makes it an ideal technique for obtaining large amounts of protein for further study with only one or two chromatographic steps.

Acknowledgments

We thank Dr. Wickner, UCLA, for the generous gift of signal peptidase.

Literature Cited

1. Sassenfeld, H. M.; Brewer, S. J. Bio/Technology 1984, 2, 76-81.
2. Germino, J.; Bastia, D. Proc. Natl. Acad. Sci USA 1984, 84, 4692-4696.

3. Moks, T.; Abrahmsen, L.; Holmgren, E.; Bilich, M.; Olsson, A.; Uhlén, M.; Pohl, G.; Sterky, C.; Hultberg, H.; Josephson, S.; Holmgren, A.; Jörnvall, H.; Nilsson, B. Biochemistry 1987, 26, 5239-5244.
4. Hopp, T. P.; Prickett, K. S.; Price, V. L.; Libby, R. T.; March, C. J.; Cerrettti, D. P.; Urdal, D. L.; Conlon, P. J. Bio/Technology 1988, 6, 1204-1210.
5. Porath, J.; Carlsson, J.; Olsson, I.; Belfrage, G. Nature 1975, 258, 598-599.
6. Sulkowski, E. BioEssays 1989, 10, 170-175.
7. Cotton, F. A.; Wilkinson, G. In Advance Inorganic Chemistry, 4th ed.; Wiley: New York, 1980; Chapters 20, 28, 31.
8. Sillén, L. G.; Martell, A. E.; Högfeldt, E.; Smith, R. M. Stability Constants of Metal-Ion Complexes; Special Publication 25; The Chemical Society: London, 1971; Suppl. No. 1.
9. Sulkowski, E.; Vastola, K.; Oleszek, D.; VonMuenchhausen, W. In Affinity Chromatography and Related Techniques; Gribnau, T. C. J.; Viser, J.; Nivard, R. J. F., Eds.; Elsevier Scientific Publishing Company: Amsterdam, 1982; pp 313-322.
10. Hutchens, T. W.; Yip, T-T.; Porath, J. Anal. Biochem. 1988, 170, 168-182.
11. Krishnan, S.; Vijayalakshmi, M. A.; Geahel, I. J. Chromatogr. 1987, 397, 339-346.
12. Smith, M. C.; Furman, T. C.; Pidgeon, C. Inorg. Chem. 1987, 26, 1965-1969.
13. Smith, M. C.; Furman, T. C.; Ingolia, T. D.; Pidgeon, C. J. Biol. Chem. 1988, 263, 7211-7215.
14. Hochuli, E.; Bannwarth, W.; Döbeli, H.; Gentz, R.; Stüber, D. Bio/Technology 1988, 6, 1321-1325.
15. Gesellchen, P. D.; Shuman, R. T. U.S. Patent 4 351 763, 1982.
16. Gesellchen, P. D.; Shuman, R. T. Tetrahedron Lett. 1976, 3369-3372.
17. Gesellchen, P. D.; Tafur, S.; Shields, J. E.; In Peptides. Structure and Biological Function. Proc. 6th Amer. Peptide Symp. Gross, E.; Meienhofer, J., Eds; Pierce Chemical Company: Rockford, IL, 1979; pp 117-120.
18. Maurer, R.; Meyer, B. J.; Ptashne, M. J. Mol. Biol. 1980, 139, 147-161.
19. Furman, T. C.; Epp, J.; Hsiung, H. M.; Hoskins, J.; Long, G. L.; Mendelsohn, L. G.; Schoner, B.; Smith, D. P.; Smith, M. C. Bio/Technology 1987, 5, 1047-1051.
20. Wickner, W.; Moore, K.; Dibb, N.; Geissert, D.; Rice, M. J. Bacteriol. 1987, 169, 3821-3822.
21. Freeman, H. C. Adv. Protein Chem. 1967, 22, 257-437.
22. Saito, Y.; Yamada, H.; Niwa, M.; Ueda, I. J. Biochem. (Tokyo) 1987, 101, 123-134.
23. Dalboge, H.; Dahl, H. H.; Pedersen, J.; Hansen, J. W.; Christensen, T. Biol. Chem. Hoppe-Seyler 1986, 367 (suppl.), 204.
24. Hsiung, H. M.; Becker, G. W. In Biotechnology and Genetic Engineering Reviews 1988, 6, 43-65.
25. Porath, J.; Olin, B. Biochemistry 1983, 22, 1621-1630.

RECEIVED February 2, 1990

Chapter 13

Site-Specific Proteolysis of Fusion Proteins

Paul Carter

Department of Biomolecular Chemistry, Genentech, Inc., 460 Point San Bruno Boulevard, South San Francisco, CA 94080

The use of gene fusions has greatly facilitated the expression and subsequent purification of heterologous proteins in *E. coli* (reviewed in 1-2, Uhlén, M.; Moks, T. Methods Enzymol., in press). A commonly used arrangement (C-terminal fusions) is one in which the gene encoding the protein of interest is fused to the 3′ end of a gene which codes for a highly expressed protein (Figure 1). This configuration is attractive because the signals which direct transcription and translation are located at the 5′ end of the gene and will function efficiently in the presence of a variety of different fusions to the 3′ end of the gene. In addition to facilitating expression, the N-terminus of such fusion (hybrid) proteins is often chosen as an "affinity handle": a polypeptide with a high affinity for a particular ligand. At least a dozen different affinity handles have been used in the purification of fusion proteins.

Although the hybrid protein may sometimes be used directly (e.g. for immunization, see 3) it is often necessary to release the protein of interest from the fusion protein and it may be desirable to use a method which generates the correct N-terminus. A variety of chemical and enzymatic methods has been used for site-specific cleavage of fusion proteins (Table I). Chemical cleavage methods have the attraction that the reagents are widely and cheaply available and that the reactions are often readily performed on a large scale. However there are several limitations intrinsic to most chemical cleavage methods. These problems are illustrated by cyanogen bromide (CNBr), which cleaves after methionine, and is the most widely used chemical for selective cleavage (reviewed in 4). Firstly, many fusion proteins will contain additional cleavage sites (~2% of residues in proteins are methionine, see 5). Secondly, the protein of interest must be able to withstand the relatively harsh reaction conditions (usually 70% formic acid in the case of CNBr). Thirdly, a limited extent of amino acid side chain modification may occur, leading to heterogeneity of the target protein. For example, oxidation of cysteine residues and bromination of tyrosine residues by CNBr may occur (4). Trace quantities of modified protein may be acceptable for biochemical studies but will preclude therapeutic use. An additional problem is the extreme toxicity of CNBr, which is a significant issue for preparative scale cleavage reactions.

Proteases are generally preferable to chemical methods for the specific cleavage of fusion proteins because corresponding cleavage sites usually occur less frequently, reaction conditions are milder and therefore compatible with a wider range of target proteins and the problem of side reactions is avoided. This review focuses on the repertoire of proteases that has been used for site-specific proteolysis of fusion proteins.

0097–6156/90/0427–0181$06.00/0

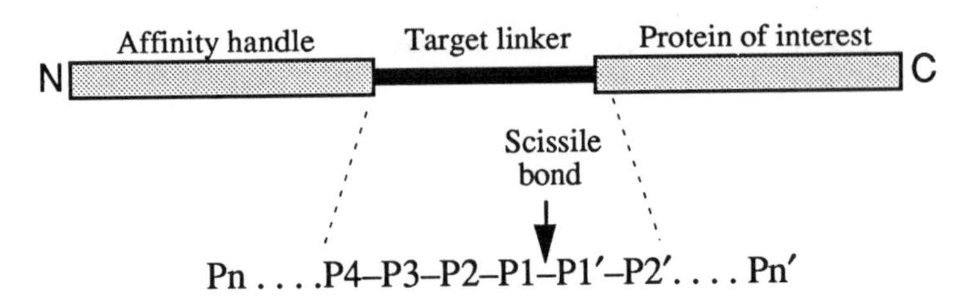

Figure 1. Design of C-terminal fusion protein for site-specific proteolysis.

Protease Substrate Specificity

In site-specific proteolysis of C-terminal fusion proteins (Figure 1) it is desirable for the protease to be exquisitely specific for residues on the N-terminal side of the scissile bond in order to restrict cleavage to the target site. However, the protease should be non-specific on the C-terminal side, to be compatible with any chosen protein. Protease specificity determinants may extend for one or more residues on either or both sides of the scissile bond, but the number and relative importance of these substrate sub-sites is highly variable between different proteases. However, for the enzymes in Table I the major specificity determinants are associated with residues on the N-terminal side of the scissile bond, which makes them well suited for specific cleavage of fusion proteins in which the C-terminal region is varied. In the case of trypsin and chymotrypsin the specificity is dominated by a single residue, (designated P1 (6) - see Table I) which limits their utility in site-specific proteolysis to a few special cases (see below). Factor X_a, thrombin, enterokinase, and kallikrein have primary substrate specificities similar to trypsin. However they have significant specificity determinants over at least 2 additional residues on the substrate, which greatly narrows their substrate specificity and consequently increases their utility for site-specific proteolysis.

The substrate specificity of the proteases shown in Table I which have extended substrate binding sites reflect known target sequences in natural (or synthetic) substrates, but are not the only sites which are efficiently cleaved by these proteases. Sub-site specificity determinants on the N-terminal side of the scissile bond are readily examined using chromogenic and fluorogenic peptide substrates, whereas analysis of sub-site specificity on the C-terminal side of the scissile bond is generally restricted to polypeptide substrates and consequently is much more limited. It is tempting to estimate the frequency of naturally occurring sites for a given "site-specific protease" by searching protein sequence databases (7, 8). However this approach is prone to inaccuracy since the protease may cleave at sites other than those previously encountered (9, 10) and the site must be accessible in order for cleavage to occur (see following section).

For candidate "site-specific proteases" it is helpful to establish a substrate specificity database. For example, it is important to identify the range of residues on the C-terminal side of the scissile bond which are compatible with efficient cleavage and to characterize any additional sites which are cleaved at a significant rate compared to the target site (Tables II - VI). Practical details on the use of these enzymes are beyond the scope of this review but may found in manufacturers' literature and from the examples cited.

Accessibility

A prerequisite for site-specific proteolysis is accessibility of the target site and inaccessibility of any additional potential cleavage sites. This is illustrated by the use of trypsin to release intact human myoglobin from a fusion protein having the N-terminus of bacteriophage λ cII protein, despite the presence of 20 potential cleavage sites (11). Accessibility

Table I. Methods for the Specific Cleavage of Fusion Proteins

	Cleavage specificity	Reference	
Chemical[a]			
Hydroxylamine	N↓G	42, 47	
Formic acid	D↓P	48, 49	
Acetic acid	D↓P	50	
Cyanogen bromide	M↓	51, 52[b]	
BNPS-skatole	W↓[c]	53	
o-iodosobenzoic acid	W↓[c]	54	
N-chlorosuccinimide	W↓[c]	*d*	
Enzyme			**Commercial Sources**[q]
Chymotrypsin	W↓ and Y↓ and F↓[e]	45	Bo Cb Cz Si,
Collagenase	P-X↓G-P[f]	15, 37-40	Bo Cb Si
Endoproteinase Lys-C	K↓[g]	46	Bo Cb
Enterokinase	D-D-D-D-K↓[h]	Table VI	Si
Factor X_a	I-E-G-R↓[i,j]	Table II	Bo Si
Kallikrein	P-F-R↓[j,k]	18	Bo Cb Si
Renin	Y-I-H-P-F-H-L↓L[l]	8	Cb Si
H64A subtilisin BPN′	A-A-H-Y↓[m]	Table IV	*r*
Thrombin	R-G-P-R↓[j,n]	Table III	Bo Si
Trypsin	R↓ and K↓[o]	11, 41-44, 77	Bo Cb Cz Pi Si
Ubiquitin protein peptidase	Ubiquitin↓[p]	Table V	N/A

[a] Chemical methods for fragmenting polypeptides are reviewed in 4. [b] Many additional examples of using cyanogen bromide to cleave fusion proteins are given in Table 2 from 1. [c] Several different methods for cleaving fusion proteins on the C-terminal side of Trp are compared in 55. [d] Nilsson, B.; Forsberg, G.; Hartmanis, M. Methods Enzymol., in press. [e] Primary and secondary sub-site specificities of α-chymotrypsin are examined in 56. [f] Consensus site in collagen which is cleaved by collagenase (57). [g] Cleavage sites for Lys-C with melittin as a substrate (Boehringer Mannheim Biochemicals: manufacturer's literature). [h] Enterokinase cleavage site in its natural substrate, trypsinogen (58). [i] Factor X_a cleavage sites in its natural substrate, thrombin (59). [j] Kinetic parameters with a large number of chromogenic and fluorogenic substrates are summarized in 25 and 26. [k] Cleavage site for plasma kallikrein in kininogen (60). [l] Minimal fragment of natural substrate, angiotensinogen, required for cleavage by renin (36). [m] Most favorable sequence identified from analysis of tetrapeptide *p*-nitroanilide substrates (7). [n] Thrombin cleavage site in its natural substrate fibrinogen (61). [o] Primary and sub-site specificities are reviewed in 62. [p] Available evidence suggests that ubiquitin as the N-terminal part of a fusion protein is necessary but not sufficient for cleavage to occur (32). [q] Commercial sources of enzyme are provided only as a guide to availability - other unnamed sources may be superior: Bo, Boehringer Mannheim Biochemicals; Cb, Calbiochem; Cz, Calzyme; Pi, Pierce; Si, Sigma Chemical Company; N/A not commercially available. [r] H64A subtilisin BPN′ variants may be obtained for non-commercial research use upon request to the author or Dr James A. Wells at the Department of Biomolecular Chemistry, Genentech Inc, 460 Point San Bruno Boulevard, South San Francisco, CA 94080. BNPS-skatole, 2-(2-nitrophenylsulfenyl)-3-methyl-3′-bromoindolenine.

of the target site may be compromised by neighboring sequences on either side of the scissile bond. One strategy for enhancing the accessibility to proteases is to flank the target sequence on both sides with short stretches of glycine residues (12) or on one side if the correct terminus of the protein of interest is required after cleavage (13). An alternative strategy to enhance substrate accessibility is to perform the digests under denaturing or reducing conditions (provided that this is compatible with the protease used) or by denaturation of the substrate prior to digestion (14). It may be possible to design a C-terminal fusion protein such that the target linker is an inherently accessible site in the affinity handle, e.g. by overlapping a known protease sensitive site (7).

Problems Encountered in Site-Specific Proteolysis

A major problem encountered in attempting site-specific proteolysis is that of non-specific cleavage at additional sites to the target site. This may be due to the "site-specific protease" itself or contaminating proteolytic activity in preparations of either the protease or fusion protein. Contaminating proteases may be overcome by the judicious addition of protease inhibitors to the digest, or by further purification of the substrate or enzyme (15). In the case where the additional cleavage events are due to the "site-specific protease", optimizing the digestion conditions to give partial proteolysis may give an acceptable yield of intact target protein (16). Failing this, one may have to adopt an alternative enzymatic (10) or chemical cleavage (17) method.

A second significant difficulty in achieving site-specific proteolysis is inefficient (or no) cleavage at the target site. This may reflect inaccessibility of the target site (see above) or the substrate specificity of the protease on the C-terminal side of the scissile bond. Many of the proteases in Table I will not cleave on the N-terminal side of proline. One strategy to circumvent this problem is to add amino acids to the N-terminus of the protein of interest and subsequently remove them using an aminopeptidase (18) or diaminopeptidase. If other methods do not enhance cleavage at the target site it may be necessary to try a different protease and fusion protein (19).

An infrequent problem is that of insolubility of the fusion protein or cleavage products, which may usually be overcome by including low concentrations of denaturant (or detergent) in the digest, provided that this is compatible with the cleavage method used (20).

Proteolysis *in vivo* is sometimes a major problem in the use of fusion proteins. It is highly desirable to obtain fusion proteins as free of proteolysis products as possible: if the fusion protein is heterogeneous (or contaminated with proteases) it may difficult to interpret the outcome of digestion experiments or to separate the protein of interest away from degradation products (P.C., unpublished data). A variety of strategies have been used to curb proteolysis *in vivo*, including the use of protease deficient host strains and optimization of growth media and temperature (reviewed in 1-2, Uhlén, M.; Moks, T. Methods Enzymol., in press). Intracellular expression of fusion proteins in *E. coli* often results in the accumulation of the fusion protein in insoluble aggregates known as "inclusion" or "refractile bodies". This has the benefit of protecting the fusion protein from proteolysis as well as allowing the fusion protein to be readily separated from most *E. coli* proteins and also nucleic acids. A disadvantage is that it necessitates refolding the protein of interest into the native conformation, although in many cases this is possible (see Kelley, R. F.; Winkler, M. E. In Genetic Engineering: Principle and Methods; Setlow, J., Ed.; Plenum: London; Vol. 12, in press).

Insulin-like growth factor II (IGF-II) is susceptible to proteolysis in *E. coli* but has been successfully expressed in a soluble form using a dual affinity fusion protein strategy (21). A tripartite fusion protein was constructed in which IGF-II was flanked on the N-terminal side by the IgG-binding domains of staphylococcal protein A and on the C-terminal side by the albumin-binding domains of streptococcal protein G. Recovery of full length IGF-II was ensured by sequential affinity purifications on IgG and albumin matrices. Furthermore this was facilitated by the observation that IGF-II in the tripartite fusion protein is less prone to proteolytic degradation in *E. coli* than when expressed as the N-terminal fusion protein lacking the albumin-binding domains. IGF-II was then excised from the tripartite fusion by the use of CNBr which also leaves a homoserine or homoserine lactone at the C-terminus. Thus one limitation of this dual affinity fusion protein strategy is that existing cleavage methods include specificity determinants of one or more residues on the N-terminal side of the scissile bond, which may preclude generation of the correct C-terminus. A potential solution of this difficulty is the use of H64A subtilisin BPN′ variants (see below). A potential limitation (not a problem for IGF-II) of dual affinity fusions is that of compromising the correct folding of the protein of interest, which may necessitate refolding *in vitro* (see Kelley, R. F.; Winkler, M. E. In Genetic Engineering: Principle and Methods; Setlow, J., Ed.; Plenum: London; Vol. 12, in press).

Survey of Enzymes Used for Site-Specific Proteolysis

Factor X_a. To date the most widely used protease for site-specific proteolysis of fusion proteins has been the blood coagulation factor X_a. This is a consequence of the pioneering work of Nagai and Thøgersen, who developed this cleavage method in conjunction with a versatile *E. coli* intracellular expression system (see 22 for a practical review). Factor X_a has been used successfully with a variety of fusion protein systems (Table II) including those in which the N-terminal region is protein A, β-galactosidase, maltose binding protein or glutathione S transferase, where efficient affinity purification methods are available (reviewed by Uhlén, M.; Moks, T. Methods Enzymol., in press).

Table II. Fusion Proteins Cleaved by Factor X_a

P4	P3	P2	P1	P1′	P2′	Fusion protein	Unique cleavage site ?	Ref.
Substrate Residues[a]								
I	E	G	R	V	H	cII / β-globin	Yes[b]	63
I	E	G	R	S	P	cII / β-globin (mutant)	Yes	22
I	E	G	R	V	L	cII / α-globin	Yes	22
I	E	G	R	G	L	cII / myoglobin	Yes	22
I	E	G	R	S	Y	cII / TMV coat protein	No	17
I	E	G	R	D	D	cII / actin	Yes	22
I	E	G	R	A	P	cII / MLC	No	22
I	E	G	R	M	A	cII / troponin C	Yes[c]	64
I	E	G	R	G	E	cII / TFIIIA	?	22
I	E	G	R	F	T	cII / MAT α1	No	16
I	E	G	R	S	S	cII / ferredoxin	Yes	65
I	E	G	R	M	N	cII / PDE γ subunit	Yes	66
I	E	G	R	?	?	cII / MLC/IGF-II receptor tail	Yes	67
I	E	G	R	D	A	cII / MOPC315V_L	Yes	68
I	E	G	R	K	E	β-gal / ribonucleaseA	No[d]	9
I	E	G	R	K	G	β-gal$_{1-375}$ / p15 or p24	No[d,e]	12
I	E	G	R	S	D	β-gal / caltrin	No[d]	20
I	E	G	R	H	G	Ubiquitin / β-gal	Yes	69
I	E	G	R	T	V	*11-1* gene / β-gal	No[d]	24
I	E	G	R	G	D	MBP / β-gal	No[d]	14
I	E	G	R	N	S	MBP / paramyosin	Yes	14
I	E	G	R	T	F	*trp E* / prorenin	Yes	70
I	E	G	R	T	?	Protein A′ / rubisco	?	71
I	E	G	R	G	I	GST / Ag 63	Yes	72
I	E	G	R	M	E	*TYA* gene / HIV-TAT	Yes	73
I	E	G	R	Y	A	Colicin$_{1-172}$ / hGRF	Yes	74
I	E	G	R	S	P	β-gal$_{1-289}$ / proAFP	No[d]	75

[a] Protease substrates are represented using the nomenclature of Schechter and Berger in which the scissile peptide bond is between the P1 and P1′ residues (6). [b] For the β-globin mutant, H143R, there is slow internal cleavage: A-L-A-R ↓K-Y (22). [c] A second site with a slower rate of cleavage was detected in the absence of divalent cations. [d] Cleavage occurs within β-gal: C-N-G-R↓W-V (9). [e] Cleavage also occurs within the C-terminal half of the fusion protein. cII, N-terminal 31 residues from bacteriophage λ cII protein; TMV, tobacco mosaic virus; MLC, myosin light chain; TFIIIA, transcription factor IIIA; MAT α1, transcription factor for regulation of α-specific genes in *Saccharomyces cerevisiae*; PDE, cGMP phosphodiesterase; IGF-II, insulin-like growth factor II; MOPC315 V_L, variable light chain sequence from MOPC315 IgA; β-gal$_{1-375}$, residues 1 to 375 from β-galactosidase; p15 and p24 are proteins derived from the HIV type I *gag* gene; *11-1* gene is from *Plasmodium falciparum*; MBP, maltose-binding protein; rubisco, small subunit of ribulose-1,5-bisphosphate carboxylase-oxygenase; GST, C-terminus of glutathione S-transferase from *Schistosoma japonicum*; Ag 63, antigen from *Plasmodium falciparum; TYA* gene is from the retrotransposon Ty; HIV-TAT, TAT protein from HIV; hGRF, human growth hormone releasing factor; proAFP, winter flounder antifreeze pro-protein.

Virtually all fusion proteins reported were cleaved at the target site by factor X_a, which suggests that substrate specificity on the C-terminal side of the scissile bond is very broad and that substrate accessibility is not a significant problem. Unfortunately cleavage (↓) by factor X_a is not limited to the sequence: I-E-G-R↓ (single letter code), as shown by additional cleavage of several fusion proteins. Only a few of these additional sites have been characterized, and they have little in common within the I-E-G-R↓ sequence: C-N-G-R↓W-V (9), S-L-S-R↓M-T (10), A-L-A-R↓K-Y (22) and A-N-F-V-K↓N-A-K↓K-Y-D-P (Luger, K., Biozentrum, Basel, unpublished data). In many cases the additional site was either in the affinity handle (9) or was cleaved much more slowly than the target site (16), which enabled the intact protein of interest to be recovered in good yield. Cleavage by factor X_a after lysine residues may be eliminated by reversible chemical modification of the substrate with 3,4,5,6-tetrahydrophthalic anhydride (Wearne, S., MRC Laboratory of Molecular Biology, Cambridge, personal communication, 1990). In a few cases however it was necessary (or desirable) to resort to an alternative cleavage method (10, 17, 23).

Factor X may be purified from bovine blood and converted to its active form (X_a) with Russell's viper venom (22). Reliable commercial sources of bovine (and human) factor X_a are available – see Table I (as are suitable chromogenic substrates such as N-benzoyl-L-Ile-L-Glu-Gly-L-Arg-*p*-nitroanilide), although problems have been noted (9, 24). The most likely contaminating activity in factor X_a preparations is thrombin, which can be selectively inhibited with the leech protein hirudin.

Thrombin. The blood clotting factor, thrombin (Table III) has been less widely used for site-specific proteolysis than has factor X_a (Table II). Thrombin is likely to be less generally useful since its substrate specificity is broader (25, 26) and there may be important specificity determinants on the C-terminal side of the scissile bond (22). Nevertheless, thrombin is a useful adjunct to factor X_a since their substrate specificities partially overlap. For example, a cII / myosin light chain fusion protein was cleaved internally by factor X_a (22), but intact myosin light chain could be obtained by cleavage of the same fusion protein with thrombin (23). Thrombin has been used to specifically and efficiently cleave a fusion protein whilst still immobilized on an affinity purification matrix (27), which obviated several steps and thus expedited product recovery. Both bovine and human thrombin are commercially available as are suitable chromogenic test substrates such as tosyl-Gly-L-Pro-L-Arg-*p*-nitroanilide (see Table I).

Table III. Fusion Proteins Cleaved by Thrombin

Substrate residues[c]								
P4	P3	P2	P1	P1′	P2′	Fusion protein	Unique cleavage site ?	Ref.
I	E	G	R	A	P	cII / MLC[a]	Yes	23
T	M	P	R	F	S	cII / GFAP[a]	(Yes)[d]	10
L	V	P	R	G	P	hGH / IGF-I[a]	No[e]	19
L	A	P	R	F	Q	β-gal$_{1-8}$ / cyt b$_5$ reductase[a]	Yes	76
L	V	P	R	G	S	GST / Ag 63[b]	Yes	71
L	V	P	R	G	S	GST / LIF	Yes	27
F	F	P	R	T	R	ZZ / bIGF-I	Yes	[f]
R	G	P	R	S	L	CAT / ANF[b]	Yes	53
L	I	A	K	G	P	Cathepsin D$_{1-50}$ /α-lactalbumin	(Yes)[g]	77

[a]Cleaved by bovine thrombin. [b]Cleaved by human thrombin. [c] Nomenclature of Schechter and Berger (6) - see Table II. Specificity determinants for human α-thrombin extend from at least P4 to P2′ (78). [d] Quantitative cleavage within GFAP (glial fibrillary acidic protein): R6↓F7 rather than at target linker I-E-G-R↓L-G. [e] Additional site(s) within hGH (human growth hormone) was proposed from absence of discrete band corresponding to hGH on a polyacrylamide gel. [f] Nilsson, B.; Forsberg, G.; Hartmanis, M. Methods Enzymol., in press. [g] Quantitative cleavage within cathepsin D: K34↓G35 rather than at target linker G-P-G-R↓G-P. (b)IGF-I, (brain) insulin-like growth factor I; cyt, cytochrome; LIF, leukemia inhibitory factor; Z, synthetic domain of *Staphylococcus aureus* protein A (79); CAT, chloramphenicol acetyl transferase; ANF, atrial natriuretic factor.

Subtilisin BPN′ Variants. The serine protease, subtilisin BPN′, (from *Bacillus amyloliquifaciens*) has been used for site-specific proteolysis (28, 29). However, its substrate specificity is far too broad for it to be generally useful. In contrast to other proteases which have wide utility for site-specific proteolysis (Table I), subtilisin has been the target for extensive protein engineering studies (reviewed in 30), which offers the unique advantage of allowing the substrate specificity to be tailored. A mutant in which the histidine in the catalytic triad was replaced by alanine (H64A) is highly specific for certain histidine-containing substrates, because the substrate histidine can substitute to some extent for the missing catalytic group - "substrate-assisted catalysis" (31). Although the activity of this H64A subtilisin variant is low, it has been possible to increase the activity of this enzyme 150-fold by further site-directed mutagenesis and by substrate optimization (7, P.C., L. Abrahmsén, J. Wells, unpublished data). Furthermore the practical utility of the H64A enzyme and an enhanced activity variant for site-specific proteolysis of fusion proteins has been demonstrated (Table IV, see 7 for practical details).

The H64A enzyme has very broad specificity on the C-terminal side of the scissile bond: at P1′ all residues (except proline and isoleucine) are compatible with cleavage and substrates with at least G, A, L, M, Y, W, R and K at P2′ (not proline) may be efficiently cleaved (Table IV and P.C., unpublished data). This suggests that it should be possible to cleave a wide range of C-terminal fusion proteins. Several short histidine-containing polypeptides have been identified which are not detectably cleaved by H64A subtilisin (31), which suggests that a histidine residue is necessary but not sufficient for polypeptide cleavage. In addition, favorable residues are apparently required for at least some of the other sub-sites, most notably at P1 (which dominates the specificity of wild-type subtilisin) but also at P4 (7). (Substrate residues are represented using the nomenclature of Schechter and Berger (6) - see Table I.)

To date 8 different fusion proteins have been specifically cleaved at the target site using H64A variant subtilisins (including 3 shown in Table IV) which suggests that these enzymes will be broadly useful. A fusion protein between synthetic domains of protein A and brain insulin-like growth factor I (bIGF-I) was quantitatively and efficiently cleaved by an H64A variant subtilisin (Nilsson, B.; Forsberg, G.; Hartmanis, M. Methods Enzymol., in press). In contrast, cleavage of very similar fusion proteins (differ only in their cleavage linkers) by thrombin and by chemical methods gave yields of only ~ 30% bIGF-I.

Table IV. Substrate Specificity of H64A[a] Subtilisin BPN′ Variants

Substrate residues[b]								S24C:H64A: E156S:G169A:	
P4[d]	P3	P2	P1	P1′	P2′	Substrate	S24C:H64A[a]	Y217L[a,c]	Ref.
F	A	H	Y	X[e]	G	Hexapeptides[f]	Yes	ND	7
F	A	H	Y	T	R	Z / AP[g]	Yes	Yes	7
F	A	H	Y	T	L	ZZ / bIGF-I	ND	Yes	*h*
F	A	H	Y	A	Y	ZZ / IGF-II	ND	Yes	*h*
I	N	H	Y	R	M	Inhibin β-chain$_{61-80}$	Yes	ND	31
M	E	H	F	R	W	ACTH$_{1-10}$	Yes	ND	31

[a] Mutants of subtilisin are designated by the single letter code for the wild-type amino acid followed by the residue number and then the amino acid replacement. Multiple mutants are identified by listing the single mutant components separated by colons, e.g. S24C:H64A represents the double mutant Ser24→Cys, His64→Ala. [b] Specificity determinants for subtilisin extend from P4 to P2′ (80). [c] This enzyme is also known as *Genenase I*. [d] Analysis of tetrapeptide *p*-nitroanilide substrates suggests that A is preferable to F at the P4 position (7). [e] X represents the 20 common amino acids except proline and isoleucine. [f] N-succinylated. [g] Very slow cleavage within the Z domain was detected with a His residue at the P1′ position (see text). [h] Nilsson, B.; Forsberg, G.; Hartmanis, M.; Methods Enzymol., in press. ND, not determined; AP, alkaline phosphatase; ACTH$_{1-10}$, residues 1 to 10 of adrenocorticotropic hormone.

Table V. Fusion Proteins Cleaved by Ubiquitin Protein Peptidase

Substrate residues[a]							Unique cleavage	
P4	P3	P2	P1	P1′	P2′	Fusion protein	site ?	Ref.
L	R	G	G	Q	L	ubiquitin / relaxin A chain	Yes	32
L	R	G	G	D	S	ubiquitin / relaxin B chain	Yes	32
L	R	G	G	M	Y	ubiquitin / *Xis*	Yes	32
L	R	G	G	G	S	ubiquitin / X[b]	Yes	32
L	R	G	G	E	F	ubiquitin / *rlx*	Yes	*c*
L	R	G	G	N	T	ubiquitin / GAP	Yes	*c*

[a] Nomenclature of Schechter and Berger (6) - see Table II. [b] X corresponds to: G - S-P-G-E-L-E-F-T-G-R-R-F-T-T-S. [c] Liu, C.-C.; Miller, H. I.; Kohr, W. J.; Silber, J. I. J. Biol. Chem., in press. *Xis*, product of *Xis* gene from bacteriophage λ; *rlx*, 38 residue peptide including 32 residues from human prorelaxin; GAP, rat gonadotropin associated protein.

Cleavage of target linkers has always occurred with the histidine residue at the expected P2 position of the substrate (F-A-H-Y↓ and A-A-H-Y↓ - see Table IV). In a few cases cleavage has also been detected at an additional site with a histidine residue at the P1′ position (P. C., J. Wells, R. Vandlen, K. Miller and S. Braxton, unpublished data). Further studies are underway to investigate the substrate requirements for proteolysis assisted by histidine at P1′. It may be possible to exploit this phenomenon for site-specific proteolysis of tripartite (21) or C-terminal fusion proteins, which both include extensions C-terminal to the protein of interest.

Ubiquitin Protein Peptidase. Several different enzymes are involved in the removal of ubiquitin from proteins in eukaryotes including ubiquitin protein peptidase, which has been cloned from yeast (32). Ubiquitin protein peptidase has been used for site-specific proteolysis of a number of fusion proteins (Table V). However its utility appears to be limited by a requirement for ubiquitin at the N-terminus of a fusion protein substrate (32). Furthermore this enzyme requires the presence of a reducing agent for activity *in vitro* and it is most efficacious against low molecular weight substrates. The C-terminal glycine (G76) in ubiquitin is apparently important since replacement by either valine or cysteine (or deletion) abolishes detectable cleavage (32). The specificity for the residue on the C-terminal side of the scissile bond (P1′) is evidently broad, with at least E, D, C, G, T and M representing favorable residues. Only proline at P1′ seems to be incompatible with cleavage.

A unique feature of ubiquitin protein peptidase compared with other proteases shown in Table I, is that suitable fusion proteins may be specifically cleaved *in vivo* in *E. coli* over-expressing ubiquitin protein peptidase (32) or by the endogenous ubiquitin-hydrolyzing activity in yeast (33). This provides a means for the heterologous expression of eukaryotic proteins and generating the correct N-termini, which may not be possible by direct expression, because a methionine residue derived from the initiation codon is often left at the N-terminus.

Enterokinase. The serine protease, enterokinase, activates trypsinogen to trypsin *in vivo* and *in vitro* (34) by specific cleavage of the peptide bond between K6 and I7. Enterokinase has been used for site-specific proteolysis of fusion proteins in which the N-terminal 8 residues includes the required cleavage sequence but also represents the epitope for a Ca^{2+}-dependent monoclonal antibody which enables affinity purification of the fusion protein (flag™ system - see Table VI). Enterokinase is capable of generating products with at least E, A, T, L and I at the N-terminus of the protein of interest. Enterokinase is commercially available (see Table I) but may require further purification to remove contaminating proteases such as chymotrypsin, trypsin and elastase (35).

Table VI. Fusion Proteins Cleaved by Enterokinase

Substrate residues[a]								
P4	P3	P2	P1	P1′	P2′	Fusion protein	Unique cleavage site ?	Ref.
D	D	D	K	A	P	flag[b] / IL-3	Yes	81
D	D	D	K	A	P	flag[b] / IL-4	Yes	81
D	D	D	K	A	P	flag[b] / IL-2	No[c]	35
D	D	D	K	A	P	flag[b] / GM-CSF	No[c]	35
D	D	D	K	A	P	flag[b] / CSF-I	No[c]	35
D	D	D	K	A	P	flag[b] / G-CSF	No[c]	35

[a] Nomenclature of Schechter and Berger (6) - see Table II. [b] Flag corresponds to the octapeptide sequence: D-Y-K-D-D-D-D-K (35). [c] A small amount of internal cleavage was detected representing ≤ 10% of the total fusion protein. IL-3, IL-4 and IL-2 are interleukins 3, 4 and 2 respectively; GM-CSF, granulocyte-macrophage colony stimulating factor; CSF-I, colony stimulating factor; G-CSF, granulocyte colony stimulating factor.

Renin. The aspartyl protease, renin, specifically cleaves angiotensinogen to release angiotensin I *in vivo* (Table I). Murine submaxillary gland renin has been used for site-specific proteolysis of a fusion protein (190 K) between Epstein-Barr virus membrane antigen and β-galactosidase (8). Quantitative cleavage at the target linker (H-P-F-H-L↓L-V) was obtained with little (if any) cleavage at additional sites. The specificity determinants for renin appear to extend for one or more residues on the C-terminal side of the scissile bond, which may limit its utility (36).

Collagenase. Although microbial collagenase was one of the first proteases used for site-specific proteolysis of fusion proteins (37), it has not yet enjoyed widespread use. A major limitation is that collagenase linkers (15, 37-39) are long (~60 residues) and contain multiple cut sites for collagenase, which often gives rise to product heterogeneity. Furthermore, even if cleavage of the linker proceeds to completion, the protein of interest is left with an extension of at least a few residues from collagen. Recently however these problems with collagenase have been largely overcome: efficient cleavage of a fusion protein between alkaline phosphatase and adrenocorticotropic hormone (ACTH) has been demonstrated at short target linkers, G-(P-X-G)$_n$-P, where $n \geq 2$ (40). Furthermore it was possible to generate ACTH with the correct N-terminus by subsequent cleavage with dipeptidyl aminopeptidase-IV (Table VII). A variety of collagenases are commercially available (see Table I) but further purification is generally (15, 37-39) but not always (40) required to remove contaminating proteases.

Trypsin and Chymotrypsin. If the protein of interest is highly resistant to proteolysis, it may be possible to achieve site-specific proteolysis using proteases such as trypsin (11, 41-44, 77) or chymotrypsin (45), which are cheap, widely available, and very well characterized. For example, intact human myoglobin was released from a fusion protein with the N-terminus of bacteriophage λ cII protein by trypsin cleavage (11). This had the side-benefit of degradation of contaminating proteins thus decreasing the need for further purification enabling myoglobin to be economically and efficiently prepared on a gram scale. The action of trypsin is readily restricted to arginine residues by citraconylation (reversible) of lysine residues (41).

Endoproteinase Lys-C. The serine endoproteinase Lys-C has primary substrate specificity similar to trypsin but narrower in that it cleaves only on the C-terminal side of lysine residues. Lys-C has been used to specifically cleave a fusion protein between *trpE* (residues 1 to 320) and epidermal growth factor, which has no lysine residues (46). This cleavage was difficult to reproduce because of variation between different batches of enzyme. Today however, reliable albeit expensive commercial sources of Lys-C from *Lysobacter enzymogenes* are available (see Table I).

Table VII. Enzymes for Modification of Amino Termini

Enzyme	Substrate Specificity[a]	Example	Substrate	Ref.
Aminopeptidase M[b]	H_2N-M↓X-P	H_2N-M↓A-P...	Met-IL2	82
Aminopeptidase P[c]	H_2N-X↓P	H_2N-A↓P-P...	Ala-IL6	18
DAP-I[d] (cathepsin C)	H_2N-X_1-X_2↓X_3	H_2N-M-V↓F-P...	Met-Asp-bGH	83
	X_2, $X_3 \neq$ P;	H_2N-M-D↓F-P...	Met-Val-bGH	83
	$X_1 \neq$ K	H_2N-M-V↓F-P...	Ala-Glu-hGH	84
DAP-IV[c]	H_2N-X_1-X_2↓X_3 X_2 = P (or A)	H_2N-G-P↓S-Y...	Gly-Pro-ACTH	40

[a] For details see 85. [b] Commercially available from Bo, Cb, Si, and Pi - see Table I. [c] Not commercially available (?). [d] Commercially available from Bo and Si. bGH and hGH are bovine and human growth hormones respectively; DAP, dipeptidylaminopeptidase.

Kallikrein. Plasma kallikrein is a serine protease that specifically cleaves kininogen to release bradykinin *in vivo*. Human plasma kallikrein has been used successfully to cleave a fusion protein containing human interleukin-2 at the N-terminus and interleukin-6 (IL-6) at the C-terminus (18). An extra alanine residue was included before the N-terminal proline of IL-6 to permit cleavage by kallikrein, which will not cleave on the N-terminal side of proline. It was then possible to generate the correct N-terminus of IL-6 by digestion with *E. coli* aminopeptidase P (Table VII).

Aminopeptidases and Dipeptidylaminopeptidases. Several enzymes have been used to cleave one or more residues from the N-terminus of proteins (Table VII) present after cleavage with some chosen endoprotease (18, 40). The use of these exoproteases also provides an alternative strategy to the use of ubiquitin protein peptidase (see above) for generating the correct N-terminus of heterologously expressed eukaryotic proteins.

Conclusion

The availability of suitable proteases for site-specific proteolysis has in the past been a major limitation to the use of fusion proteins. This problem has been extensively eroded by more extensive characterization and wider availability of known proteases, by the discovery of new highly specific proteases, and by tailoring the substrate specificity of existing proteases by protein engineering.

Acknowledgments

The author thanks Drs Björn Nilsson, Mathias Uhlén, Steven Wearne and Chung-Cheng Liu for generously communicating manuscripts prior to publication and helpful discussions, and Dr Jim Wells for continued support.

Legend of Symbols

The single letter amino acid code has been used: A, Ala; C, Cys; D, Asp; E, Glu; G, Gly; H, His; I, Ile; K, Lys; L, Leu; M, Met; N, Asn; P, Pro; Q, Gln; R, Arg; S, Ser; T, Thr; V, Val; W, Trp.

Literature Cited

1. Harris, T. J. R. In Genetic Engineering; Williamson, R., Ed.; Academic: London, 1983; Vol. 4, p 127.
2. Marston, F. A. O. Biochem. J. 1986, 240, 1.
3. Kleid, D. G.; Yansura, D.; Small, B.; Dowbenko, D.; Moore, D. M.; Grubman, M. J.; McKercher, P. D.; Morgan, D. O.; Robertson, B. H.; Bachrach, H. L. Science 1981, 214, 1125.

4. Fontana, A.; Gross, E. In Practical Protein Chemistry – A Handbook; Darbre, A., Ed.; John Wiley and Sons: London, 1986; Chapter 2.
5. Klapper, M. H. Biochem. Biophys. Res. Commun. 1977, 78, 1018.
6. Schechter, I.; Berger, A. Biochem. Biophys. Res. Commun. 1967, 27, 157.
7. Carter, P.; Nilsson, B.; Burnier, J. P.; Burdick, D.; Wells, J. A. Proteins: Structure, Function and Genetics 1989, 6, 240.
8. Haffey, M. L.; Lehman, D.; Boger, J. DNA 1987, 6, 565.
9. Nambiar, K. P.; Stackhouse, J.; Presnell, S. R.; Benner, S. A. Eur. J. Biochem. 1987, 163, 67.
10. Quinlan, R. A.; Moir, R. D.; Stewart, M. J. Cell Sci. 1989, 93, 71.
11. Varadarajan, R.; Szabo, A.; Boxer, S. G. Proc. Natl. Acad. Sci. USA 1985, 82, 5681.
12. Ellinger, S.; Glockshuber, R.; Jahn, G.; Plückthun, A. J. Clin. Microbiol. 1989, 27, 971.
13. Georgiou, G.; Schmidt, C.; Baneyx, F. Fermentor 1989, 9, 92.
14. Maina, C. V.; Riggs, P. D.; Grandea, A. G. III; Slatko, B. E.; Moran, L. S.; Tagliamonte, J. A.; McReynolds, L. A.; di Guan, C. Gene 1988, 74, 365.
15. Scholtissek, S.; Grosse, F. Gene 1988, 62, 55.
16. Tan, S.; Ammerer, G.; Richmond, T. J. EMBO J. 1988, 7, 4255.
17. Turner, D. R. Ph. D. Thesis, Cambridge University, 1987.
18. Tonouchi, N.; Oouchi, N.; Kashima, N.; Kawai, M.; Nagase, K.; Okano, A.; Matsui, H.; Yamada, K.; Hirano, T.; Kishimoto, T. J. Biochem. 1988, 104, 30.
19. Nishikawa, S.; Yanase, K.; Tokunaga-Doi, T.; Kodoma, K.; Gomi, H.; Uesugi, S.; Ohtsuka, E.; Kato., Y.; Suzuki, F.; Ikehara, M. Protein Engineering 1987, 1, 487.
20. Heaphy, S.; Singh, M.; Gait, M. J. Protein Engineering 1987, 1, 425.
21. Hammarberg, B.; Nygren, P.-Å, Holmgren, E.; Elmblad, A.; Tally, M.; Hellman, U.; Moks, T; Uhlén, M. Proc. Natl. Acad. Sci. USA 1989, 86, 4367.
22. Nagai, K.; Thøgersen, H. C. Methods Enzymol. 1987, 153, 461.
23. Reinach, F. C.; Nagai, K.; Kendrick-Jones, J. Nature 1986, 322, 80.
24. Sieg, K.; Kun, J.; Pohl, I.; Scherf, A.; Müller-Hill, B. Gene 1989, 75, 261.
25. Lottenberg, R.; Christensen, U.; Jackson, C. M.; Coleman, P. L. Methods Enzymol. 1981, 80, 341.
26. Kawabata, S-i.; Miura, T.; Morita, T.; Kato, H.; Fujikawa, K.; Iwanaga, S.; Takada, K.; Kimura, T.; Sakakibara, S. Eur. J. Biochem. 1988, 172, 17.
27. Gearing, D. P.; Nicola, N. A.; Metcalf, D.; Foote, S.; Willson, T. A.; Gough, N. M.; Williams, R. L. Bio/Technol. 1989, 7, 1157.
28. Jacobsen, H.; Klenow, H.; Overgaard-Hansen, K. Eur. J. Biochem. 1974, 45, 623.
29. Richards, F. M.; Vithayathil, P. J. J. Biol. Chem. 1959, 234, 1459.
30. Wells, J. A.; Estell, D. A. Trends Biochem. Sci. 1988, 13, 291.
31. Carter, P.; Wells, J. A. Science 1987, 237, 394.
32. Miller, H. I.; Henzel, W. J.; Ridgway, J. B.; Kuang, W.-J.; Chisholm, V.; Liu, C.-C. Bio/Technol. 1989, 7, 698.
33. Sabin, E. A.; Lee-Ng, C. T.; Shuster, J. R.; Barr, P. J. Bio/Technol. 1989, 7, 705.
34. Craik, C. S.; Largman, C.; Fletcher, T.; Roczniak, S.; Barr, P. J.; Fletterick, R.; Rutter, W. J. Science 1985, 228, 291.
35. Hopp, T. P.; Prickett, K. S.; Price, V. L.; Libby, R. T.; March, C. J.; Cerretti, D. P.; Urdal, D. L.; Conlon, P. J. Bio/Technol. 1988, 6, 1204.
36. Skeggs, L. T.; Lentz, K. E.; Kahn, J. R.; Hochstrasser, H. J. Exp. Med. 1968, 128, 13.
37. Germino, J.; Bastia, D. Proc. Natl. Acad. Sci. USA 1984, 81, 4692.
38. Hanada, K.; Yamato, I.; Anraku, Y. J. Biol. Chem. 1987, 262, 14100.
39. Snouwaert, J. N.; Jambou, R. C.; Skonier, J. E.; Earnhardt, K.; Stebbins, J. R.; Fowlkes, D. M. Clin. Chem. 1989, 35 (suppl.) B7.
40. Daum, J.; Donner, P.; Geilen, W.; Hübner-Kosney, G.; Isernhagen, M.; Scheidecker, H.; Seliger, H.; Boidol, W.; Siewert, G. Eur. J. Biochem. 1989, 185, 347.
41. Shine, J.; Fettes, I.; Lan, N. C. Y.; Roberts, J. L.; Baxter, J. D. Nature 1980, 285, 456.

42. de Geus, P.; van den Bergh, C. J.; Kuipers, O.; Verheij, H. M.; Hoekstra, W. P. M.; de Haas, G. H. Nucleic Acids Res. 1987, 15, 3743.
43. Latta, M.; Knapp, M.; Sarmientos, P.; Bréfort, G.; Becquart, J.; Guerrier, L.; Jung, G.; Mayaux, J.-F. Bio/Technol. 1987, 5, 1309.
44. Moser, R.; Frey, S.; Münger, K.; Hehlgans, T.; Klauser, S.; Langen, H.; Winnacker, E.-L.; Mertz, R.; Gutte, B. Protein Engineering 1987, 1, 339.
45. Dahlman, K.; Strömstedt, P.-E.; Rae, C.; Jörnvall, H.; Flock, J.-I.; Carlstedt-Duke, J.; Gustafsson, J.-Å. J. Biol. Chem. 1989, 264, 804.
46. Allen, G.; Paynter, C. A.; Winther, M. D. J. Cell Sci. Suppl. 1985, 3, 29.
47. Moks, T.; Abrahmsén, L.; Holmgren, E.; Bilich, M.; Olsson, A.; Uhlén, M.; Pohl, G.; Sterky, C.; Hultberg, H.; Josephson, S.; Holmgren, A.; Jörnvall, H.; Nilsson, B. Biochemistry 1987, 26, 5239.
48. Nilsson, B.; Holmgren, E.; Josephson, S.; Gatenbeck, S.; Philipson, L.; Uhlén, M. Nucleic Acids Res. 1985, 13, 1151.
49. Szoka, P. R.; Schreiber, A. B.; Chan, H.; Murthy, J. DNA 1986, 5, 11.
50. Huston, J. S.; Levinson, D.; Mudgett-Hunter, M.; Tai, M.-S.; Novotny', J.; Margolies, M. N.; Ridge, R. J.; Bruccoleri, R. E.; Haber, E.; Crea, R.; Oppermann, H. Proc. Natl. Acad. Sci. USA 1988, 85, 5879.
51. Itakura, K.; Hirose, T.; Crea, R.; Riggs, A. D.; Heyneker, H. L.; Bolivar, F.; Boyer, H. W. Science 1977, 198, 1056.
52. Goeddel, D. V.; Kleid, D. G.; Bolivar, F.; Heyneker, H. L.; Yansura, D. G.; Crea, R.; Hirose, T.; Kraszewski, A.; Itakura, K.; Riggs, A. D. Proc. Natl. Acad. Sci. USA 1979, 76, 106.
53. Knott, J. A.; Sullivan, C. A.; Weston, A. Eur. J. Biochem. 1988, 174, 405.
54. Villa, S.; De Fazio, G.; Donini, S.; Tarchi, G.; Canosi, U. Eur. J. Biochem. 1988, 171, 137.
55. DiMarchi, R.; Long, H.; Epp, J.; Schoner, B.; Belagaje, R. In Synthetic Peptides: Approaches to Biological Problems; Tam, J. P.; Kaiser, E. T., Eds.; Alan R. Liss: New York, 1989, p 283.
56. Keil, B. Protein Seq. Data Anal. 1987, 1, 13.
57. Harper, E. In Collagenase; Mandl, I., Ed.; Gordon and Breach: New York, 1972; p 19.
58. Maroux, S.; Baratti, J.; Desnuelle, P. J. Biol. Chem. 1971, 246, 5031.
59. Magnusson, S.; Petersen, T. E.; Sottrup-Jensen, L.; Claeys, H. In Proteases and Biological Control; Reich, E.; Rifkin, D. B.; Shaw, E. Eds.; Cold Spring Harbor, New York, 1975, p 123.
60. Claeson, G.; Friberger, P.; Knös, M.; Eriksson, E. Haemostasis 1978, 7, 76.
61. Blombäck, B.; Blombäck, M.; Edman, P.; Hessel, B. Biochim. Biophys. Acta 1966, 115, 371.
62. Wilkinson, J. M. In Practical Protein Chemistry – A Handbook; Darbre, A., Ed.; John Wiley and Sons: London, 1986; Chapter 3.
63. Nagai, K.; Thøgersen, H. C. Nature 1984, 309, 810.
64. Reinach, F. C.; Karlsson, R. J. Biol. Chem. 1988, 263, 2371.
65. Coghlan, V. M.; Vickery, L. E. Proc. Natl. Acad. Sci. USA 1989, 86, 835.
66. Brown, R. L.; Stryer, L. Proc. Natl. Acad. Sci. USA 1989, 86, 4922.
67. Glickman, J. N.; Conibear, E.; Pearse, B. M. F. EMBO J. 1989, 8, 1041.
68. Baldwin, E.; Schultz, P. G. Science 1989, 245, 1104.
69. Chau, V.; Tobias, J. W.; Bachmair, A.; Marriott, D.; Ecker, D. J.; Gonda, D. K.; Varshavsky, A. Science 1989, 243, 1576.
70. Imai, T.; Cho, T.; Takamatsu, H.; Hori, H.; Saito, M.; Masuda, T.; Hirose, S.; Murakami, K. J. Biochem. 1986, 100, 425.
71. Landry, S. J.; Bartlett, S. G. J. Biol. Chem. 1989, 264, 9090.
72. Smith, D. B.; Johnson, K. S. Gene 1988, 67, 31.
73. Braddock, M.; Chambers, A.; Wilson, W.; Esnouf, M. P.; Adams, S. E.; Kingsman, A. J.; Kingsman, S. M. Cell 1989, 58, 269.

74. Geli, V.; Baty, D.; Knibiehler, M.; Lloubès, R.; Pessegue, B.; Shire, D.; Lazdunski, C. Gene 1989, 80, 129.
75. Peters, I. D.; Hew, C. L.; Davies, P. L. Protein Engineering 1989, 3, 145.
76. Shirabe, K.; Yubisui, T.; Takeshita, M. Biochem. Biophys. Acta 1989, 1008, 189.
77. Wang, M.; Scott, W. A.; Rao, K. R.; Udey, J.; Conner, G. E.; Brew, K. J. Biol. Chem. 1989, 264, 21116.
78. Bode, W.; Mayr, I.; Baumann, U.; Huber, R.; Stone, S. R.; Hofsteenge, J. EMBO J. 1989, 8, 3467.
79. Nilsson, B.; Moks, T.; Jansson, B.; Abrahmsén, L.; Elmblad, A.; Holmgren, E.; Henrichson, C.; Jones, T. A.; Uhlén, M. Protein Engineering 1987, 1, 107.
80. Robertus, J. D.; Kraut, J.; Alden, R. A.; Birktoft, J. J. Biochemistry 1972, 11, 4293.
81. Prickett, K. S.; Amberg, D. C.; Hopp, T. P. BioTechniques 1989, 7, 580.
82. Nakagawa, S.; Yamada, T.; Kato, K.; Nishimura, O. Bio/Technol. 1987, 5, 824.
83. Hsiung, H. M.; MacKellar, W. C. Methods Enzymol. 1987, 153, 390.
84. Dalbøge, H.; Dahl, H.-H. M.; Pedersen, J.; Hansen, J. W.; Christensen, T. Bio/Technol. 1987, 5, 161.
85. McDonald, J. K.; Barrett, A. J. In Mammalian Proteases: a Glossary and Bibliography; Academic Press: London, 1986; Vol. 2, p 132.

RECEIVED January 30, 1990

Chapter 14

Purification Alternatives for IgM (Human) Monoclonal Antibodies

G. B. Dove, G. Mitra, G. Roldan, M. A. Shearer, and M.-S. Cho

Cutter Biological Laboratory, Miles, Inc., Berkeley, CA 94704

Methods suitable for purification of IgM monoclonal antibodies from serum-free tissue culture supernatants are described in this case study. We review techniques found to be useful in purifying proteins and the techniques applied to establish utility. Partitioning techniques include polyethylene glycol (PEG) precipitation, size exclusion chromatography, anion and cation exchange chromatography, hydroxylapatite chromatography, and immunoaffinity. Modification techniques include the use of enzymes (e.g. DNAse).

Protein purity is achieved primarily with precipitation, size exclusion chromatography, and immunoaffinity. DNA removal is greatest with anion exchange, immunoaffinity,and a combination of DNAse and size exclusion chromatography. Virus is partitioned most effectively through hydroxylapatite and immunoaffinity. A cascade of several appropriate steps provides contaminant protein clearance of >100x (purity greater than 99%), DNA clearance of >1,000,000x, and virus clearance of >100,000x.

This paper presents a case study of the definition of purification processes for monoclonal IgMs produced by tissue culture fermentation. We review techniques that we have found to be useful and the approaches taken to establish their utility. The monoclonal antibodies are specific to various bacterial antigens for use as a therapeutic product.

Serum-containing and serum-free (supplemented with proteins) broths are purified, although the purification methods are optimized for serum-free media. Purification is necessary to remove contaminants introduced by the media and the cells. Contaminants include media components (albumin, transferrin, insulin, and many serum components), nucleic acids, viruses, and other cellular products.

0097–6156/90/0427–0194$06.00/0

Polymers, high molecular weight aggregates and fragments of IgM and albumin must be removed as well.

There are several approaches to purification in terms of properties (Table 1). The common properties are size, charge and ionic interaction, and affinity. Separation is optimized by exploiting (e.g. size between IgM and albumin) or creating differences (e.g. DNAse) between the product IgM and contaminants.

Techniques discussed in this study include filtrations, precipitation with PEG and salts (1, 2, 3), size exclusion chromatography (41, 5, 6), anion (4, 5)and cation exchange (7), contact with hydroxylapatite (8, 9), immunoaffinity, addition of reagents (e.g. DNAse (10)) and combinations of these techniques (11).

Materials and Methods.

Monoclonal (human) antibodies of class M (m-IgM) were derived from human B lymphocyte cell lines, designated A, B, C, D, and E. Antibody from each line was directed toward a different, specific bacterial antigen. Cells were grown in suspension culture or hollow fiber. Polyclonal plasma-derived IgM was obtained from Cohn fraction III (ethanol fractionation of human plasma (12).

Chemicals were reagent-grade. Salts were obtained from Mallinckrodt, Paris, Ky. Enzymes were obtained from Sigma, St. Louis, Mo. Chromatography resins and equipment, unless noted otherwise, were obtained from Pharmacia, Uppsala, Sweden. Hydroxylapatite (DNA-Grade Bio-Gel HTP) was obtained from Bio-Rad Laboratories, Richmond, Ca.

Proteins were characterized by SDS-polyacrylamide gels using a 2-10% agarose gradient, stained with either Coomasie Blue or silver. General product and contaminant analysis were quantitated by Pharmacia FPLC (Fast Protein Liquid Chromatography) system with a column matrix of Superose 6, utilizing size exclusion chromatography. Buffers defined the state as native, reduced, or denatured. Absorbance at 280 nm. was used for approximate measurements.

Various ELISA were developed to provide a consistent basis for loss of IgM as well as denaturation during processing. An antigen ELISA using anti-u chain IgG detected the Fc region and is useful in monitoring yield. A functional ELISA indicated antibody binding efficiency to antigen.

Epstein-Barr virus (EBV) was derived from B95-8 cells (13). EBV-specific nuclear antigen (EBNA) was demonstrated by an anticomplementary immunofluorescence (ACIF) assay with modifications (14). Residual native DNA was assayed by dot blot hybridization analysis (15, 16). P32-labeled DNA was prepared by nick-translation of host cell DNA isolated from culture harvests of cell line C. The DNA was spiked into various process steps and recovered (17). Samples were spotted to filter paper, precipitated and washed with 10% trichloroacetic acid (TCA) and measured by scintillation counter (LKB, model #1217 Rack Beta).

Table 1. Isolation Parameters for Purification

Component	Size (x1000 MW)	Isoelectric Point (Neutral Charge)	Other Characteristics
IgM	800	pH 6-6.5	
Contaminants:			
Albumin	69	pH 5	
DNA	variable	pH 5	HAPT DNAse
Viruses	>1000	pH 4-6	
<u>Basis of Separation</u>:			
Albumin	SEC		
DNA		IEC	HAPT DNAse
Viruses	SEC		Inactivation
All			Immunoaffinity

SEC: size exclusion chromatography
IEC: ion exchange chromatography
HAPT: hydroxylapatite
DNAse: degradation by DNAse
Inactivation: elimination of biological activity

Clearance values were calculated by a ratio of concentrations:

$$\frac{\frac{(\text{Contaminants})}{(\text{IgM})} \text{ before purification step}}{\frac{(\text{Contaminants})}{(\text{IgM})} \text{ after purification step}} = \text{Clearance}$$

For example, 99% removal of contaminants with 100% yield generates a factor of 100x. Contaminants included proteins, DNA, and viruses.

Results and Discussion.

Precipitation. Clarified harvests were concentrated on 100,000 MW membrane, adjusted to specific pH and precipitated with PEG (Table 2, cell line B). Low pH and high PEG concentration resulted in the highest yield of IgM and the lowest purity. It should be noted that, as the purity of the initial material increased, the yield from precipitation with PEG increased. Stability data substantiate this observation; IgM is stabilized in solution by proteins (e.g. albumin). Further, IgM at a concentration of less than 50 ug/ml precipitated poorly.

Precipitation of high molecular weight aggregates with low concentrations of PEG were studied. 1% and 2% PEG produced only slight reductions of aggregates, as measured by FPLC-Superose 6.

Precipitation with other agents and reprecipitation of an initial precipitation were examined, utilizing ammmonium sulfate, dextran sulfate, ethanol, and boric acid (Table 2). With the exception of ammonium sulfate, these agents did not precipitate IgM at the concentrations tested. Reprecipitation gave low yields and unremarkable purity with all agents.

Size Exclusion Chromatography (Gel Filtration). Initial experiments with plasma-derived IgM demonstrated good separation of IgM from contaminants on Sephacryl S-300, an acrylic based gel. However, separation and yield were poor with monoclonal IgMs. Sepharose CL-6B, an agarose based gel, produced excellent resolution from albumin.

A typical analytical chromatogram of a PEG precipitate on FPLC-Superose 6 (30 x 1 cm. dia.) is shown in Figure 1, compared to a typical large-scale run (80 x 37 cm. dia.). The leading (left side) peak consisted of aggregates of IgM and albumin, the second (largest) peak was IgM, and the third peak was primarily albumin with minor low molecular weight fragments. Clearance of DNA was approximately 10x. Yields were improved by high salt concentration, which reduced non-specific binding and increased stability. However, several workers have demonstrated improved separation with the use of both low and high ionic strength buffers (4, 18).

Degradation of DNA by DNAse. Endogenous or exogenous DNAses degrade the contaminant DNA by enzymatic cleavage, changing the size and charge of the DNA and thereby altering the efficiency of the separation cascade between the product (IgM) and contaminant (DNA).

Table 2. Precipitation by Various Agents

Agent	Concentration of agent (%)	pH	Yield (%)	Purity (% IgM)
Samples and agent were mixed for 1 hr. at 4 C.				
Line B, 200 ug/ml.				
PEG	10	7.4	84	80-95
PEG	8	7.4	75	95+
PEG	5	7.4	20	95+
PEG	5	5.5	66	95+
PEG	5	7.4	10	98+
PEG	10	5.5	95	80-95
PEG	10	7.4	86	80-95
Line E, 100 ug/ml.				
Amm. Sulfate	18	6.5	7	45
Amm. Sulfate	18	7.2	0	--
Amm. Sulfate	24	7.2	49	90
Amm. Sulfate	31	7.2	57	38
Line C, 100 ug/ml.				
Amm. Sulfate	24	7.2	79	80
PEG	12	5.5	96	60
The PEG precipitate of line C was reprecipitated by each of the following:				
PEG	10	7.2	39	85
PEG	12	5.5	80	80
Amm. Sulfate	11	7.2	0	--
Amm. Sulfate	24	7.2	71	90
Dextran Sulfate	10	7.2	2	--
Ethanol	25	7.2	0	--
Boric Acid	5	5.0	0	--

Amm: Ammonium

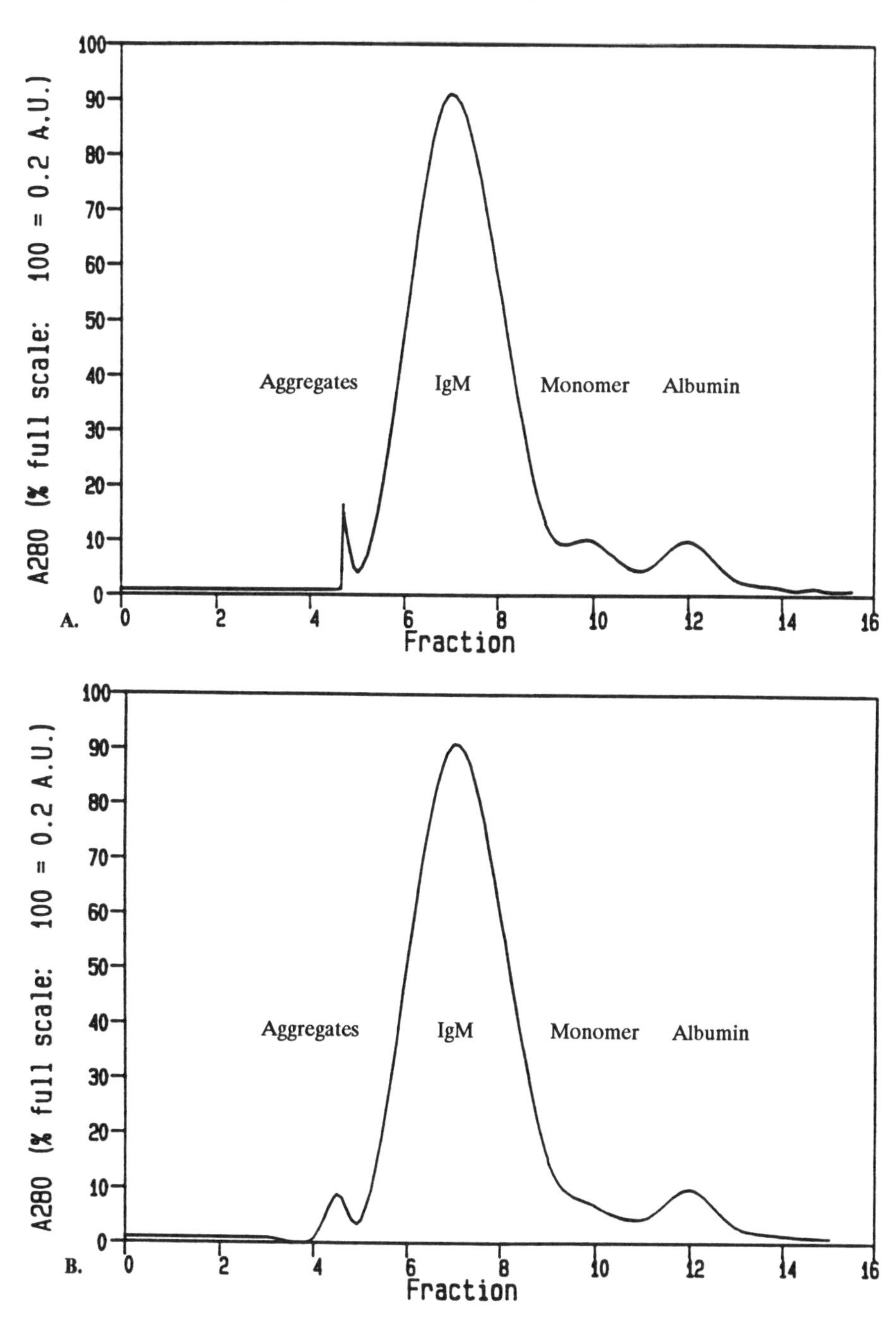

Figure 1. Chromatograms of size exclusion (SEC).
A. FPLC-Superose 6, 30 x 1.0 cm. dia.
B. Sepharose CL-6B, 80 x 37 cm. dia.
Load: precipitate by PEG.

Exogenous DNAses were added to the purification processes to accelerate the effect of the endogenous enzymes, and degraded virtually all of the DNA quickly to allow separation. Specifically, in precipitations or size exclusion chromatography, reduction of the molecular weight of the contaminanting DNA improved separation from a high molecular weight product (e.g. IgM). Bovine pancreas DNAse was immobilized on an agarose matrix to characterize and demonstrate controlled degradation of a purified DNA preparation passed through a column of the matrix. The purified DNA preparation consisted initially of molecular weight 1,000,000. Two column passes resulted in an approximate bell distribution with a broad range of 1,000,000 to 10,000 daltons and median at molecular weight 100,000. Many passes resulted in a narrow range of approximately 10,000 daltons. DNA degradation was monitored by size exclusion chromatography and SDS-PAGE (Figure 2). In a separate experiment, a harvest of IgM was passed through the DNAse column and then fractionated on FPLC-Superose 6. DNA clearance was 10,000x, compared to 10x without DNAse digestion (10).

Anion Exchange. Anion exchange on DEAE-Sepharose was optimized for IgM purification from DNA. The following buffer conditions were studied: Tris, pH 8.0; phosphate, pH 6.5; sodium acetate, pH 6.5; and no buffer. Elution was achieved by a linear gradient of sodium chloride. Separation from other proteins was marginal, but removal of albumin was slightly superior in the phosphate buffer. The salt buffers gave comparable reduction in DNA (10,000x), with the unbuffered system giving 1,000x.

To demonstrate removal of DNA more clearly, several preparations were bound to DEAE-Sepharose: a) a partially purified IgM preparation, b) a purified preparation of DNA, and c) a combination of both. Preparations were bound in 0.05 M Tris, 0.05 M NaCl, pH 8.0 and eluted by linear gradient of sodium chloride. Referring to Figure 3A, IgM was recovered in the first elution peak at 0.15 M NaCl. The second and third elution peaks at 0.2 and 0.34 M NaCl contained native DNA. Figure 3B shows elution of the DNA preparation at 0.3 and 0.35 M NaCl. Mixing the purified IgM and DNA preparations and repeating the elution yielded a precise superimposition of the two chromatograms (Figure 3C), indicating the two entities eluted independently.

Typical chromatograms are shown in Figure 4, with elution by linear gradient and step elution. A small amount of protein did not bind, indicated by the peak on the left. A high salt strip after elution of the IgM produced a peak of similar magnitude, which contained IgM, DNA, and albumin.

A step elution at 0.15 M NaCl gave DNA clearances listed in Table 3. Native DNA was assayed by dot blot hybridization. Clearance studies with P32 labeled DNA gave substantially lower clearance factors. P32 labeled DNA derived from harvests of cell line C gave approximately twice the clearance of viral DNA.

Other anion exchange resins gave comparable or poorer results for IgM. Use of Q-Sepharose resulted in slightly tighter binding of both IgM and DNA. Repetitive chromatography on DEAE did not result in increased removal of DNA from IgM.

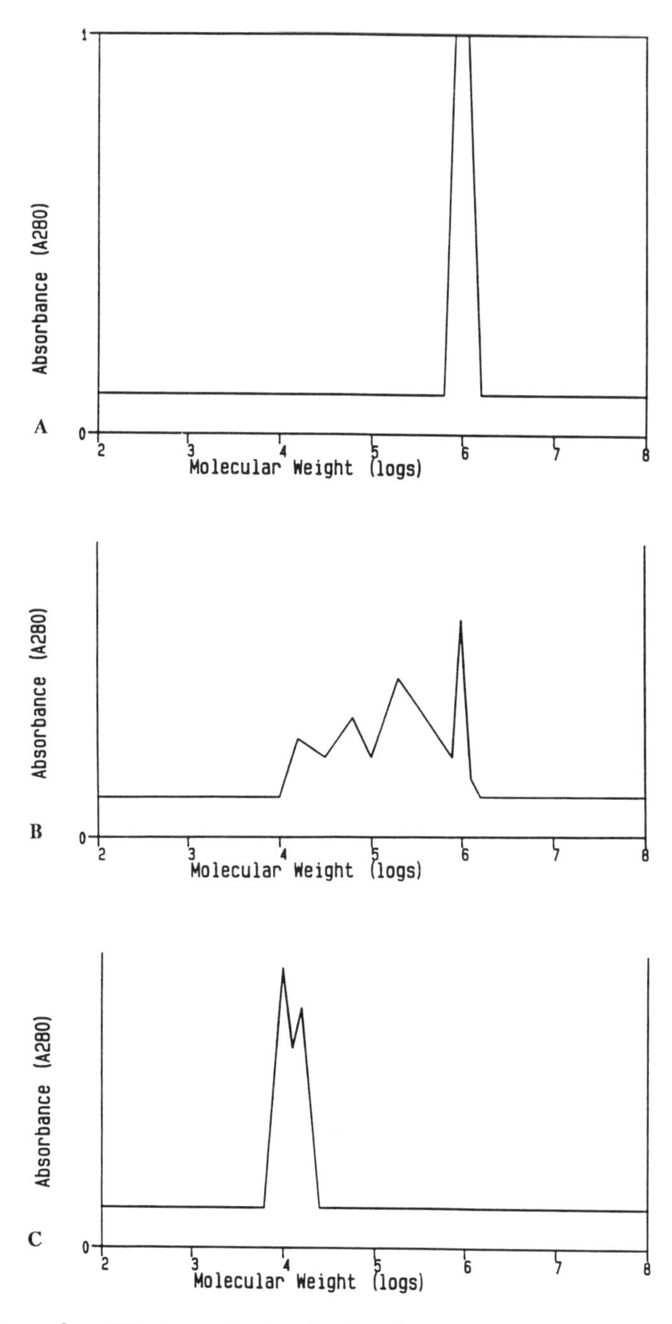

Figure 2. DNA degradation by immobilized DNAse. Chromatograms (FPLC-Superose 6) of:
A. 0 passes across DNAse column(initial DNA prep.).
B. 2 passes. C. 14 passes.

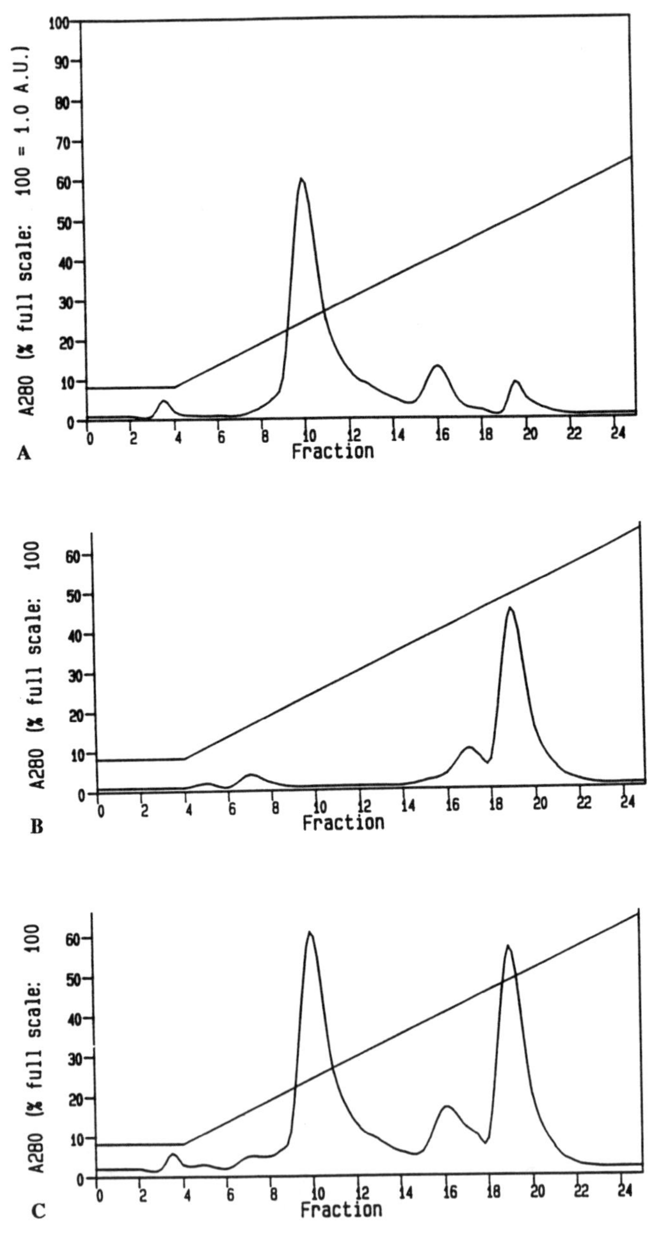

Figure 3. Chromatograms (DEAE Sepharose) of:
A. Purified IgM. B. Purified DNA. C. Combination.

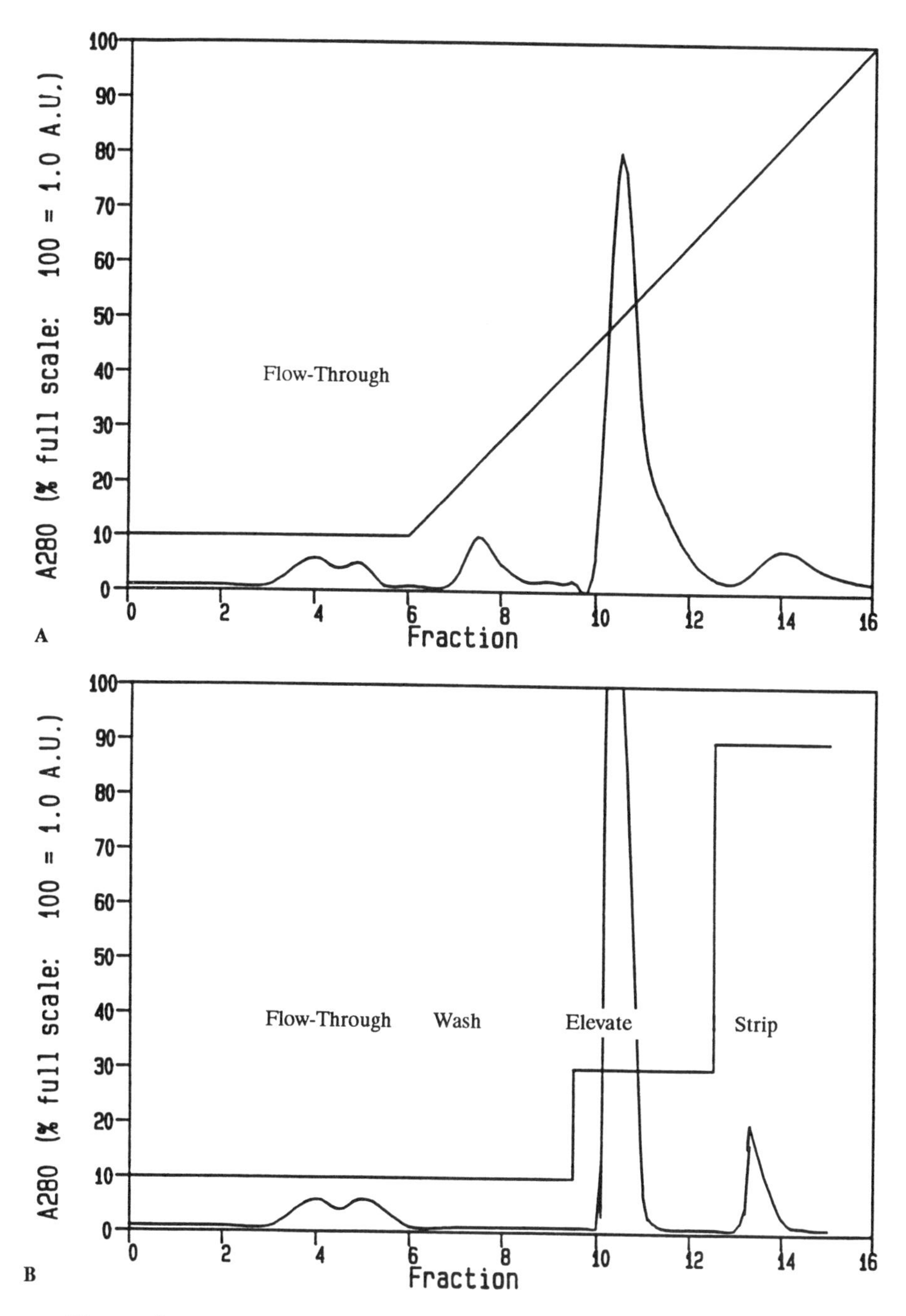

Figure 4. Chromatograms (DEAE Sepharose, load ppt. by PEG).
A. Linear gradient elution. B. Step elution.

Cation Exchange. Cation exchange provides a reverse pH dependency compared to anion exchange. In anion exchange, contaminants with a low isoelectric point (e.g. albumin IEP = pH 4.8) bind tightly at neutral pH as they are anionic. In cation exchange, these low IEP contaminants do not bind. Unfortunately, purification of these IgMs (IEP = 6-6.5) was poor, with incomplete binding and poor separation from contaminants (Table 4). Experience with other monoclonal antibodies (e.g. several IgGs) indicated higher yields and purity.

Hydroxylapatite. Contact with hydroxylapatite results in binding of DNA and proteins (8, 9). Binding is reduced with higher concentrations of phosphate in the same way as higher ionic strength reduces binding on ion exchange. Chromatograms generated with a linear gradient of phosphate resembled those of DEAE-Sepharose (Figure 3). Because the material proved difficult to scale-up, further experiments were conducted in batch mode. An optimum in phosphate concentration was established by binding DNA at a concentration where the IgM would not bind. Table 5 demonstrates high yield and clearance at 0.05-0.1 M phosphate buffer. Clearance values were done with several labels to determine if the molecular weight of the label affected clearance. HMW, MMW, and LMW labels refer to high, medium and low molecular weights produced by degradation of the DNA. Native DNA was assayed by dot blot hybridization. The small label gives values somewhat higher than the other labels. As the amount of hydroxylapaptite is increased, the IgM yield decreases and DNA clearance increases.

Immunoaffinity. Immunopurification of IgM by an anti-u chain IgG was defined. A murine IgG was produced in stirred tank tissue culture fermentation. The impure IgG was purified by precipitation with PEG and anion exchange (DEAE-Sepharose). The IgG was bound to solid supports of control pore glass, Bio-Rad Affigel, FMP, and AH-Sepharose. Matrices were prepared as follows:

A) Control Pore Glass (CPG): Activation is achieved by periodate oxidation of polysaccharides forming aldehyde groups. Primary amine groups of the ligand are then covalently bound.

B) Affi-Gel (Biorad): Gel consists of N-hydroxysuccinimide esters of a derivatized cross-linked agarose support. Free aryl amino groups displace the N-hydroxysuccinimide and form a stable amide bond.

C) FMP-Activated Hydrophilic Gel (Bioprobe Int.): Modified Trisacryl GF 2000 contains 2-alkloxypyridinium salts. These salts react with amino groups to displace pyridinium groups at neutral pH.

D) CNBr activation of AH-Sepharose CL4B: Hydroxyl groups on the agarose react with CNBr and form cyanate esters, which in turn bind primary amines.

Table 6 lists the IgG coupling efficiency and the amount of IgM (cell line B) that bound to the respective anti-IgM matrices. AH Sepharose demonstrated the greatest net binding of IgM at both 2 and 18 hours (16.2 mg/ml). Substantially higher IgG-IgM binding

Table 3. Anion Exchange Chromatography
DEAE-Sepharose: Clearance of DNA

Expt.	Cell Line	Process Step	Concn. of DNA (pg DNA/mg IgM)	Clearance Factor (Pre/Post DEAE)
With native DNA:				
1	C	Pre-DEAE	2,000,000	
		Post-DEAE	60	33,000
2	C	Pre-DEAE	1,700,000	
		Post-DEAE	6	283,000
3	C	Pre-DEAE	2,000,000	
		Post-DEAE	24,000	83
4	D	Pre-DEAE	2,000,000	
		Post-DEAE	35	57,000
With P32 labels:				
5	C			278
6	C			294
7	Viral			176
8	Viral			142

Table 4. Cation Exchange Chromatography
S-Sepharose: Yield and Protein Purity

Salt Concn. (M)	pH	IgM Yield (%)	Purity (% IgM)
Initial		100	10
0.01	4.0	100	20
0.05	4.0	100	20
0.1	4.0	80	40
0.05	4.8	90	20
0.1	4.8	75	50
0.05	5.5	50	30
0.1	5.5	30	30

Table 5. Hydroxylapatite.
Effect of Phosphate Concentration

Phosphate Concn. (M)	Concentration of HAPT (%)	Yield (%)	DNA Clearance (pre/post HAPT)	
			HMW label	LMW label
0.0	1	45	--	--
0.05	1	80	24	139
0.1	1	95	69	145
0.15	1	97	39	--
0.20	1	97	5	--
0.25	1	100	1	--
			MMW label	native
0.07	1	93	9	--
0.07	2	93	--	22
0.07	2	93	20	--
0.07	10	82	25	--

HMW: high molecular weight label
LMW, MMW: low and medium molecular weight labels respectively.
native: DNA assay by dot blot hybridization.

Table 6. Immunoaffinity.
Coupling Efficiencies and Binding of IgM

Support	IgG Bound (mg/ml)	IgM Bound* (mg/ml gel)	Ratio Bound IgM/IgG
CPG	2.77	1.27	0.46
Affi-Gel	9.82	6.64	0.68
FMP	2.14	3.77	1.76
Seph.	10.72	16.2	1.51

*: Contact for 18 hours.

efficiencies were demonstrated by FMP (1.76) and Sepharose 1.51) compared to the other two.

Elution conditions were established by testing three buffers of increasing strength on CPG: a)2 M MgCl2, pH 7, b) 0.05 M NaAcetate, pH 4.0, c) 0.2 M glycine, pH 2.8. Only the last buffer yielded significant recovery of IgM. Batch elution of IgM from all four supports using 0.2 M glycine, pH 4 was conducted repeatedly on each support. Values for IgM elution and IgG leaching are given in Table 7. Affigel is different from the other supports in that it contains a carbon spacer arm. This spacer may account for the very high elution results. While Affigel may not have the optimum ratio of IgM bound to IgG, the percentage of IgG coupled and the elution values identify this as a feasible support.

Purification was very high, with greater than 500x clearance of contaminating proteins and DNA.

Table 7. Immunoaffinity.
Elution of IgM from IgG matrices

Support	Batch Elution(#)	IgG (ng/ml)	IgM (mg/ml)	IgM (mg)	Recovery (% load)
CPG	1	<240	0.494	0.502	40
	2	<240	0.004		
	3	<240	<0.004		
Affigel	1	266	5.136	5.264	80
	2	<240	0.124		
	3	<240	<0.004		
FMP	1	279	0.670	0.716	19
	2	<240	0.041		
	3	<240	<0.004		
Sepharose	1	399	4.968	7.002	43
	2	<240	1.988		
	3	<240	0.046		

Summary of Methods.

Yields and clearance factors are summarized for various steps in Table 8. Yields of IgM were typically 75-90% which demonstrates comparable purification potential from each step. Protein purity was demonstrated by FPLC-Superose 6 profiles (Figure 5), illustrating the increased purity of tissue culture harvest purified through a cascade of several steps.

Protein purity is achieved primarily with precipitation, size exclusion chromatography, and immunoaffinity. DNA removal is observed with anion exchange, immunoaffinity,and a combination of

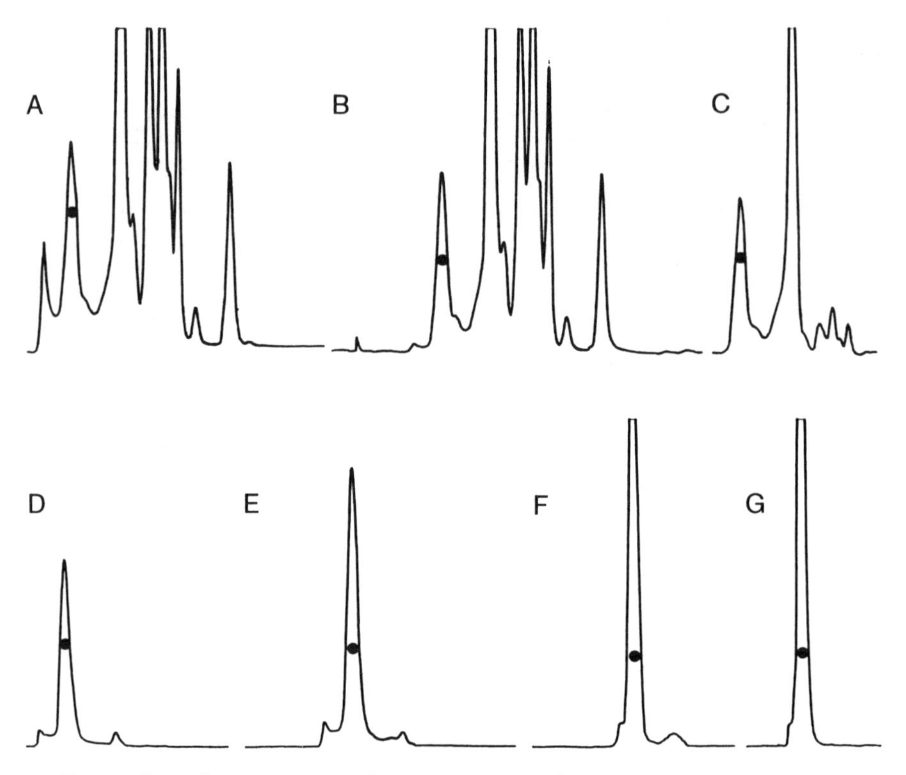

Figure 5. Chromatograms (FPLC-Superose 6) of several steps:
A. TCF, filtered through 0.2 um filter.
B. (A), filtered through high MW membrane.
C. (B), concentrated by ultrafilter.
D. (C), precipitated by 5% PEG. E. (C), precipitated by 10% PEG.
F. (E), eluted on DEAE. G. (E), fractionated on Sepharose CL-6B.
● indicates IgM peak.

Table 8. Summary of Purification Factors

Process Step	Yield (%)	Clearance Factor (x) Protein	DNA	Virus
Filter TCF	100	1	1	0
Precipitation	80	5	1	<10
Size Exclusion	85	100	10	100
with DNAse	85	100	10,000	100
Anion Exchange	75-85	5	>10,000	20
Cation Exchange	75	2	2	20
Hydroxylapatite	90	2	20	1,500
Immunoaffinity	80	500	10,000	5,000

DNAse and size exclusion chromatography. Virus is partitioned through hydroxylapatite and immunoaffinity.

A cascade of several appropriate steps provides contaminant protein clearance of >100x (purity greater than 99%), DNA clearance of >1,000,000x, and virus clearance of >100,000x. Viruses may also be removed by inactivation as compared to partitioning.

Literature Cited.

1. Hasko, F.; Vaszileva, R. Biotech. Bioeng. 1982, 24, 1931.
2. Polson, A.; Ruiz-Brava, C. Vox Sang. 1972, 23, 107.
3. Wickerhauser, M.; Hao, Y.L. Vox Sang. 1972, 23, 119.
4. Clezardin, P.; Bougro, G; McGregor, J.L. Chrom. 1986, 18, 261.
5. Jehanli, A; Hough, D. J. Immun. Meth. 1981, 44, 199.
6. Wichman, A.; Borg, H. Biochim. Biophys. Acta 1977, 490, 363.
7. Prior, C.P.; Doyle, K.R.; Duffy, S.A.; Hope, J.A.; Moellering, G.J.; Prior, G.M.; Scott, R.W.; Tolbert, W.R. Bio/Technology 1989, 43, no. 1, 15.
8. Brooks, T.L.; Stevens, A. American Laboratory, 1985, October, 54.
9. Stanker, L.H.; Vanderlaan, M.; Juarez-Salinas, H. J. Immun. Meth. 1985, 76, 157.
10. Dove, G.; Hall, E. Characterization and Utility of DNAses in Tissue Culture Fluids. 1987, Engineering Foundation.
11. Mariani, M.; Bonelli, F.; Tarditi, L.; Calogero, R.; Camagna, M.; Spranzi, E.; Seccamani, E.; Deleide, G.; Scassellati, G. Biochromatography 1989, 4, 149.
12. Cohn, E.J.; Strong, L.E.; Hughes, W.L.; Mulford, D.J.; Ashworth, J.N.; Melin, M.; Taylor, H.L. J. Amer. Chem. Soc. 1946, 68, 459.
13. Miller, G.; Lipman, L. Proc. Natl. Acad. Sci. 1973, 70, 190-4.
14. Reedman, B.M.; Klein, G. Int. J. Cancer 1973, 11, 499-520.
15. Southern, E.M. J Mol Biol 1975, 98, 503-517.
16. Wahl, G.M.; Stern, M.; Stark, G.R. Proc. Natl. Acad. Sci. 1979, 76, 3683-87.
17. De Rie M.A.; Zeijlemaker, W.P.; Von dem Borne, A.E.G.K. J. Immun. Meth. 1987, 102, 187.
18. Bouvet J.P.; Pires, R.; Pillot, J. J. Immun. Meth. 1984, 66, 299.

RECEIVED January 18, 1990

Chapter 15

Analytical, Preparative, and Large-Scale Zone Electrophoresis

Cornelius F. Ivory[1], William A. Gobie[1], and Tri P. Adhi[2]

[1]Department of Chemical Engineering, Washington State University, Pullman, WA 99164–2710
[2]Department of Chemical Engineering, University of Idaho, Moscow, ID 83843

Electrophoresis is capable of resolving biological molecules on the basis of differences in their molecular weights, mobilities, isoelectric points, or various combinations of these properties. At laboratory scale, this includes some of the most powerful techniques available for the purification and characterization of biologically active molecules. Because of this, electrophoresis is routinely used for analytical and bench-scale preparative work.

Recent advances in equipment design are dramatically changing the speed, resolution and scale of zone electrophoresis. For instance, capillary electrophoresis provides extraordinarily high plate efficiencies at loadings well below 1 μg. In the range, 1 μg-100 mg, electrochromatography can combine the resolving power of electrophoresis with chromatography while retaining many of the superior scaling properties of chromatography. At loadings greater than 1 gm, zone electrophoresis may offer significant advantages over other processing methods.

We shall begin with simplified descriptions of capillary zone electrophoresis and electrochromatography. Then, after a brief discussion of the obstacles which had hindered development of large-scale electrophoresis equipment, we shall describe the device we have developed and illustrate the performance of this apparatus using several examples of protein fractionation.

0097–6156/90/0427–0210$09.50/0

Gel electrophoresis (**GE**) is used primarily in the characterization of small amounts, e.g. milligrams, of protein or nucleic acids and, to a lesser extent, in preparative work involving roughly 1-100 mg of sample. The major advantage that electrophoresis has enjoyed over other bench techniques has been the high resolution attainable in a simple, inexpensive apparatus.

The primary competitor to electrophoresis at all scales is high performance liquid chromatography (**HPLC**). During the past ten years improvements in resin properties and operating protocols coupled with automation have greatly increased the resolution, ease-of-use and reliability of **HPLC** columns. At bench scale, **HPLC** is competitive with one-dimensional **GE**, e.g. SDS-PAGE or starch **GE**, but at the time of writing **HPLC** offers somewhat lower resolution than two-dimensional **GE** (1,2).

At preparative and industrial scales, e.g. 1-100 grams, **HPLC** has outstripped electrophoresis for purifying peptides and small proteins. The main reason for this is that most types of chromatography are readily scaleable while, for many decades, electrophoresis has resisted efforts at scale-up. This situation is changing and it appears likely that several electrophoresis instruments will be developed in the very near future which may be competitive with **HPLC** in terms of resolution, capacity and/or economics.

We will begin with brief descriptions of both capillary zone electrophoresis and electrochromatography emphasizing the mechanisms governing separation and the problems hindering further development of each technique. Then, after reviewing the events which preceded invention of the multi-gram **RCFE**, we will describe this apparatus and several of the experiments Bill and I have performed using it. Persons interested in a more detailed account of this and related work are directed to recent reviews (3,4) which contain references to many of the seminal papers on this subject.

CAPILLARY ZONE ELECTROPHORESIS

Capillary zone electrophoresis (**CZE**) offers important advantages over conventional **GE** in the characterization of proteins and nucleic acids. It is much faster than **GE** and can be fully automated and instrumented to include sample injection, detection and collection. In addition, **CZE** can separate solutes with a resolution which is one to two *orders of magnitude* greater than **HPLC** at sample loadings several orders of magnitude smaller than **HPLC**.

While electrophoresis in "microbore" capillaries, inner diameter (ID) less than 1 mm, was demonstrated several decades ago (5-7), the introduction of capillary *zone* electrophoresis (**CZE**) in 25-100 μm ID capillaries using voltages up to 30 kV is generally attributed to Jorgenson (8-10). The enormous electric fields used in **CZE** allow resolution of biomolecules approaching the ideal limit

(Equation 1 assumes no electroosmosis, an infinitesimally thin pulse load and no anomalous sources of dispersion) in plate efficiency, N_p, which is over an order of magnitude greater than the efficiency of most high performance liquid chromatography columns (**HPLC**).

$$N_p = \frac{\mu EL}{2\,D_m} \approx \frac{10^{-8}\;30\times10^{+3}\;1}{2\;10^{-10}} = 1.5\times10^{6}\,\frac{\text{plates}}{\text{meter}} \qquad [1]$$

Small capillaries can efficiently dissipate Joule heat generated by the enormous power densities, P_w, used in these columns without suffering significantly from natural convection or anomalous dispersion (11-13). As a result, plate efficiencies in excess of 500,000 plates per meter can be obtained in practice.

$$P_w = \frac{ELI}{\vartheta} \approx \frac{30\times10^{+3}\;1\;100\times10^{-6}}{\pi\left(25\times10^{-6}\right)^2 1} = 15.3\times10^{18}\;\frac{\text{Watts}}{m^3} \qquad [2]$$

Researchers have extended capillary electrophoresis to include isoelectric focusing (14), isotachophoresis (15-17), SDS-PAGE (18) and micellar electrophoresis (19-21). To encompass all of these applications and reflect the power of these techniques, the term **CZE** was broadened to high-performance capillary electrophoresis (**HPCE**). Compared with **GE** (22,23), **HPCE** is fast and it is readily amenable to automation including sample loading, detection and sample collection.

Because of its speed and simplicity, analytical **HPCE** is rapidly being adopted in the laboratory, e.g. a fully instrumented **HPCE** column takes less than fifteen minutes to automatically process a set of samples from initial loading to final analysis. There are now more than a half-dozen commercial units available with prices ranging from under $10,000 to over $50,000.

Band Spreading in **CZE**. If the fluid is driven through the capillary by an external pressure field, it adopts a parabolic velocity profile (b) which distorts the sample,

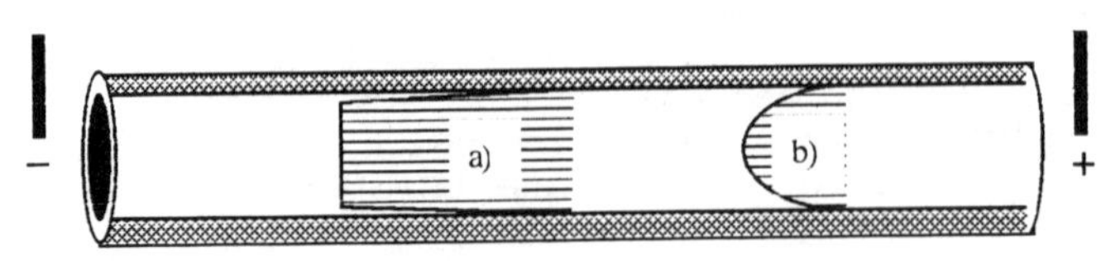

Figure 1. Velocity profiles in a capillary resulting from a) pure electroosmotic flow and b) pressure-driven flow. The electro-osmotic flow (a) is flat except in a region 10-100 Å thick near the capillary surfaces. As a result, sample dispersion is purely diffusive. The pressure-driven flow (b) is parabolic and gives rise to Taylor dispersion which is several orders of magnitude larger than diffusion.

leading to Taylor dispersion (24-26). As illustrated in equation 3, even at low flowrates Taylor dispersion dominates molecular diffusion. However, if the fluid is driven by a purely electroosmotic flow (a), its velocity profile remains flat and dispersion in the tube is governed solely by molecular diffusion, D_m. As a result, the eluting peaks are sharp and plate efficiencies are high.

$$D = D_m + \frac{R_L^2 U^2}{48\,D_m} \approx 10^{-10} + \frac{(25\times10^{-6})^2(10^{-4})^2}{48\;10^{-10}} \approx 14\times10^{-10}\;\frac{m^2}{s} \qquad [3]$$

In **CZE**, heat dissipation and solute dispersion are coupled so *temperature* profiles in the capillary must be carefully evaluated to assess their influence on dispersion. A cross section through a typical fused-silica capillary (figure 2) shows four phases involved in heat transfer: the fluid in the lumen, usually aqueous buffer with 10-50 mM electrolyte, a structural "glass" wall, a stabilizing polymer film and the environment outside the capillary. The fluid in contact with the outer surface is most often air but forced liquid cooling is becoming more popular (27-31).

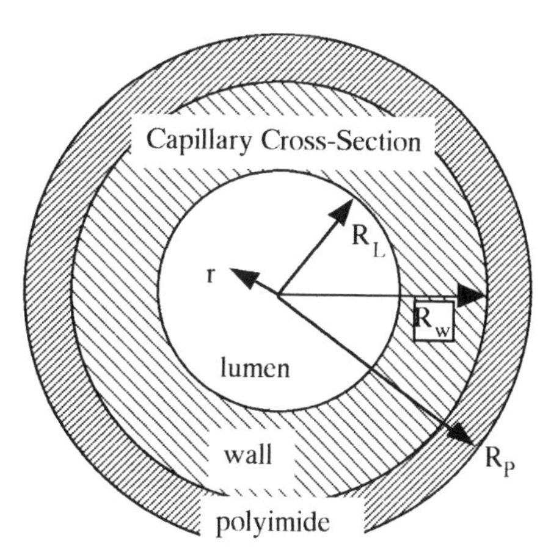

Figure 2. Cross section of a typical fiber showing lumen, fiber wall and the polyimide coating found on many commercially available capillaries. Joule heat is generated in the lumen and dissipated to air.

Heat Transfer.

Temperature profiles in single capillaries are estimated using a standard model for heat transfer with electrical generation. Heat is produced by passing current through the lumen of the fiber and is dissipated to air or liquid coolant through the outer surface of the capillary. Our model includes the autothermal effect in addition to the four phase resistances and predicts a dimensionless temperature profile given by equation 4 with the autothermal coefficient, λ, and the dimensionless heat transfer coefficient, Bi, defined below.

$$\Theta = \frac{1}{\lambda^2}\left(\frac{\mathrm{Bi}\,J_o(\lambda\eta)}{\mathrm{Bi}\,J_o(\lambda) - 2\lambda J_1(\lambda)} - 1\right) \qquad [4]$$

$$\lambda = \sqrt{\sigma_1 \frac{\sigma_o E^2 R_L^2}{k_L}}$$

$$\frac{1}{Bi} = \frac{k_{air}}{R_P h_{air}} + \frac{\log_e (R_P/R_W)}{k_P/k_{air}} + \frac{\log_e (R_W/R_L)}{k_W/k_{air}}$$

Because of the autothermal effect, our model indicates that the lumen temperature profile will be nearly, but not exactly, parabolic as had been predicted by Grushka et al. (31). If the capillary is air-cooled the major resistance to heat transport out of the lumen is external transfer to air. Therefore, if the heat transfer coefficient to air is not accurately calculated, the capillary temperature may be underestimated by more than 10°C.

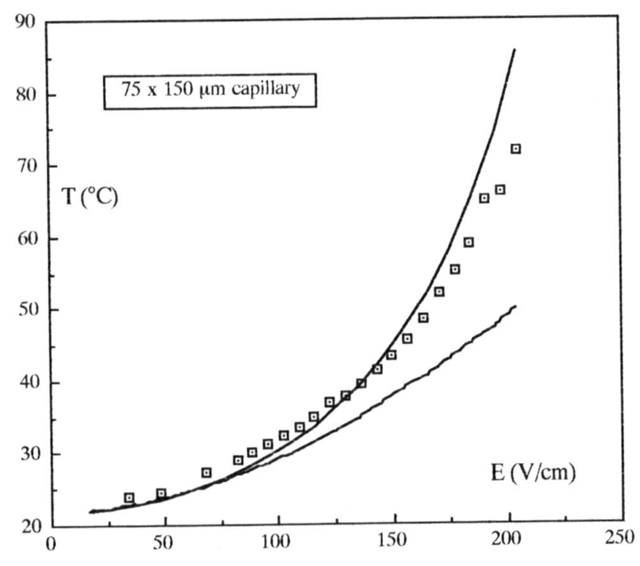

Figure 3. Comparison of data (⊡) with theory (—) for the average lumen temperature in a glass fiber with a 75 μm ID and 150 μm OD. The lower line is calculated using the theory of Grushka et al. (31). The upper line was derived using equation 4 and takes into account the autothermal effect.

A common omission made even by experts in this field is to ignore the autothermal effect (32) on the temperature rise. This effect arises because as electrolyte in the capillary lumen becomes warmer it becomes more conductive. More current is carried through the lumen causing it to heat up further and conduct even more current. The end effect of this positive "feedback" is to amplify Joule heating, eventually leading to thermal runaway. Ignoring the autothermal effect leads to underprediction of the temperature excursion by about 50% when the predicted temperature rise is 50°C (figure 3). This can lead to significant errors in the calculation of the dispersion coefficient and hamper efforts to counteract dispersion.

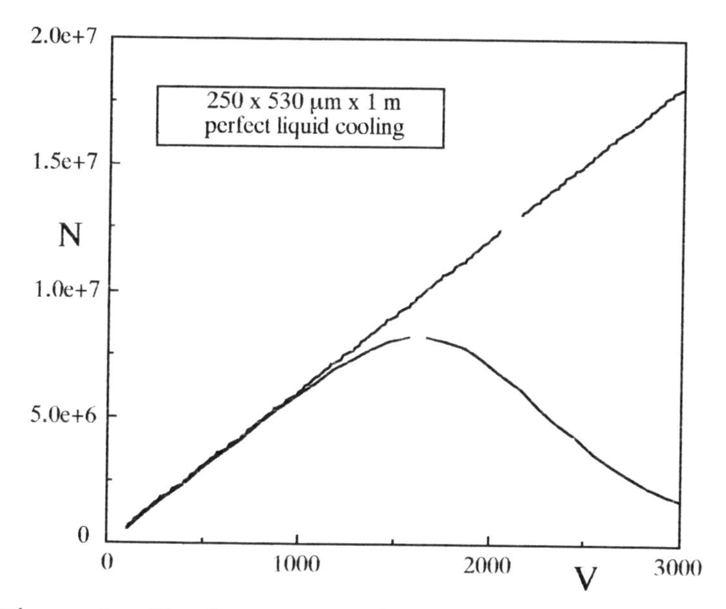

Figure 4. The lower curve illustrates the deterioration of the plate efficiency with increasing applied voltage in a 1 meter long capillary. The upper is computed using an idealized isothermal theory. Very high efficiencies are allowed in these calculations we have not accounted for the thickness of the initial pulse.

The Effect of Heat Transfer on Band Spreading. In air cooled capillaries the temperature difference between the wall and the center of the lumen is 1°C or less and is proportional to the power dissipation rate. This temperature difference alters the buffer viscosity and leads to a radial variation in the electrophoretic mobility, i.e. fastest at the centerline and slowest at the wall. Because of this radial variation in the electrophoretic velocity, solute disperses more rapidly than would be expected by diffusion alone. This is the primary reason for the deterioration of **HPCE** resolution at high power levels.

$$D = D_m + \frac{R_L^2 \mu_o^2 E^2 \lambda^4 Bi}{3072 D_m \left[Bi J_o(\lambda) - 2\lambda J_1(\lambda)\right]} \qquad [5]$$

The relationship between applied power and plate efficiency, N_p, can be determined by applying Taylor dispersion theory to the non-isothermal capillary. This yields the formula for the dispersion coefficient given by equation 5 (it is important to note that a similar formula was first derived by Grushka et al. (31) using the method of Gill and Sankarasubramanian (33). They did not take into account the autothermal effect but did show deterioration in N_p with increasing voltage) for N_p which makes several important predictions. First, it predicts that under normal operating conditions, N_p will go through a maximum (lower curve in figure 4) as the voltage is increased. This type of behavior has been observed by several researchers (30,31) and results from competing effects, i.e. N_p is directly proportional to the electric field (c.f. equation 1) but inversely proportional to applied power which goes as the *square* of the electric field strength. Once the source of dispersion was identified, steps were taken to minimize it.

Optimization of **HPCE**. The second important prediction of equation 5 is that thermally-induced dispersion can be compensated by applying a pressure drop across the capillary. Recalling that the temperature profile in the lumen is nearly parabolic, it follows that the thermal distortion of the solute is also parabolic. Since the velocity profile under an applied pressure is also nearly parabolic, it follows that the parabolic portions of the electrophoretic and hydrodynamic velocities should superimpose in such a way as to very nearly cancel.

$$D = D_m + \frac{R_L^2}{3072\, D_m}\left[\frac{\mu_o E \lambda^2\, Bi}{Bi\, J_o(\lambda) - 2\lambda J_1(\lambda)} - \frac{R_L^2}{\rho\nu}\frac{dP}{dx}\right]^2 \qquad [6]$$

Again applying Taylor dispersion theory but including the applied pressure drop this time yields an approximation (equation 6) for the dispersion valid for air-cooled capillaries. To maximize N_p, the pressure drop is adjusted until the term in square brackets vanishes. At this point, the dispersion coefficient reduces to the molecular diffusion coefficient and the plate efficiency returns to the ideal predicted by equation 1 and illustrated by the straight line in figure 4.

We are now in the process of extending this prediction to liquid cooled capillaries and considering whether this approach will significantly improve the *resolution* of single and multiple-capillary electrophoresis devices.

ELECTROCHROMATOGRAPHY

Early attempts at preparative electrochromatography (**EC**) suffered from field-induced dispersion (34,35). This problem was not effectively addressed until the invention (36) of an **EC** column capable of focusing proteins at the interface between two granular gel phases. In this apparatus, field-induced dispersion is virtually eliminated by careful manipulation of the electric field and flowrate. Purified, concentrated product can then be withdrawn from the column after the field is turned off.

O'Farrell's method of **EC** provides a way to overcome the problem of electric field-induced dispersion. A jacketed column is packed with gel filtration media in two contiguous *regions* which have distinct K_{av} for a targeted protein. The region nearest the solvent inlet is packed with a gel which tends to exclude (K_{av}~0)a target protein while the other region is packed with a gel which tends to include (K_{av}~1) that protein.

When buffer is pumped *slowly* through the excluding region into the including region in the absence of an electric field, solutes run through the excluding region at a relatively high velocity but slow down as they move into the region containing the including gel. This decrease in the apparent migration velocity occurs because the solute spends a greater fraction of its time inside the stationary gel phase unaffected by convection in the continuous phase.

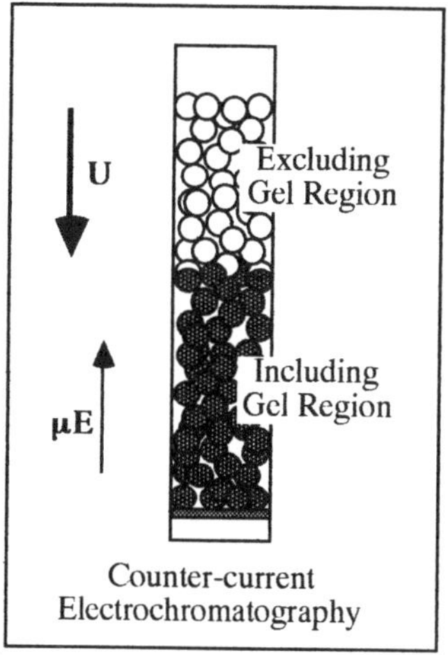

*Figure 5. Countercurrent Electrochromatography (**CACE**) apparatus drawn after O'Farrell (36). (Adapted from reference #38. Copyright 1990 ACS).*

If a longitudinal current is then applied across the bed so that the electrophoretic migration of the target protein is in the direction opposite to the flow of carrier, i.e. counter-current, then a range of values of the electric field can be found which allows the protein to migrate up to the interface between the two gels and focus nearby. If the electric field is increased beyond the upper limit of this range, the protein migrates away from the interface and moves upstream through the excluding region of the column. If the electric field is below this range then the solute migrates downstream from the interface.

Proteins whose electrophoretic mobilities fall inside this range are retained at the interface while all other solutes elute upstream or downstream of the interface. A discrete range of electric field will not only stop specific solutes but will also *concentrate* them at the interface between the gel regions. Purified protein is then removed from the interface and the column can then be reloaded.

This is possible because, unlike the convective flow which acts only in the continuous phase, the electric field acts on the solute in both the continuous phase and in the stationary phase. So while a solute moves down through the column with the convective flow, its net migration may be in the opposite direction due to the cumulative effects of electrophoresis in both phases.

This technique has the important advantage of being able to focus protein into a thin band *in native electrolyte* (As opposed to isoelectric focusing (**IEF**) which is performed near the protein's pI where solubility is often lowest. **IEF** is usually carried out in the presence of ampholytes which are expensive and may bind to protein). The large scale dispersion exhibited in other **EC** devices is reduced in this apparatus, thus removing one of the most important limitations on **EC**. Furthermore, this process can be carried out in well-defined, inert media using common laboratory equipment and can be extended to batch capacities approaching one gram per run.

O'Farrell described several experiments which he conducted to illustrate the simplicity and versatility of this technique. In one experiment he used BioGel P-10 (exclusion limit: 20 kd) as the excluding gel and BioGel A-50m (exclusion limit: 50,000,000 kd) as the including gel in a **CACE** column. He then loaded 10 mg of ferritin onto the column and adjusted the electric field and solvent flow rate so that the ferritin focused near the interface between the gels.

Under the conditions listed in that paper, ferritin was focused into a band about 2 mm thick below the interface and thus apparently concentrated to 13% protein, i.e. 130 mg/ml.

Modelling **CACE**. Our efforts in this area have centered on reproducing O'Farrell's original experiments, developing a predictive mathematical model of **CACE** and extending this technique to continuous operation in order to increase the processing rate. The details of our work are contained in papers (37,38) published elsewhere but the salient features of that work are summarized here.

To model a solute focusing near the gel interface, one must use the nonlinear mass conservation equations for each of the ions in the buffered electrolyte in addition to each of the macro-ions in the feed. A set of these equations must be included for each continuous and each stationary phase in each *region* of the column. Thus, a minimal system consisting of a single protein, one co-ion and one counter-ion in a column poured using one excluding gel and one including gel would require 3(2 electrolytes plus 1 macro-ion)×4(1 continuous and 1 stationary phase per region)=12 second order ordinary differential equations of the form

$$\varepsilon\frac{\partial c_i}{\partial t} + \mathbf{K}_{o,i}\mathbf{a}_o\left(c_i - \frac{s_i}{K_{av,i}}\right) = -\frac{\partial J_{c,i}}{\partial x}$$
$$(1-\varepsilon)\frac{\partial s_i}{\partial t} + \mathbf{K}_{o,i}\mathbf{a}_o\left(\frac{s_i}{K_{av,i}} - c_i\right) = -\frac{\partial J_{s,i}}{\partial x} \qquad [7]$$

where c_i and s_i are the phase-average concentrations in the continuous and stationary phases, respectively. The fluxes on the right-hand side of equation 7 have the form

$$J_{c,i} = \varepsilon \left[\left(U + \mu_{c,i} E \right) c_i - D_{c,i} \frac{\partial c_i}{\partial x} \right]$$
$$J_{s,i} = \left(1 - \varepsilon \right) \left[\left(V + \mu_{s,i} E \right) s_i - K_{av,i} D_{s,i} \frac{\partial \left[s_i / K_{av,i} \right]}{\partial x} \right] \qquad [8]$$

where D are dispersion coefficients and U and V are the solvent velocities in the continuous and stationary phases, respectively. In addition to equations 7 and 8, local continuity (left, below) and electroneutrality (right, below) constraints must be satisfied at every point along the x axis.

$$\varepsilon \frac{\partial U}{\partial x} + \left(1 - \varepsilon \right) \frac{\partial V}{\partial x} = 0; \qquad \sum_{i=1}^{N} z_i c_i \approx 0; \qquad \sum_{i=1}^{N} z_i s_i \approx 0$$

Four conditions must be specified on each interior phase interface and two on each exterior boundary. At the exterior boundaries we typically specify the concentrations of the electrolyte(s) and either set the concentration to zero or set no-flux conditions for the macro-ion.

$$c_i^U = c_i^L \qquad \int_{\vartheta} \left(\varepsilon\, c_i + \left(1 - \varepsilon \right) s_i \right) A_c \, dx = M_i$$
$$\frac{s_i^U}{K_{av,i}^U} = \frac{s_i^L}{K_{av,i}^L} \qquad J_{c,i}^U + J_{s,i}^U = J_{c,i}^L + J_{s,i}^L \qquad [9]$$

There are several sets of interior boundary conditions which yield unique solutions of the governing equations and, as yet, there is not enough information available to determine which is the "best" set. All of the sets we have considered give similar results and one admissible set is given above. The two boundary conditions on the left represent continuity of the concentrations across the interface in both the continuous and stationary phases, the condition on the lower right specifies continuity of the mass fluxes and the integral constraint (applied only to the macro-ion) specifies the amount of protein loaded onto the column. Once the current is specified, this system of equations is well posed and may be solved numerically.

Using this model, we have been able to predict (a) the existence and extent of the focusing 'window', (b) the concentration, location and breadth of the focused band and (c) the effect of interphase mass transport on dispersion of the focused band. A simulation of our ferritin focusing experiment is shown below in figure 6 together with a photograph of the simulated experiment.

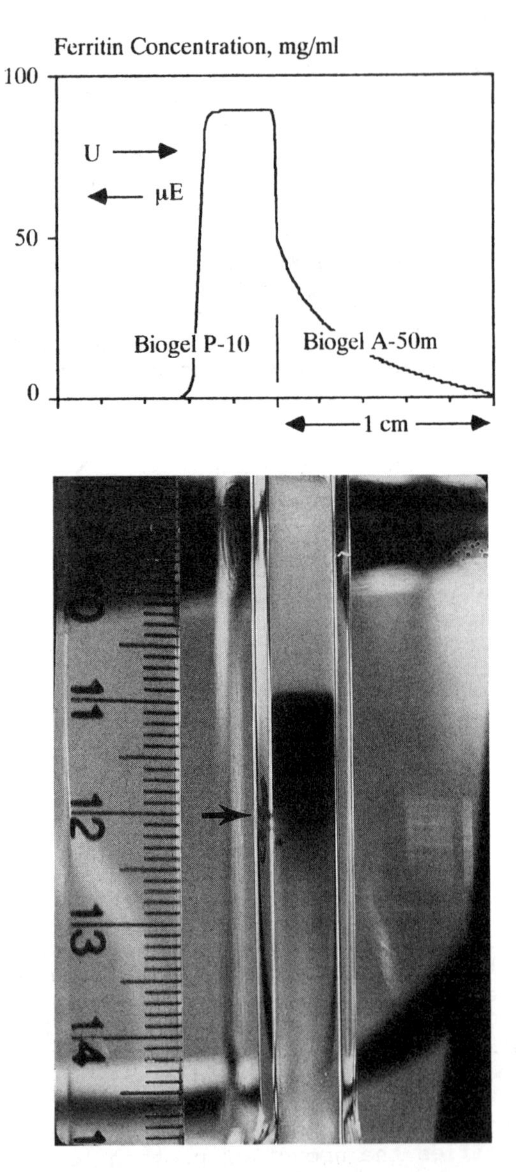

Figure 6. A comparison of experiment and theoretical simulation of ***CACE.*** *The gross appearance of the simulation is comparable to the features observed in the experimental apparatus. The gel interface is located at the 12 mm mark. The major discrepancy between theory and experiment is between the predicted (8%) and apparent (25%) concentration in the focussed band. (Adapted from reference #38. Copyright 1990 ACS).*

Experimental Results. Our attempts to repeat O'Farrell's experiments using ferritin have met with limited success. After building a column similar to the one described in his paper, we were easily able to focus the protein into a tight band near the interface between the gels. However, we observed several important differences between phenomena reported in his paper and our results.

First, we were never able to focus ferritin in the lower (including) gel but were routinely able to focus it just above the interface, i.e. in the excluding gel region. Because of this, the apparent ferritin concentration in our apparatus was on the order of 25% protein rather than 13% as reported by O'Farrell. Second, the ratio of the electric field strength to the hydrodynamic velocity required to focus ferritin in our apparatus was roughly ten times higher than that reported by O'Farrell. This is consistent with results communicated by other researchers, notably Jean Hunter at Cornell and Bruce Locke out of Carbonell's group at North Carolina, but at this time there is no obvious explanation for this discrepancy.

As figure 6 shows, we obtain good qualitative agreement between simulation and experiment. In particular, ferritin has focused above the interface with a sharp leading edge and a diffuse tail extending about 1 cm downstream of the interface. The major source of disagreement between theory and experiment is the concentration in the focused band. The predicted concentration is about 8% protein in the most concentrated portion of the band while the apparent concentration observed in the experiment is about 25%.

At the present time we do not know why this discrepancy exists. The most plausible explanation put forward to date (39) is that the electric field polarizes excluded protein against the surface of the gel in much the same way that concentration polarization occurs in ultrafiltration. This could easily lead to apparent concentrations in excess of the 8% predicted by our model.

Continuous Electrochromatography. While O'Farrell's batch apparatus is relatively easy to assemble from common laboratory equipment, sample loading and product extraction are cumbersome and the rate of product recovery is low. To address these points, we decided to adapt batch **CACE** for continuous processing and, it turns out, these problems can be resolved in continuous **CACE**. Although in principle this can be done by simply adding feed and offtake streams to the column, in practice this interferes with operation of the column.

Apparatus. Two concessions are made in our design of a continuous **CACE** column. First, the feed is introduced at a single port in the center of the column and withdrawn continuously from two points. Because of this, our column cannot fractionate feed into more than two product streams. However, it is important to note that most strictly continuous processes split product into only two streams.

In choosing to operate the column as a continuous, one-dimensional process, one concedes the possibility of one-step fractionation of multicomponent feeds, but retains fine control over resolution. Furthermore, a one-dimensional process can be readily incorporated into a two-dimensional cascade when multicomponent fractionation is required.

Second, in order to simplify continuous loading and extraction of sample, a small volume called a *port* (figure 8), is opened between gel regions. Because it contains no gel, solutes can be withdrawn from or injected into a port without disturbing the gel beds. A photograph of our apparatus is shown in figure 9. As indicated, the column is divided into four regions which are separated from each other by three ports: one feed port and two offtake ports.

The central, bulb-shaped vessel acts as feed port, the bulb having ample volume to serve as a reservoir in semi-batch experiments (approximately 15 ml). Two tubular glass arms rise vertically out of the bulb. These hold the separating gels whose primary function is to fractionate the feed into two product streams, one directed upstream and the other directed downstream. These arms are made vertical to simplify bed packing. The bulb is segregated from the separating gels in the right and left-hand arms using glass frits which serve to support and isolate the gels. The gels which fill the separating arms of this apparatus are chosen with distinct exclusion limits to optimize separation.

Above each of the separating gels is an offtake port, another glass frit and another gel which is used to retain and concentrate products at the offtake ports.

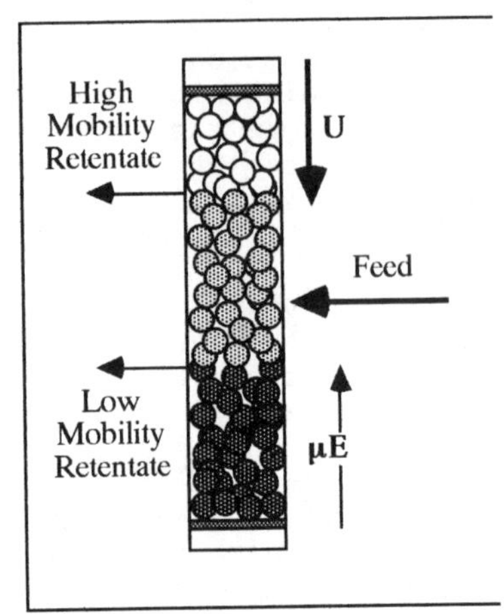

*Figure 7. Illustrative drawing of continuous **CACE** showing the location of the feed and offtake streams. Compare this with the photograph of the device shown in figure 9. (Adapted from reference #38. Copyright 1990 ACS).*

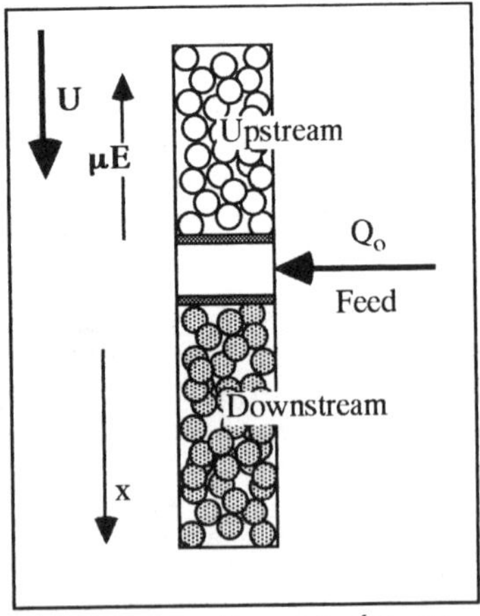

Figure 8. A port is an open volume formed between two gels. This facilitates the loading and extraction of sample. The feed port is shown here. (Adapted from reference #38. Copyright 1990 ACS).

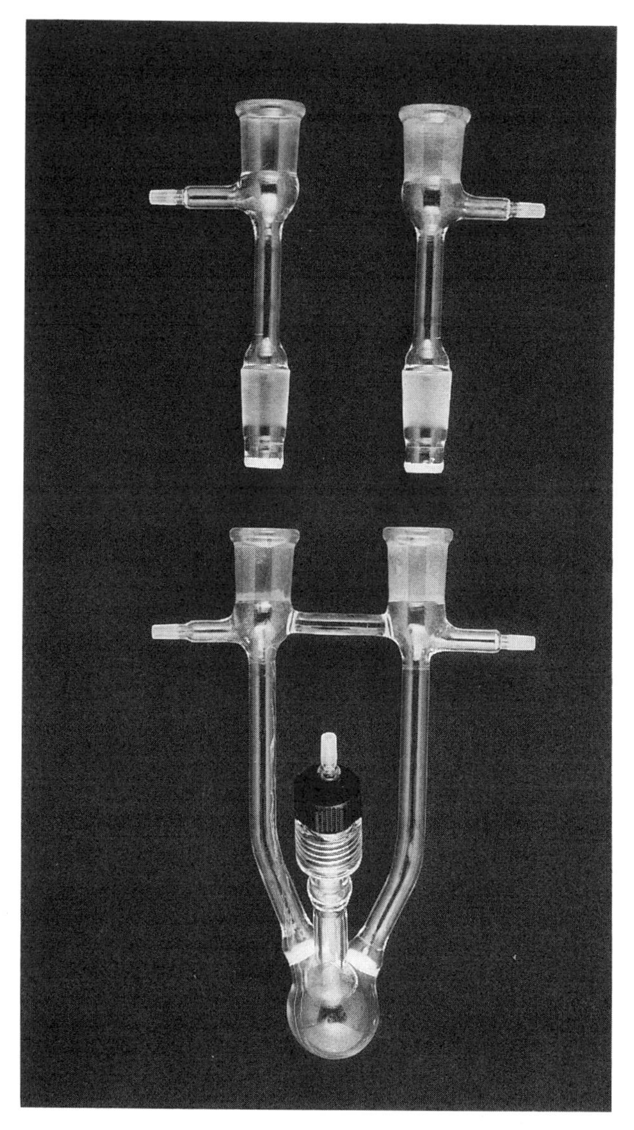

Figure 9. Photograph of our continuous **CACE** *showing the four sections of the apparatus which hold the gel regions and the three ports used to introduce and remove solutes. The bulb at the center of the apparatus is the feed port which connects the left and right separating arms of the chamber. The offtake ports are visible at the top of these arms. The disconnected upper portions of the apparatus hold the focussing gels and the flow control ports are visible at the top of these sections. (Reprinted from reference #38. Copyright 1990 ACS).*

Above each of these gels sits another port which is used to control the buffer flowrate through the apparatus. Above this last port is a polyacrylamide plug which isolates the column from the electrodes.

Thus we see that four gels and five ports are required for continuous bi-component fractionation. Assuming convective flow from left to right, it is customary for the gel on the upper left to have the lowest value of K_{av} of the four gels, typically zero in this position. The K_{av} is increased stepwise, up to a value of 1.0, as one moves sequentially through the gels.

The U-shaped tube and tube sections with glass frits facilitate pouring of the beds and eliminate intermingling of different gels so, if the bed is disturbed for any reason, it can be readily restored. If a gel must be replaced, only that section of the column containing the gel need be disturbed. Also, the U-shape allows the column to be dipped into a cooling bath, maintained at 4°C by a Lauda circulator, which replaces the cumbersome jacket otherwise required to remove Joule heat.

The tube dimensions are 6 mm OD, 4 mm ID. The upper tubes are approximately 70 mm long and the lower tubes 90 mm long. The tubes flare to 10 mm ID to accommodate the coarse glass frits and to 10-12 mm to join the taper fittings. The side arms are equipped with male Luer fittings.

Continuous Experiments. For the first experiment, the semi-batch processing of hemoglobin:azo-casein solutions, the **CACE** column was packed with four gels. They were, in order of increasing K_{av}, Sephadex G-50, G-100, G-200 and Bio-Gel A-50m. The column was rinsed overnight with running buffer, 10 mM phosphate at pH 7.0, and a 1% mixture of equal parts Hb and azo-casein was charged into the feed port. A potential drop of 1-2 kV was established across the electrodes and the buffer flow rate adjusted to about 0.1 ml/hr through the column. The electric field and flow rate were then adjusted until azo-casein migrated upstream against the flow and hemoglobin (Hb) ran downstream with the flow. By individually manipulating the flows across the two offtake ports, the separated proteins were recovered with no apparent mass losses. When checked by electrophoresis on 1% agarose gels with Coomassie R-200 stain, no cross-contamination of products was detected.

A second set of semi-batch experiments was performed with dye-labelled BSA and Hb under similar operating conditions. In this case the molecular weights of the proteins are close enough that it is impossible to separate them by conventional gel filtration. Our apparatus readily separated BSA from Hb on the basis of differences in their electrophoretic mobilities, again yielding products that were essentially pure and with no detectable mass losses.

Continuous operation with this apparatus has proved to be difficult, chiefly because the very low flow rates force us to use gravity-driven flow. This arrangement does not give stable flow rates over times scales of hours, because the gravity head is so small that the Mariotte bottle used as the buffer reservoir functions as a barometer, causing the flow rates respond to changes in atmospheric pressure during the course of a run. The semi-batch experiments required frequent monitoring and adjustment. We intend to overcome these problems with a new column which will operate at flow rates about 10x higher, the flow being provided by a small peristaltic pump.

Downstream Processing. The maximum processing rate which can be achieved in our apparatus is less than 0.01 $gm/cm^2/hr$ of protein. This rate is low because **CACE** operates under two severe constraints. First, electrophoretic mobilities are small, generally on the order of 10^{-4} $cm^2/V\text{-}sec$, so reasonably high electrophoretic velocities require large electric fields. Second, electric field strengths are limited by constraints on the temperature which increases roughly as $O(\sigma E^2 d^2)$ where d is a characteristic length scale which pertains to cooling, i.e. in our apparatus it is the column radius. Large electric fields imply thin columns, effectively limiting the throughput which can be obtained.

Before batch, semi-batch or continuous **CACE** can become useful preparative techniques, research will have to be undertaken to improve our understanding not only of the mechanisms involved in separating and focusing solute bands, but also those involved in solute dispersion. Carefully controlled experiments are needed to test the predictions of the theory and to measure empirical coefficients. New equipment designs and improved packings are needed to allow scale-up by two or more orders of magnitude into the gram per hour range.

It is our belief that, while the details of the physics involved in **CACE** are poorly understood, the existing theoretical and empirical framework is sufficiently well-developed to allow important advances in modelling and to spur improvements in equipment design. As experiments furnish more information about relevant mechanisms and verify advanced theoretical developments, **CACE** should evolve into an indispensable preparative technique.

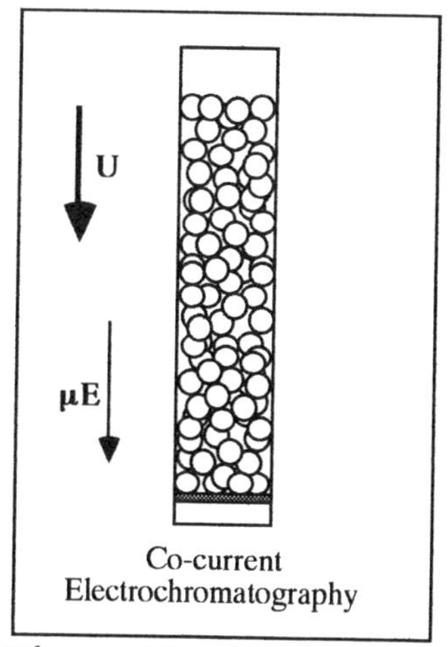

Figure 10. In co-current electrochromatography, the hydrodynamic and electrophoretic velocities are in the same direction.

Co-Current Electrochromatography. Early on in our investigation of O'Farrell's **CACE**, we realized that if a solute's electrophoretic velocity in the stationary phase were in the same direction as the flow of the continuous phase, the behavior of the column would be very different from that observed in **CACE**. In counter-current **EC** solutes move in opposite directions in the stationary and mobile phases while in co-current **EC** they move in the same direction.

In formulating a model for **EC**, we felt it essential that both co-current and counter-current operation be treated using the same equations. The only difference we allowed in our analysis is that, in the absence of focusing, it is not necessary to retain the nonlinear coupling terms between equations. Instead, the electric field, hydrodynamic and electro-osmotic velocities (E, U & V) are assumed constant. According to an idealized analysis based on the linearized form of equations 7-9, this has a dramatic effect on dispersion.

When the net motion of the solute is in the same direction as the hydrodynamic flow, the elution time, t_e, and dispersion, σ, *in long columns* having negligible mechanical dispersion, i.e. Pe $\rightarrow \infty$, are

$$t_e = \frac{\{1+\alpha\}}{\{1+\alpha\gamma\}}\left[\frac{L}{U+\mu_c E}\right] \quad [10]$$

$$\sigma^2 = \frac{2}{\beta}\frac{[\alpha(1-\gamma)]^2}{[1+\alpha\gamma]^3}\left[\frac{L}{U+\mu_c E}\right]^2 \quad [11]$$

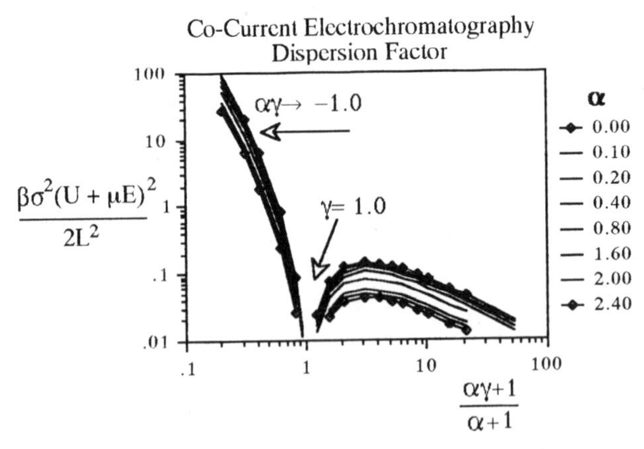

Figure 11. In co-current chromatography dispersion due to mass transfer goes through a minimum near γ=1. This occurs because solutes in the continuous and stationary phases move at the same velocity.

From equation 10 it is seen that our theory predicts that peak separation can be increased by adjusting γ to change the elution times of various components. Moreover, equation 11 for the dispersion indicates that the peak width varies in an unexpected way which strongly influences the resolution of the column.

As seen in figure 11 it is apparent that as $\alpha\gamma \rightarrow -1^+$, dispersion increases rapidly making conventional pulse-load operation unsuitable. However, near γ=1.0 column dispersion due to interphase mass transfer resistance dips dramatically. Theory predicts that, under proper operating conditions, some peaks could be made up to an order of magnitude sharper while eluting in less than half the time that would be required without the field.

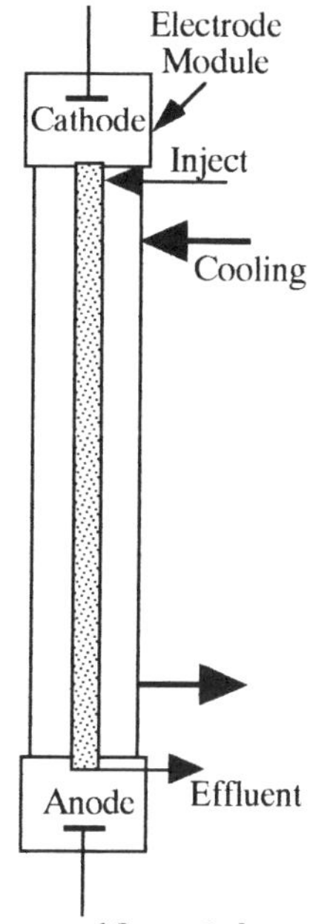

*Figure 12. Schematic of our co-current **EC** column showing electrode header and cooling jacket.*

This minimum occurs when solute in the stationary phase moves at the same velocity as its counterpart in the mobile phase. Although mass transfer still proceeds at a finite rate, the solute in the stationary phase does not lag behind the solute in the mobile phase so smearing due to interphase mass transport, which is unavoidable in conventional gel filtration (i.e. γ=0), is eliminated near γ=1.

We have recently completed construction of an apparatus for co-current **EC** column and run a number of experiments comparing results with and without electric fields. Our results so far are inconclusive: while we have observed improved resolution of easily separable proteins and dyes, the observed enhancement is somewhat less than predicted by our theory. The main culprit, as usual, appears to be the effect of Joule heating on the viscosity of the carrier fluid and we are currently modifying both our theory and our apparatus to account for this effect.

In closing this section we should note that our theoretical and early experimental results are in conflict with results reported by Ladisch and Rudge this past year. At the present time it is our belief that discrepancies between theory and observation stem from differences in column construction and operation, choice of granular gels, and mechanisms accounted for in the models. Eventually, we expect that differences in theory and experiment will be resolved as more is learned about co-current **EC**.

MULTIGRAM ELECTROPHORESIS.

The history of large-scale electrophoresis begins with Philpot (40) who first described an apparatus for continuous protein processing in 1940. Although his design anticipated many of the difficulties which would hinder research in this field for years to come, he was unable to get it working properly before the Second World War interrupted his work.

It was not until the beginning of the next decade that Grassmann and Hannig (41) in West Germany and Svensson and Brattsten (42) in Sweden independently announced development of equipment for continuous electrophoresis. Although these devices were different in many respects, they possessed a number of similarities, e.g. both chambers consisted of slits mounted vertically with electrodes on either side. In addition to this, both used "co-current" (i.e. in continuous flow chambers co-current refers to buffer flow in the same direction as gravity. Counter-current flow is in the direction opposite to gravity) buffer flow and were packed with anti-convectant media. Both devices could process roughly 1 ml per hour of sample, about 0.01 gram protein per hour for a 1% protein solution.

From 1970 to 1980 important advances were made in our understanding of fluid dynamics and heat transfer in free-flow electrophoresis. The anticonvectant was removed from the chamber slit and the slit thickness reduced to less than a millimeter (43), the origin and minimization of "crescent" dispersion was described by Strickler and Sacks (44,45), the onset of unstable natural convection was predicted by Saville (46) and the autothermal effect was discussed (47). Even with these advances, throughput in Hannig's "thin-film" apparatus remained below 5 ml/hr (0.05 grams protein per hour). In addition, a number of alternative approaches did yield significant advances in throughput.

The Biostream. Philpot (48) eliminated natural convection from his earlier apparatus. He did this by passing carrier fluid axially (counter-current) through the annulus (3-5 mm thick) formed between two concentric cylinders with the outer one serving as anode and rotating in much the same manner as the outer cylinder of a Couette viscometer (figure 13). In this device, which is known as the Biostream Separator, the outer cylinder rotates at speeds approaching 150 rpm, generating a stabilizing radial velocity gradient.

Feed is injected through a ring-shaped opening just below the electrodes. Current passes radially across the annulus separating charged solutes as they pass between the electrodes. These are then collected through discrete ports located near the top of the chamber. This configuration allows operation at throughputs in excess of three liters per hour, over three orders of magnitude greater than the "thin-film" devices.

The Biostream uses counter-current flow to discourage unstable convection with the radial stress field giving added stability. In addition, the parabolic velocity profile of the carrier fluid is used to sharpen the solute bands during separation. Finally, because the electric field in the Biostream runs perpendicular to the transverse annular surfaces, there is no electroosmosis and dispersion is, in principle, controlled by diffusion alone (49). This apparatus is commercially available from CJB Developments, Inc. It costs about $250,000 FOB and is rated for processing 2-4 liters per hour of sample.

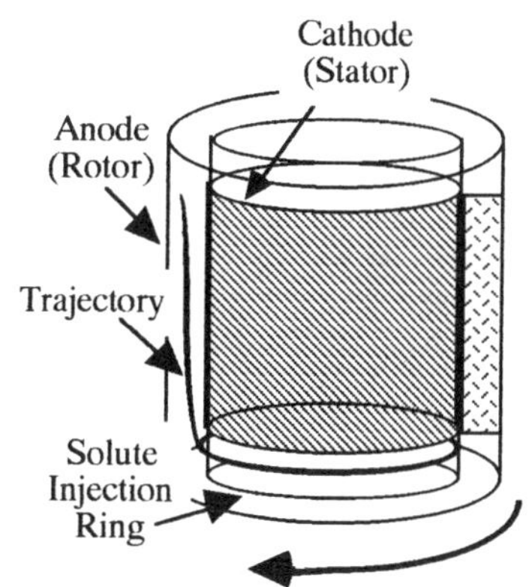

Figure 13. Illustration of the Biostream. Solute enters through a ring at the base of the device and flows up through the electric field with the carrier fluid. Separated fractions are collected through offtake ports at the top of the chamber.

While the Biostream processes protein rapidly, it is a relatively low resolution separator as compared with "thin-film" devices. As a rule of thumb, clean separation of components requires mobilities on the order of 10^{-4} cm^2/V-s with a 50% difference in the mobilities of key components (50). A number of pilot studies (49, 51) conducted with the Biostream have indicated that there is an anomalous source of dispersion which keeps the unit from operating up to its theoretical optimum. Roughly speaking, the dispersion measured with proteins is about three times greater than would be expected from molecular diffusion alone and this detracts considerably from the Biostream's resolving power.

Microgravity. **NASA**, in collaboration with McDonnell-Douglas, conducted a number of low-**g** electrophoresis experiments, most notably in Skylab and on several Shuttle flights. The basic idea was to eliminate natural convection from the thin-film device by reducing the influence of gravity in the Rayleigh number,

$$Ra \equiv \frac{g\omega\,|\nabla T|\,h^4}{\kappa\nu} \qquad [12]$$

thus dropping the chamber below the critical Ra for unstable natural convection. In micro-**g** the chamber thickness could be increased several-fold (52,53). However, the majority of the 500× scale-up reported for electrophoresis in space (54) resulted primarily from the use of a 25% protein feed mixture, not from an increase in device thickness. Although this work contributed significantly to our appreciation of the effects of gravity on electrophoretic separations, the increase in throughput obtained in space does little to mitigate the prohibitive expense of boosting equipment and materials into orbit.

Field-Step Focusing. An idea which makes excellent use of commercial thin-film electrophoresis technology was put forward by Wagner (55,56) working in collaboration with Bender & Hobein. This process, known as Field Step Focusing (**FSF**, figure 14), introduces sample into a chamber such as Bender & Hobein's Elphor VaP 22 as a broad, low conductivity curtain of buffer spanning the central portion of the chamber and running between two curtains of buffer which have substantially higher conductivity. Typically, the central curtain is 3-5 cm across and 0.5-1 mm thick, its conductivity is 10^{-5} mho/cm and the conductivities of the outer curtains are some 100-1000 times higher.

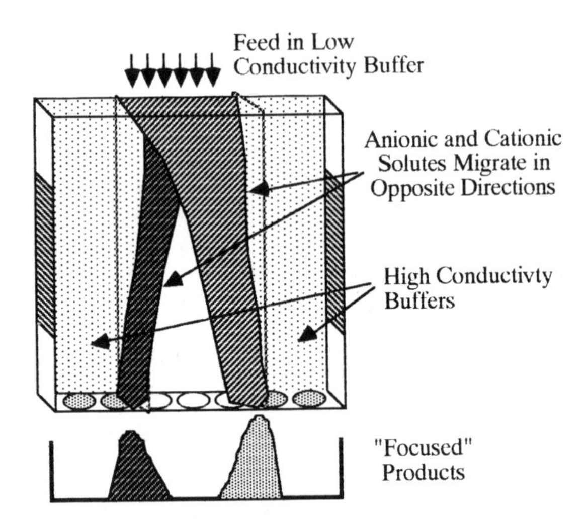

Figure 14. Illustrative drawing of Wagner's apparatus for Field-Step Focusing. The feed is injected with the entire central curtain with the pH adjusted so that solutes below their pI run toward the cathode while solutes above their pI run toward the anode. When solutes reach the high conductivity buffer their electrophoretic mobilities decrease and the solutes "focus" at the interface.

The pH of the central curtain is adjusted so that it is intermediate to the pIs of the two "key" components and, when the electric field is applied, these components migrate to opposite sides of the central curtain. On contact with the high conductivity fluids their electrophoretic velocities decrease significantly so the solutes slow down and "focus" at the interfaces between the two buffer curtains. In his papers Wagner has demonstrated clean separations of synthetic binary mixtures of proteins and a number of other biomaterials. The four advantages of **FSF** indicated by Wagner are that (a) focusing is achieved without ampholytes, (b) throughput is high, (c) separation is optimized by adjusting pH and (d) separation is conducted away from pI so precipitation is avoided.

We do not believe that Wagner has ever suggested or claimed that **FSF** could be scaled substantially beyond one gram per hour but we have heard reports that 10-20 grams per hour processing has been achieved. In our opinion, to exceed the 1 gram/hr benchmark in the Elphor VaP, **FSF** would require concessions which would limit its utility in separating proteins. Specifically, (a) the low conductivity curtain fluid would have low protein solubility, (b) the device would exhibit "crescent" dispersion, (c) the curtain buffering capacity would be extremely low, (d) the solutes would need to be oppositely charged and have electrophoretic mobilities in excess of ±0.75 μm-cm/V-s to traverse the central curtain. Finally, the thickness of the chamber and the conductivity ratio in the curtain would have to be carefully selected to avoid (e) overheating of the protein or (f) convective dispersion associated with the conductive-dielectric instability (57).

RECYCLING

The development which most heavily influenced my work in scale-up was the concept of recycling. The basic idea behind recycling in electrophoresis is that solutes do not have to be completely separated in a single pass through the apparatus since partially purified component streams could be repeatedly cycled through the device until separation is complete.

Two obvious advantages would result from this approach. First, the quality of the separation could be carefully controlled. If 90% product purity were desired, so many cycles would be used; if 99% purity were needed, more cycles would be added by increasing the number of recycle *stages*, N_R, until the desired purity was achieved.

Second, the electric field strength could be reduced since solutes do not have to be separated in a single pass. Noting that the electric field could be reduced as $1/N_R$, a rough calculation indicated that recycling would allow us to scale the process roughly as N_R^2.

RIEF (Recycle Isoelectric Focusing). The application of recycle to continuous electrophoretic processing was pioneered by Bier and coworkers (58-60) for isoelectric focusing (figure 15). The heart of this apparatus is an adiabatic, multichannel slit partitioned into compartments by closely spaced, fine porosity nylon screens which damp intercompartmental convection but freely transmit macro-ions as they migrate in the electric field.

During a run each of the compartments develops a characteristic pH which is highest at the cathode and lowest at the anode and which changes sharply at the nylon mesh screens. Once a pH gradient is established, polyampholines are circulated continuously through the chamber until all components have migrated to their isoelectric points. The electric field is then switched off and the apparatus and reservoirs drained of product. Later designs have eliminated the screens but use a thinner slit to discourage lateral mixing.

Because the system is modular, heat generated by electrical dissipation is removed from the process streams via a multichannel heat exchanger located downstream. The process streams are monitored externally for temperature, pH, UV absorbance, etc.

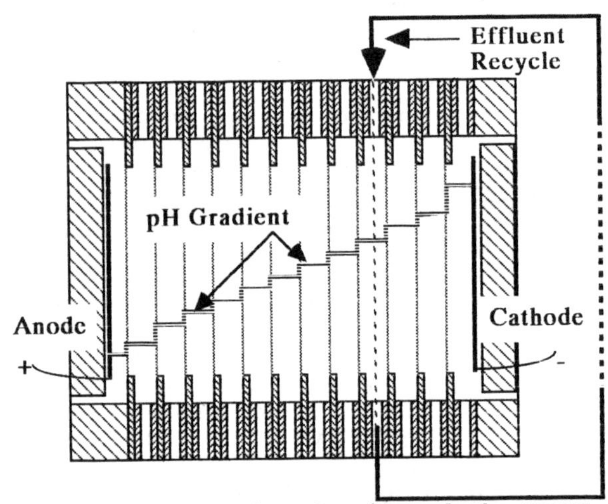

*Figure 15. Illustrative drawing of the **RIEF**. Solutes focus in the compartment(s) closest to their pI and are withdrawn from the chamber after it is turned off. Note that effluent is always recycled directly back to the compartment from which it was withdrawn.*

Bier and coworkers have produced several generations of **RIEFs**. Early models operated in the range of 2 gram/hr by processing about two liters of feed containing 0.1% protein. Later models have been operated in semi-batch mode at throughputs in the neighborhood of *two liters per hour* and should be considered competitive with the Biostream in terms of volumetric throughput but superior in terms of resolution.

Application of the **RIEF** does have several drawbacks, but none of them are particularly severe. First, there is the cost and use of the ampholyte system needed to establish a stable pH gradient. Assuming that the ampholytes are used only once and are purchased from a commercial supplier, they would introduce a base operating cost of $100-1,000 per gram of protein purified. However, if the ampholytes are manufactured in-house or are reprocessed on-site, **RIEF** operating costs associated with ampholytes could drop below $5 per gram of protein purified.

If ampholytes, which bind to some proteins, are used it may be difficult to gain federal approval for application of this process in downstream processing of therapeutics until safe levels of ampholyte contamination are established. However, it is possible to run the **RIEF** using an ampholyte-free buffer systems (61) and this would remove consideration of cost and contamination from application of the **RIEF**. This apparatus is available from Ionics, Inc., as the ISOPREP System.

Despite its power, commercial development of **RIEF** has been slow for reasons which are currently not clear. What makes this difficult to understand is that, while the full-scale unit has not done well, a scaled-down version of the **RIEF** marketed by Bio-Rad and known as the Rotofor has been doing very well as a benchtop preparative device.

RCFE (Recycle Continuous Flow Electrophoresis). I first proposed recycling *zone* electrophoresis in 1980. At that time conventional wisdom held that recycling would not work with zone electrophoresis as it had with isoelectric focusing. The reason for this is that the two processes are fundamentally different: isoelectric focusing is an "equilibration" process (62) in which solutes migrate to a point of dynamic equilibrium. In this case it migrates to the point in the system where the pH is equal to the solute's pI and then remains at that point. Separation occurs because the solutes have distinct pIs. By contrast, zone electrophoresis is a "rate-governed" process in which solutes are continuously displaced by the applied electric field. Separation occurs because solutes migrate at different velocities.

This problem is circumvented in recycle zone electrophoresis the effluent streams are shifted by a distance Δ, before reinjection at the inlet ports (figure 16). This gives rise to an effective *counterflow* of solvent in the direction opposite to the electrophoretic migration of the solutes. This counterflow is adjusted so that it splits the feed components into two product streams: one containing the components which migrate faster than the counterflow and the other containing components which migrate slower than the counterflow. There is a region near the feed port where the two solutes overlap, but theory predicts (63,64) that both will separate and can then be collected in essentially pure form.

In principle, any two solutes with different mobilities could be separated by proper selection of Δ. Theoretical analyses (63,64) and experiments had shown us that

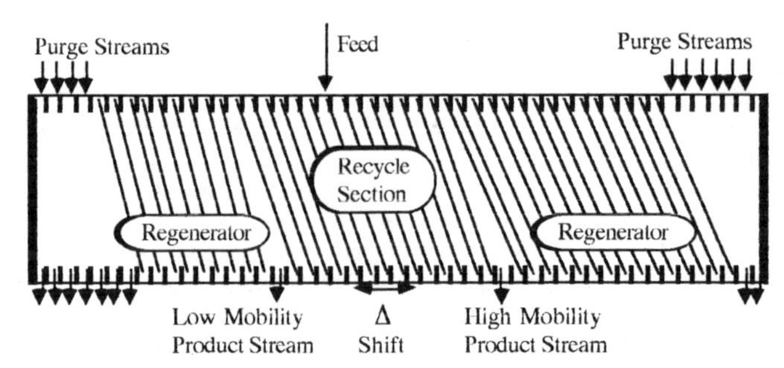

Figure 16. The ***RCFE*** *with regenerators. The regenerators keep solute from migrating past the outlet ports and this helps keep solute losses to a minimum. Note that opening up offtake ports automatically creates the regenerators and that the magnitude of the shift and the strength of the countercurrent flow is altered at each regenerator. (Reprinted from reference #63. Copyright 1990 ACS).*

continuous flow electrophoresis with recycle could simultaneously overcome the problems of natural convection, overheating and crescent dispersion.

Solutes migrate at their "shifted" velocities, i.e. electrophoretic plus convective, until they reach the regenerator units on either side of the recycle section. In the regenerators the counterflow of solvent is altered by changing Δ to keep solute from migrating past the outlet port (figure 16). With proper design complete separation can be achieved with virtually no loss of biological activity.

Separation in the **RCFE** is carried out in native buffer with background conductivities as high as 0.01 mho/cm (≈ 50 mM buffer) so low protein solubility is not a problem. Power requirements are kept low by using small electric fields, e.g. less than 40 V/cm, so the temperature rise in the chamber is typically less than 5°C. Besides this the **RCFE** is hydrodynamically stable when operated adiabatically with upflow and, because of the low electric fields used in the device and broad concentration gradients which accompany separation, it is not susceptible to the conductive-dielectric instability encountered in **IEF**.

RCFE Problems. In continuous operation the **RCFE** will split a multicomponent feed into only two product streams so, if there are more than two solutes in the feed, a "key" component must be chosen at which to split the solutes and offtake streams containing more than one product must be sent for further processing.

For instance, imagine a four component mixture in which two components are difficult to separate from one another while the others are relatively easy to separate. Three chambers would be needed. In the first chamber one of the "easier to separate" components would be taken off while the other three were sent on for further processing. Next, the properties of the buffer would be changed by diafiltration and next simplest component fractionated from the three-component mixture. The remaining bi-component mixture could be processed in a final chamber optimized for that separation.

Cascading the **RCFE** would allow independent optimization of each step in a difficult, multicomponent separation. In the chemical process industry many batch processes, e.g. centrifugation, ion exchange and affinity chromatography, and virtually all continuous separations , e.g. extraction, distillation, ultrafiltration, etc., are designed to split product into two streams. When several products must be isolated from a multicomponent feed, this is often done using a cascade of separating units (65). A cascade can also be used to conduct a particularly difficult separation, e.g. purification of uranium isotopes (66).

RCFE Advantages. Having noted these disadvantages, there are a number of important advantages to be gained by substituting zone electrophoresis for conventional technology. These are:

a. high resolution,
b. negligible mass losses, e.g. by adsorption
c. negligible activity losses, e.g. by denaturation,
d. continuous, semi-batch or batch processing,
e. broad particle size range, and
f. low cost.

The ultimate resolution (a) of solutes in the **RCFE** depends largely on the number of recycle *stages* designed into the chamber. Since the design engineer has complete control of this variable, any degree of product purity can, in principle, be attained.

The **RCFE** is operated without anticonvectant or adsorbent so the surface area available for adsorption is six to eight orders of magnitude lower than in packed beds. Because of this, non-specific adsorption (b) is negligible. In addition, solutes are forced by the regenerators to exit the chamber only at the two offtake ports so, unless there is a great deal of cross contamination between the ports, solute losses will be very small.

No special solvents (e.g. acetonitrile), solubilizing agents (e.g. SDS or urea), or modifiers (e.g. ampholytes), are needed to perform zone electrophoresis in the **RCFE**. Although any solvent which does not attack the chamber interior surfaces could be used, our approach has been to use the simplest buffer system which will allow good separation while maintaining protein (c) viability.

The **RCFE** was designed to operate continuously (d) with single or multiport injection feed and two offtake ports. However, theory indicates that it can also be run semi-batch or batch with multiple offtake ports to fractionate product into more than two product streams. Ongoing experiments indicate that, in batch operation, a single component can be simultaneously purified from cell homogenate *and* concentrated beyond 1% protein.

The **RCFE** can simultaneously accommodate solutes ranging in size from several Angströms to hundreds of microns in diameter (e). Bill and I have routinely run dyes through the apparatus to test or debug the chamber. On the other end of the size spectrum we have processed dilute whole blood and, in an experiment designed to test the limitations of our chamber, we processed a mixture of P-10 (BioRad) and G-25 (Pharmacia) gel filtration media. Although we were not able to fractionate the gels according to composition, we succeeded in keeping these rapidly sedimenting particles suspended in the chamber for several hours.

RCFE Economics. The operating costs of the **RCFE** are low (f) and capital costs show a dramatic economy of scale. The capital cost for a chamber which processes 1 kg/hr of protein is less than three times that of a chamber which processes 1 gm/hr. We generally run with total power to the chamber less than 100 W and the pump is rated 1/8 hp, roughly 100 W. Assuming power is available at $0.07/kW-hr, the power costs for running the chamber continuously are about 1¢ per hour. The make-up electrolyte is cheap: we usually run 10 mM phosphate buffer which, even at laboratory prices, costs less than 5¢ per hour in the prototype. At 1 gm/hr processing rates these expenses contribute about 7¢ per gram to the overall processing costs.

By comparison, the operating costs incurred for protein processing by large-scale reversed phase **HPLC** or dye-ligand affinity chromatography are at least $100 per gram of protein if one takes into account only the packing, solvent and solvent disposal costs. Furthermore, there is little economy of scale associated with scale-up of chromatography.

EXPERIMENTS

Experiments on proteins using a chamber with a 2 mm gap and 50 recycle ports indicate that the principles delineated in Gobie et al. (63) are closely followed in practice. In one experiment, a model feed solution composed of nearly equal weights of Hb and BSA in 10 mM phosphate buffer at 5% total protein was processed continuously at 30 ml/hr (1.5 grams protein per hour) with a holdup time of about one hour and power dissipation less than 25 watts. The recycle flowrate was set at approximately 120 ml/hr so each of the products was diluted by a factor of roughly four from its feed concentration (with a slight modification of the chamber, dilution can be entirely eliminated).

Product was withdrawn from one port on the left and one port on the right hand side and neither product showed any measurable cross contamination with either Hb or BSA according to our analysis by agarose gel electrophoresis. Agarose gel electrophoresis is a simple test that we routinely use to give a quick check on our results. It is not as powerful a technique as SDS-PAGE or isoelectric focusing. However, we have presented the results in this form because (a) the gels are very easy to interpret and (b) the results are consistent with all the results we have obtained from our tests and from the ones run later at Boehringer-Mannheim.

We have also processed several other synthetic protein mixtures in the **RCFE** including azoCasein-Hb (figure 17). In general these studies have progressed well and high-purity products have been recovered from both offtake ports.

The most important fractionation we have conducted to-date was the purification of monoclonal antibodies (MoAb, kindly supplied to us by Boehringer-Mannheim) secreted by hybridoma into a complex medium containing BSA from fetal calf serum. This separation is important because antibodies produced by continuous culture are used in diagnostic tests and the market demand for pure MoAbs is expected to increase dramatically as they are used more often for therapeutic purposes.

We began with an overnight dialysis of 60 mls raw broth to adjust feed conductivity and pH. This was followed by continuous electrophoresis for a period of three hours at 20 mls/hr feed with a total recovery of about 94 mg IgG. We kept a small amount of the product to run our own tests and sent the remainder back to Boehringer-Mannheim for complete biochemical analysis. Our in-house analysis using SDS-PAGE and agarose gel electrophoresis showed that both IgG and BSA were recovered with negligible cross-contamination as illustrated in figure 18.

Boehringer-Mannheim's more thorough battery of tests included SDS-PAGE, isoelectric focusing, hydroxyapatite **HPLC** and an antibody-enzyme performance assay. Our contact at Boehringer-Mannheim, Roger Sokoloff, concluded that the purity of products obtained by zone electrophoresis is comparable with the product obtained in their commercial process which consists of an ammonium sulfate precipitation (to concentrate protein, remove albumin and fetal IgG) followed by resuspension and either Protein-A affinity or cation exchange chromatography.

It is difficult to see in the figure above, but both MoAb products contain a very small amount of contamination. The precise nature of this contamination is not known although it appears to be slightly different in the two cases. In the absence of a rigorous biochemical analysis, my best guess at this time is that the contamination in the commercially-purified solution is the result of proteolytic degradation while the electrophoretically purified solution is contaminated with a very small amount of IgG from the fetal calf serum.

In closing this section it should be noted that, at the time these experiments were run, the **RCFE** did not have any

*Figure 17. Separation of azoCasein from hemoglobin at 1 gm/hr in the **RCFE**. Looking directly into the front face of the **RCFE**, one can easily see the low mobility Hb on the left-side of the chamber but the azoCasein, which is light yellow and elutes at the top on the right, is difficult to see in this B&W photograph. The right-most edge of the azoCasein is marked at the tip of the arrow.*

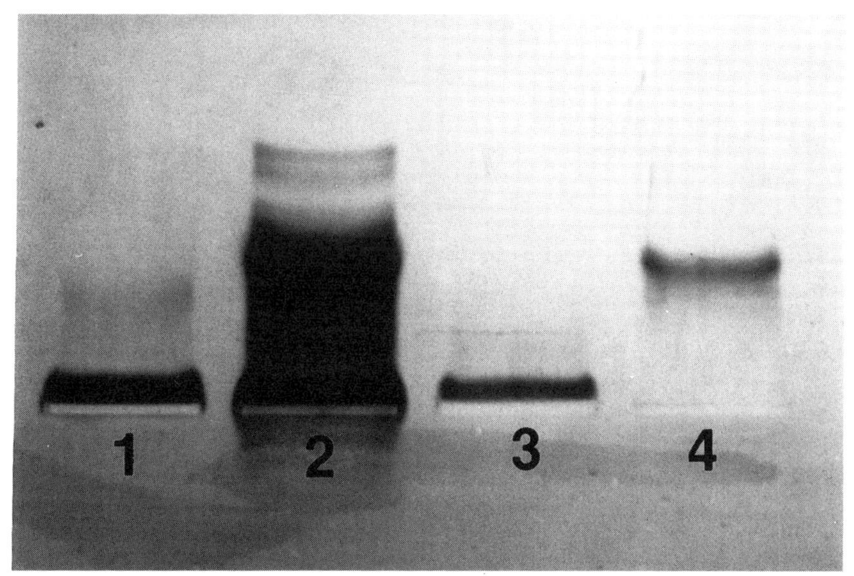

*Figure 18. Agarose gel electrophoresis of the raw feed and products from the purification of Boehringer-Mannheim's MoAb culture supernatant. Track #1: Product from protein A affinity chromatography column. Track #2: Raw feed. Track #3 Low mobility fraction (MoAb) from **RCFE**. Track #4: High mobility product from the **RCFE** (BSA). Further analysis by SDS-PAGE, isoelectric focusing and hydroxyapatite **HPLC** confirmed that the recovered MoAb stream was comparable in purity to commercially produced protein.*

equipment installed to allow for monitoring or control of the separation process. That is, unless the operator could see the materials being separated and adjust column operating conditions, he had no way of knowing whether the separation was going well. This was especially true of the MoAb separation where, since none of the proteins involved are visible, we were forced to run the chamber "blind." To alleviate this problem, we have fitting the chamber with a scanning UV spectrophotometer and are planning a new battery of test separations to shake down the spectrophotometer.

CONCLUSION

The last two decades have witnessed important advances in analytical, preparative and large-scale electrophoretic processing. The field has reached a point where many of the problems hindering development of suitable equipment have been partially or completely solved and, in many cases, advanced instrumentation for bench, pilot and industrial-scale zone electrophoresis is either commercially available now or in the later stages of development.

The challenge still remains to convincingly demonstrate that *zone* electrophoresis can provide high resolution at scales ranging from μg to kg. If successful, the speed and resolution of the new analytical devices and the economic advantages of preparative zone electrophoresis should persuade industry to consider implementing this technology initially in pilot studies and then, as the technology matures, in full-scale bioprocessing.

ACKNOWLEDGMENTS

The authors would like to thank Dr. Roger Sokoloff of Boehringer-Mannheim for supplying the raw materials for the MoAb separation and then providing us with a detailed analysis of the electrophoretically purified and chromatographically purified products. This work was supported in part by NSF Grant CBT-8813864 and by the Washington Technology Center.

NOTATION

A_c	Cross-sectional area
Bi	Biot number
c	Continuous phase conc.
D	Dispersion coefficient
D_m	Molecular diffusivity
E	Electric field strength
g	Gravity acceleration
h	Chamber slit thickness
I	Current/current density
J_i	ith component molar flux
J_o	Bessel fct order zero
J_1	Bessel fct order one
k	Thermal conductivity
K_{av}	Distribution coefficient
K_o	Mass transfer coeff.
L	Capillary/column length
M_i	Macro-ion loading
N_p	Plate efficiency
N_R	Recycle stages
P	Pressure
P_w	Power
Ra	Rayleigh number
R_L	Lumen radius
R_P	Polymer outer radius

R_W Wall outer radius
s Stationary phase conc.
t_e Elution time
T Temperature
U Continuous phase velocity
V Stationary phase velocity
x Axial coordinate
z_i ith component charge

Greek Symbols

α Dim. distribution coeff.
β Dim. mass transfer coeff.
γ **EC** velocity ratio
Δ Recycle shift
ε Gel bed void fraction
η Dim. radial coordinate
Θ Dimensionless temperature
ϑ Capillary volume
κ Thermal diffusivity
λ Autothermal coefficient
μ Electrophoretic mobility
ν Kinematic viscosity
ρ Density
σ_e Electrical conductivity
σ_1 Conductivity coeff.
ω Coefficient of expansion

Mathematical Symbols

∇ Grad operator

LITERATURE CITED

1. O'Farrell, P., J. Biological Chem., **250** 4007 (1975).
2. Scopes, R. K., *Protein Purification, Principles and Practice,* Springer-Verlag, New York (1982).
3. Ivory, C. F., Separ. Sci. Technol. **23**(8&9) 875 (1988).
4. Mosher, R. A., W. Thormann, N. B. Egan, P. Couasnon and D. Sammons, in *New Directions in Electrophoretic Methods,* J. W. Jorgenson and M. Phillips, eds., ACS Symp. Series No.**335**, American Chemical Society, Washington, D.C. 1987.
5. Hjerten, S., Chromatogr. Rev, **9** 122 (1967).
6. Hjerten, S., J. Chromatogr., **270** 1 (1983).
7. Mikkers, F. E. P., F. M. Everaerts and Th. P. M. Verheggen, J. Chromatogr., **169** 11 (1979).
8. Jorgenson, J. W. and K. D. Lukacs, Anal. Chem., **53** 1298 (1981).
9. Jorgenson, J. W. and K. D. Lukacs, J. High Res. Chromatogr. & Chromatogr. Commun., **4** 231 (1981).
10. Jorgenson, J. W. and K. D. Lukacs, Science, **222** 266 (1983).
11. Roberts, P. H., Quart. J. Mech. Appl. Math., **22**(2) 211 (1969).
12. Rhodes, P. H. , R. L. Snyder and G. O. Roberts, J. Colloid Interface Sci., **127** (1989).
13. Turk, R. S., Ph.D. Thesis, University of Notre Dame, Notre Dame, Indiana (1990).
14. Hjerten, S. and M-D. Zhu, J. Chromatogr., **346** 265 (1985).
15. Thormann, W., Separ. Sci. Technol., **19** 455 (1984).
16. Thormann W., R. A. Mosher and M. Bier, in *Chemical Separations, Vol. I: Principles*, p. 153, C. J. King and J. D. Navratil (eds), Proceedings of the First International Conference on Separations Science and Technology, Litarvan Literature, Denver 1986.

17. Everaerts, F. M. Mikkers, F. E. P. and Th. P. M. Verheggen, in *Isotachophoresis: Theory, Instrumentation and Applications*, J. Chromatogr. Libr., Vol. 6, Elsevier Publ. Co., Amsterdam, 1976.
18. Cohen, A. S., A. Paulus and B. L. Karger, Chromatographia, **24** 15 (1987).
19. Burton, D. E., M. J. Sepaniak and M. P. J. Maskarinec, J. Chromatogr. Sci., **24** 347 (1986).
20. Walbroehl, Y. and J. W. Jorgenson, Anal. Chem., **58**(2) 481 (1986).
21. Grossman, P. D., J. C.Colburn, H. H. Lauer, R. G. Nielsen, R. M. Riggin, G. S. Sittampalam and E. C. Rickard, Anal. Chem., **61** 1186 (1989).
22. Hames, B. D. and D. Rickwood, eds., *Gel Electrophoresis of Proteins. A Practical Approach*, IRL Press, Washington, D.C. 1981.
23. Andrews, A. T., *Electrophoresis: Theory, Techniques, and Biochemical and Clinical Applications*, Clarendon Press, Oxford 1986.
24. Taylor, G. I., Proc. Roy. Soc., **A219** 186 (1953).
25. Taylor, G. I., Proc. Roy. Soc., **A223** 446 (1954).
26. Aris, R., Proc. Roy. Soc., **A235** 67 (1956).
27. Ewing, A. G., R. A. Wallingford and T. M. Olefirowicz, Anal. Chem., **61** 292A (1989).
28. Wallingford, R. A. and A. G. Ewing, Adv. Chromatogr., 1 (1989).
29. Jorgenson, J. W., Anal. Chem., **58** 743A (1986).
30. Jorgenson, J. W., in *New Directions in Electrophoretic Methods*, J. W. Jorgenson and M. Phillips, eds., ACS Symposium Series **335**, p. 199 Wash. D.C. 1987.
31. Grushka, E., R. M. McCormick and J. J. Kirkland, Anal. Chem., **61** 241 (1989).
32. Lynch, E. and D. A. Saville, Chem. Engng. Commun., **9** 201 (1981).
33. Gill, W. N. and R. Sankarasubramanian, Proc. Roy. Soc., **A316** 341 (1970).
34. Vermeulen T., L. Nady, J. Krochta, E. Ravoo and D. Howery, Ind. Engng. Chem. Process Des. Dev., **10**(1) 91 (1971).
35. Hybarger, R. M., C. W. Tobias and T. Vermeulen, Ind. Engng. Chem. Process Des. Dev., **2**(1) 65 (1963).
36. O'Farrell, P. H., Science, **227**(4694) 1586 (1985).
37. Gobie, W. A. and C. F. Ivory, AIChE J. (1990) *submitted.*
38. Gobie, W. A. and C. F. Ivory, Biotech. Prog., **6**(1) 21 (1990).
39. Rudge, S. R. and M. Ladisch, presented in the session on *Separation of Bioactive Compounds* at the Annual Meeting of the AIChE (1989).
40. Philpot, J. St. L., Trans. Faraday Soc., **39** 38 (1940).
41. Grassman, W. and K. Hannig, Naturwissenschaften, **37** 397 (1950).
42. Svensson, H. and I. Brattsten, Arkiv for Kemi., **1**(47) 401 (1949).

43. Barrollier, J., E. Watzke, and H. Gibian, Z. Naturforschung, **13b** 754 (1958).
44. Strickler, A. and T. Sacks, Prep. Biochem., 3(3) 269 (1973).
45. Strickler, A. and T. Sacks, Ann. N. Y. Acad. Sci., **209** 497 (1973).
46. Saville, D. A., PhysicoChemical Hydrodyn., **1** 297 (1980).
47. Lynch, E. D. and D. A. Saville, Chem. Engng. Commun., **9** 201 (1981).
48. Philpot, J. St. L., in *Methodological Developments in Biochemistry*, Vol. 2. Preparative Techniques, E. Reid, ed., Longmans, England 1973.
49. Mattock, P., G. F. Aitchison and A. R. Thompson, Separ. Purif. Methods, **9**(1) 1 (1980).
50. Beckwith, J. B. and C. F. Ivory, Chem. Engng. Commun., **54**(1-6) 301 (1987).
51. Noble, P. T., Biotech. Progress, **1**(4) 237 (1985).
52. Saville, D. A. and S. Ostrach, Final Report, Contract #Nas-8-31349 Code 361 (1978).
53. Rhodes, P. H., NASA Tech. Memo. NASA TM-78178 (1979).
54. Hymer, W. C., et al., Cell Biophysics, 10(1) 61 (1987).
55. Wagner, H. and R. Kessler, p.303 in *Electrophoresis '82*, D. Stathakos, ed., Walter de Gruyter & Co., Berlin 1983.
56. Wagner, H., V. Mang, R. Kessler, A. Heydt and R. Manzoni, in *Electrophoresis '83*, Walter de Gruyter & Co., Berlin 1984.
57. Turk, R. S., *A Coulombic-Dielectric Instability*, Ph.D. Thesis, University of Notre Dame 1990.
58. Bier, M. and N. Egan, in *Developments in Biochemistry*, vol. 7, Haglund, Westerfield and Ball, eds., Elsevier-North Holland, Amsterdam 1979.
59. Bier, M., N. B. Egan, G. E. Twitty, R. A. Mosher and W. Thormann, in *Chemical Separations, Vol. I,* C. J. King and J. D. Navratil, eds., Litarvan Literature, Denver, Colorado 1986.
60. Bier, M., in *Separation, Recovery, and Purification in Biotechnology*, Asenjo, J. and J. Hong, eds., ACS Symposium Series 314, American Chemical Society, Washington, DC 1986.
61. Bier, M., R. A. Mosher and O. A. Palusinski, J. Chromatogr., **211** 313 (1981).
62. King C. J., *Separation Processes*, Second Edition, McGraw-Hill Book Co., New York 1980.
63. Gobie, W. A., J. B. Beckwith, and C. F. Ivory, Biotechnology Progress, **1**(1) 60 (1985).
64. Gobie, W. A. and C. F. Ivory, in *Separation, Recovery, and Purification in Biotechnology*, J. A. Asenjo and J. Hong, eds., ACS Symposium Series No. 314, 169 (1986).
65. Wankat, P., Separ. Sci. Technol., **19**(11&12) 801 (1985).
66. Olander, D. R., Scientific American, **239**(2) 37 (1978).

RECEIVED January 18, 1990

Chapter 16

Applied Electric Fields for Downstream Processing

Scott R. Rudge and Paul Todd

National Institute of Standards and Technology, Center for Chemical Technology, Boulder, CO 80303–3328

Electric potentials at interfaces are governed by the distribution of ions between the phases. The fundamental thermodynamics of ionic partitioning are reviewed, and the implications of the resulting double layer for transport processes in electric fields are discussed. The special considerations of electric field applications, concerning heat and mass transport, are presented. Electric fields have been applied to aqueous biphasic demixing and to cell separations by density gradient electrophoresis and free flow electrophoresis, for the purpose of producing gram quantities of protein. Analytical applications of electric fields, for the purpose of monitoring downstream processes, have also been successful. These applications include capillary electrophoresis, electrospray mass spectrometry, and pulsed field electrophoresis.

Electric fields may be applied as external force fields to drive or enhance rate processes in downstream processing equipment. In applications which currently employ gravitational or centrifugal fields, or energy or chemical gradients, electric fields may often be substituted to achieve similar process effects. The purpose of this chapter is to review the fundamental basis for the existence of electric potentials in chemical systems, to show how these fundamentals lead to working electrokinetic systems, and finally to discuss selected applications of electric fields to enhance rates, transport solutes, and/or to monitor downstream (non–reaction) processes.

THERMODYNAMICS of CHARGED INTERFACES

When two distinct and stable phases are contacted, solutes in each phase distribute between the phases according to their chemical potentials. This is true, for example, for organic salts in oil–water emulsions, salts and proteins in aqueous two phase systems, ions in certain membranes or films, or electrons at metal/metal interfaces. The affinity of one phase for one solute over another results in different concentrations of each solute in each phase. This affinity is expressed as a chemical potential, defined as

$$\mu_i^\alpha = \mu_i^o + RT \ln x_i^\alpha \quad (1)$$

for a dilute ideal solute i in phase α, where μ is the chemical potential, μ^o is the standard or reference potential, x is the concentration, R is the gas constant and T is temperature. The differences in concentration between phases are expressed as partition coefficients

$$K_i = \frac{x_i^\alpha}{x_i^\beta} = \exp\left[\frac{\mu_i^{o\beta} - \mu_i^{o\alpha}}{RT}\right] = \exp\frac{\Delta\mu_i^o}{RT} \quad (2)$$

Several methods are available for estimating chemical potentials and activity coefficients for uncharged compounds (1). For uncharged compounds, such as electrolytes, the chemical potential is given by

$$\bar{\mu}_i^\alpha = \bar{\mu}_i^{o\alpha} + RT \ln x_i^\alpha + z_i F\phi^\alpha \quad (3)$$

where $\bar{\mu}_i^\alpha$ is the electrochemical potential of solute i in phase α, $\bar{\mu}_i^o$ is the standard or reference "uncharged" chemical potential, z_i is the valence of solute i, F is Faraday's constant (96459 C/equiv), and ϕ^α is the potential on phase α. The last term of equation (3) represents the work required to bring z charges into phase α. The resulting distribution coefficient is given by

$$K_i = \exp\left[\frac{\Delta\bar{\mu}_i^o + z_i F\Delta\phi}{RT}\right] \quad (4)$$

where $\Delta\phi = \phi^\beta - \phi^\alpha$ and $\Delta\bar{\mu}_i^o = \bar{\mu}_i^{o\beta} - \bar{\mu}_i^o$. Standard potentials and activities for electrolytes may be found in Newman (2) and Robinson and Stokes (3). In the case of electrolytes, $\Delta\bar{\mu}^o$ refers to a single ion, while $\Delta\phi$ is a common force for all species. For example, for a hypothetical electrolyte A^+B^-, where partitions of $K_a = 2.0$ and $K_b = 0.2$ can be observed in the absence of charge, 2.5 times as much B^- as A^+ would exist in phase β, while 4 times as much A^+ as B^- would exist in phase α. However, electroneutrality requires that the ions exist in equal amounts in each phase, or

$$x_a^\alpha = x_b^\alpha = x^\alpha \ , \ x_a^\beta = x_b^\beta = x^\beta \ , \ K_a = K_b \quad (5)$$

and

$$\Delta\phi = \frac{\Delta\bar{\mu}_a^o - \Delta\bar{\mu}_b^o}{(z_b - z_a)F} \quad (6)$$

In the case of the electrolyte A^+B^-, the potential difference is –30 mV at 27°C, and the resulting partition coefficient for both ions is 0.63.

It may be seen, then, that an electric potential gradient can support an "uncharged" chemical potential gradient between two charged solutes in systems which are treated ideally. These results may be extended to nonideal systems. Extensive activity data and correlations for electrolytes may be found in Robinson and Stokes (3), and the measurement and correlation of electrolyte thermodynamics remains an active and challenging area of research (see, for example,4–9). The extension to non–ideal mixtures introduces more stringent conditions for a zero electric potential between two phases. Excellent and more extensive developments of electrochemical thermodynamics may be found in the books by Newman(2) and Hunter(10).

SURFACE CHARGE and the DOUBLE LAYER

Thermodynamically, an electric potential supports gradients in chemical potential across an interface. The gradients of chemical and electrical potential must balance each other for all ions in the system away from the interface as

$$\frac{d\bar{\mu}_i}{dy} = -z_i F \frac{d\phi}{dy} \qquad (7)$$

Substituting for $\bar{\mu}_i$

$$RT\,\frac{d\ \ln\ x_i}{dy} = -z_i F \frac{d\phi}{dy} \qquad (8)$$

which, upon integration, gives the Boltzmann distribution of ions away from the interface

$$x_i = x_{ib}\ \exp -\left[\frac{z_i F \phi}{RT}\right] \qquad (9)$$

where x_{ib} is the bulk concentration of x_i ($y = \pm \infty$, $\phi = 0$, $x_i = x_{ib}$). The charge density, defined as the number of equivalents of each ion, per volume,

$$\rho_e = \sum_i F z_i x_i \qquad (10)$$

is 0 in the bulk, electroneutral solution, but varies near the interface according to equation (9). The Poisson equation of electrostatics relates the electric potential to the spatial variation in charge density

$$\nabla^2 \phi = \frac{\rho_e}{\epsilon\epsilon_0} \qquad (11)$$

where ϵ is the dielectric constant of the medium and ϵ_0 is the permittivity of free space, 8.854 x 10^{-12} F/m. Equations 9, 10, and 11 combine to form the Poisson–Boltzmann equation, which gives the electric potential gradient in terms of the properties of the system

$$\nabla^2 \phi = -\frac{1}{\epsilon\epsilon_0} \sum_i F z_i x_{ib}\ \exp -\left[\frac{z_i F \phi}{RT}\right]) \qquad (12)$$

The Poisson–Boltzmann equation (eq. 12) is usually solved in linearized form by taking the series expansion for the exponential term ($e^{-x} = 1 - x + \frac{x^2}{2!} - \cdots$) and ignoring second order and higher terms

$$\nabla^2\phi = -\frac{1}{\epsilon\epsilon_0}\left(\Sigma\, F x_{ib} z_i - \Sigma\, F^2 z_i^2 x_{ib}\phi/RT\right) \tag{13}$$

The first term on the right hand side of equation 13 must be zero so that electroneutrality is satisfied in the bulk solution. The linearized equation for a flat interface then, is

$$\frac{d^2\phi}{dy^2} = \kappa^2\phi \tag{14}$$

where κ is the inverse Debye length

$$\kappa = \left[\frac{F^2\, \Sigma z_i^2 x_{ib}}{RT\epsilon\epsilon_0}\right]^{0.5} \tag{15}$$

The solution to the Poisson–Boltzmann equation for planar geometry is

$$\phi = \phi_0 \exp\left(-\kappa y\right) \tag{16}$$

where ϕ_0 is the potential at the interface compared to the potential in the bulk phase. Figure 1 shows the distribution of electric potential, and the ions A^+ and B^- for the example cited above. In calculating the potential and ionic distribution, a value for dielectric constant of 80 (corresponding to dilute aqueous environment) for the α phase, and 2 (corresponding to organic environment) for the β phase were chosen. The calculated values for the Debye length were 4.9 and 0.62 Å, respectively. The truncation of the Taylor series requires that ϕ_0 be small, the limiting value for which the approximation is valid is 25.7 mV/z at 25°C (10). For higher surface potentials and different geometries, integration of the non–linearized Poisson–Boltzmann equation is possible (for details, see (10)).

ELECTROKINETICS in TRANSPORT PROCESSES

Electric fields act on charged particles as a body force. If there is an imbalance of charge in a differential volume, as there is near an interface, an electric field will act on that charged volume as if it were a charged particle. The charge density near the interface for salt A^+B^- is shown in Figure 2. A force balance on the electrical and mechanical stress tensors at a planar interface within a charged double layer was derived by von Smoluchowski for a fixed interface

$$E_z\rho_e = -\eta\frac{d^2v_z}{dy^2} \tag{17}$$

where y is the direction perpendicular to the interface, z is the direction of the electric field (parallel to the interface), η is the viscosity of the medium (assumed Newtonian) and v_z is the velocity of the fluid. Substituting for the charge density, ρ_e, from Poisson's equation

$$E_z\epsilon\epsilon_0\frac{d^2\phi}{dy^2} = \eta\frac{d^2v_z}{dy^2} \tag{18}$$

The result is obtained by integrating between $y = \infty$ ($\phi = 0$, $v_z = v_e$) and $y = 0$ ($\phi = \zeta$, $v = 0$),

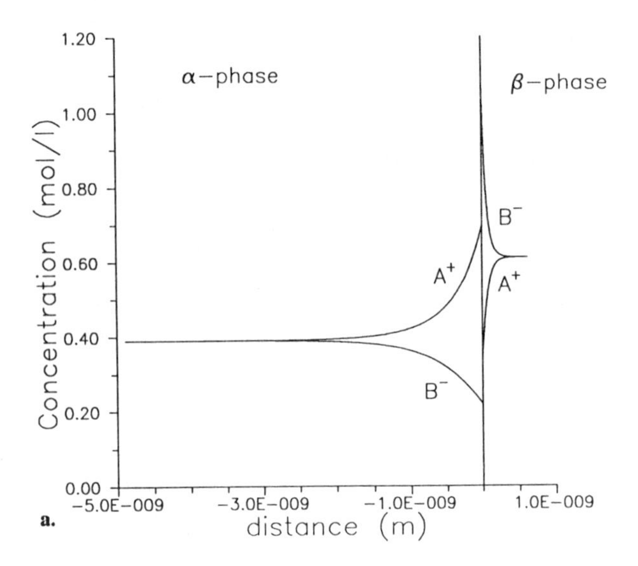

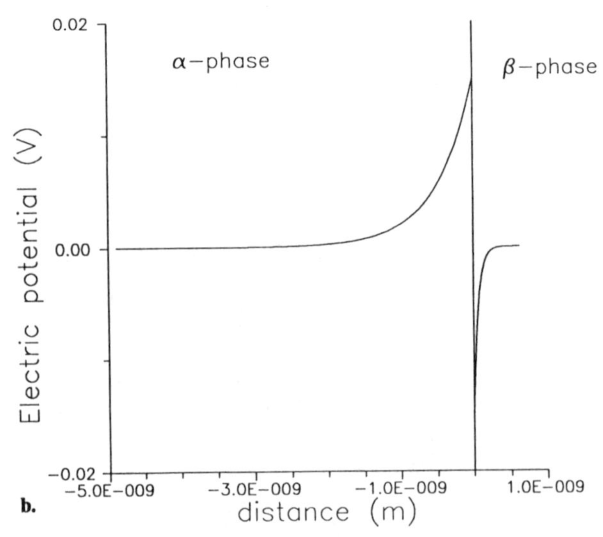

Figure 1.
The distribution of the ions A^+B^- near the interface of the α and β phases (a) and the resulting electric potential distribution (b). Dielectric constants of 80 (water) and 2 (oil) have been used for the α and β phases, respectively.

$$\frac{v_e}{E_z} = u_e = -\frac{\epsilon_0 \epsilon \zeta}{\eta} \tag{19}$$

where u_e is defined as the electrophoretic mobility. There is a subtle relationship between the integration limits resulting in equations 19 and 16. When determining the potential distribution, the integration is made from the interface itself, where the potential is ϕ_0. When calculating the velocity of the fluid away from the interface, the integration must be made from the slipping plane, or surface of shear, where the fluid velocity is zero and the potential is ζ (the zeta potential). The exact relationship between the thermodynamic surface potential difference, $\Delta\phi$, and the electrokinetic zeta potential, ζ, is system dependent, and considered indeterminant for all practical purposes. It is an important distinction because the length scale over which the electric potential drops ($1/\kappa$) is on the order of nanometers or less, so layers of hydration become significant in the attenuation of potential.

Whenever charged double layers move with respect to one another, an electrokinetic process is said to occur, and the zeta potential is important. The four electrokinetic processes, electrophoresis, electroösmosis, streaming potential and sedimentation potential, are shown schematically in Figure 3. They are four permutations on the possibility that a charged interface is fixed or free, and that the force applied is an electrical gradient or a pressure gradient. If a pressure gradient is applied, a macroscopic potential may be measured, if an electric field is applied, migration is observed, or a pressure gradient results.

A streaming potential (Figure 3a) results when electrolyte fluid is pumped through a tube where a surface potential exists on the tube walls. There are many considerations in the evaluation and measurement of streaming potentials (10) such as flow profile, and surface and tubing conductivity, which detract from the usefulness of streaming potentials for measuring surface potentials.

A sedimentation potential (Figure 3b) may be generated when a colloidal suspension settles under gravity. In this case, the movement of the particles with respect to the continuum is responsible for the potential. Both streaming and sedimentation potential are useful methods for measuring zeta potential in idealized systems. Their processing implications appear to be small.

When an electric field is applied across a tube with a surface potential in an electrolyte, the electrolyte will begin to flow (Figure 3c) in a process called electroosmosis, or electroendoosmosis. The direction and magnitude of the flow is given by equation (19). Electroosmotic flow generally has a flat velocity profile (compared to parabolic flow profiles found in Poiseuille flow), since the flow is driven from the walls of the tube, rather than from the center. All the flow variation occurs within the region of the double layer. This is valid when the radius of the tube is much larger than the Debye length ($\kappa r >> 1$), but is not generally true for pores and capillaries. For very small pores ($\kappa r \simeq 1$) a different solution to the Poisson–Boltzmann equation is required, because there is no distance from the interface at which $\phi = 0$, and the curvature of the interface is important (11). A flow profile closer to parabolic is found in these cases. Electroosmotic flow in closed tubes also results in parabolic flow, since conservation of mass requires that flow along the walls be compensated by back flow through the center of the tube.

When an electric field is applied to a dispersed system, the dispersed particles migrate in a process known as electrophoresis, as shown in Figure 3d. There is a significant body of literature addressing the physics of this process, the most recent advances in theory being made by O'Brien and White (12). Equation (19) gives the magnitude and reverse sign for electrophoretic

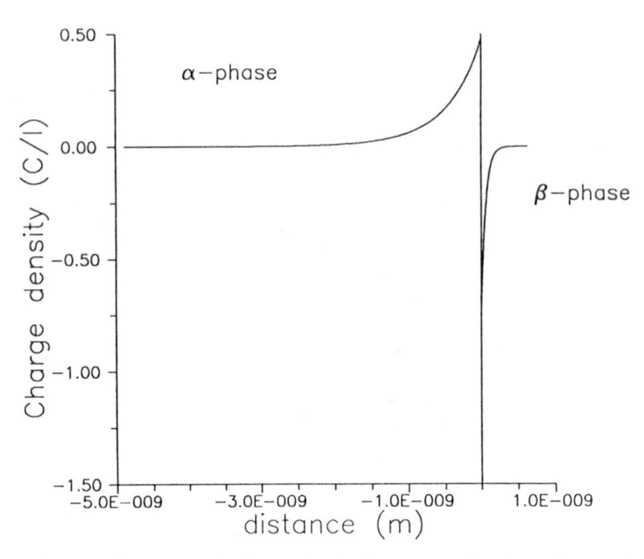

Figure 2. The charge density of the salt $A^{+}B^{-}$ near the interface, derived from Figure 1.

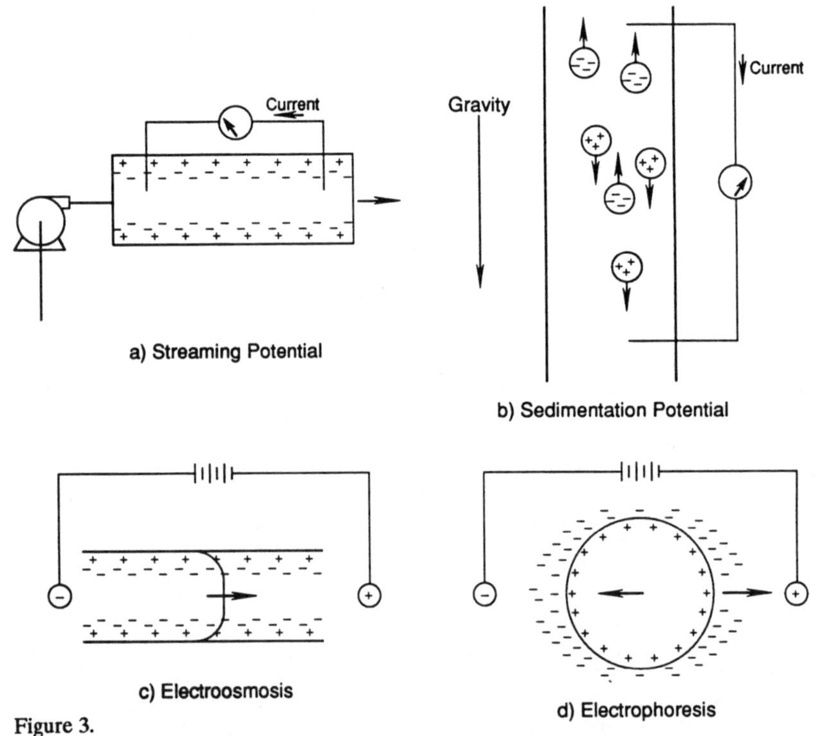

Figure 3.
The four electrokinetic processes: a) streaming potential, b) sedimentation potential, c) electroosmosis and d) electrophoresis.

migration of a large, non–conducting sphere with a small double layer ($\kappa r >> 1$). Hückel obtained another limiting solution for the mobility of a small sphere with a large double layer ($\kappa r << 1$)

$$u_e = \frac{2\epsilon\epsilon_0\zeta}{3\eta} \tag{20}$$

When a particle and double layer are on the same order of magnitude, distortion of the double layer with respect to the particle, a phenomenon known as relaxation, becomes important. Good correlations for the electrophoretic mobility of particles of all sizes and potentials have been found by O'Brien and White (12), and Wiersema et al. (13).

In electrokinetics as applied to bioprocessing (0.001M to 0.1 M buffers), equation (19) would apply to cell sized particles, bacteria and larger. Equation (20) applies to protein molecules and smaller solutes. Double layer relaxation phenomena occur in intermediate cases, such as viruses, small vesicles, and small subcellular particles such as ribosomes. Ionic strength is a strong modulator of these categories, since $\kappa \propto (z_i^2 x_{ib})^{0.5}$. The relationship between the Debye length, κ, and the molarity of salts of different valences (equation 15) is shown in Figure 4.

A distinction should be made between electrophoresis and electrolyte conduction. Electrophoresis technically only occurs when an interface causes a charge imbalance. Electrolyte solutions will conduct an electric current without the presence of an interface, simply because the salts are dissociated. The distinction appears to be made based on the origin of the charge, and on the length scale over which one may assume that continuum properties still hold. Ions, such as sodium and chloride, have a charge and a mobility in solution which is unaffected by the other ions or their concentrations (2,3). Colloidal particles, on the other hand, are affected thermodynamically by the properties of the continuum. Proteins, theoretically, belong to the former group. However, proteins exhibit many of the properties associated with colloidal particles. Their charge is a complicated function of pH due to their diversity of ionizable groups, and they are presumed to have an interior and a surface which differ distinctly in properties. It is difficult to justify extension of continuum derivations of potential and mobility to proteins, however. Their surfaces are not defined, and do not have an evenly distributed potential, and they are not necessarily spherical (Haggerty and Lenhoff, J. Chromatog., submitted 1989). Nevertheless, proteins are discussed in terms of electrophoresis and zeta potential in most instances. Hunter (14), Rudge and Ladisch (15,16), and Bier et al. (17–19) have recognized the importance of the ionic character of proteins in electrophoretic processes in their models.

SPECIAL CONSIDERATIONS in ELECTROSTATIC PROCESSES

A major problem in the scale–up of electrokinetic processes is heating and mixing (20). A major advantage of electrokinetic processes is the speed at which they attain equilibrium. The relevant scaling laws are given here.

HEATING and MIXING. Electric energy is dissipated as heat according to the equation

$$W = IE \tag{21}$$

In systems in which Ohm's law applies

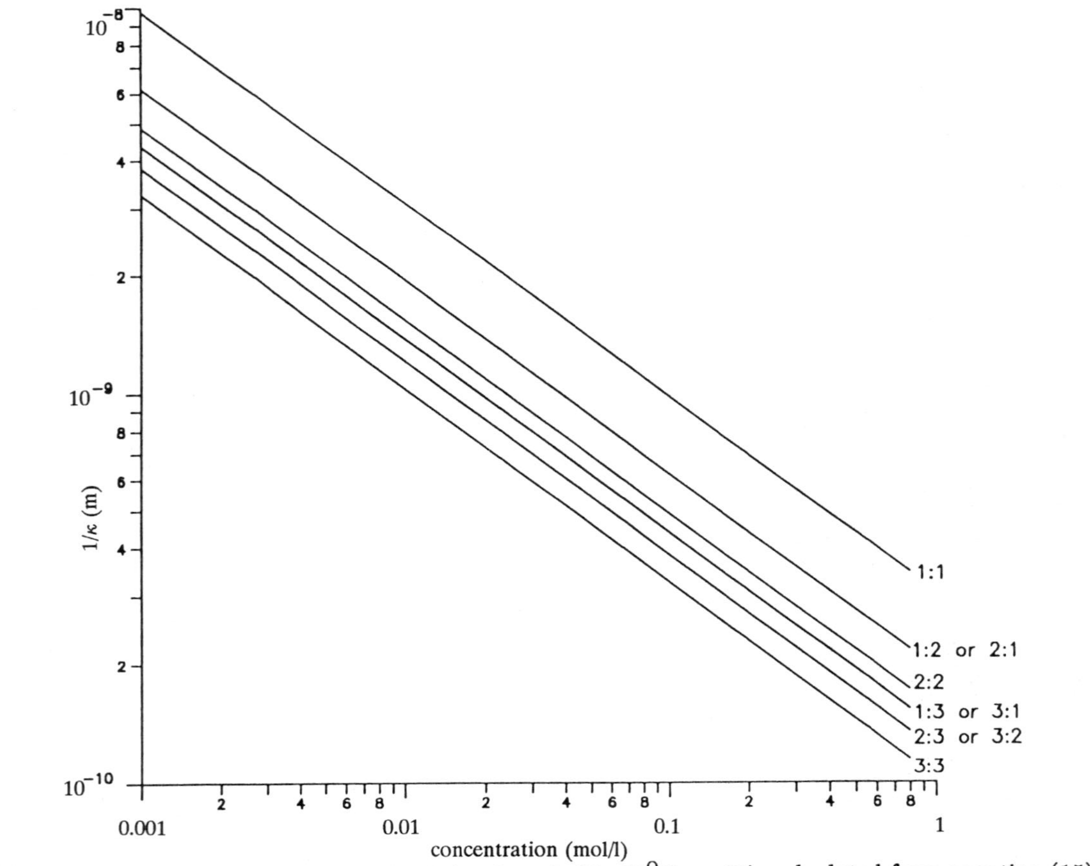

Figure 4. The effect of valence and molarity on the Debye length ($T=27^{\circ}C$, $\epsilon=80$), calculated from equation (15).

$$W = I^2/k \quad (22)$$

where k is the conductivity of the system, the heat generation increases as the square of the current passed. For this reason, nearly all electrokinetic applications are in the most resistant media compatible with the unit operation (16). In an adiabatic, stagnant system, the temperature increases uniformly with time as

$$\Delta T = \frac{IE}{C_p \rho V} t \quad (23)$$

where C_p is its heat capacity of the media carrying the current, ρ is the density of the material and V is the volume of the system. This equation represents the maximum temperature increase in an electrophoretic system. Presuming that the system boundaries remain at the initial temperature, the Rayleigh number

$$Ra = \frac{g\beta \ \Delta T d_c^3 \bar{\rho} C_p}{k_t \eta} \quad (24)$$

and the Grashof number

$$Gr = \frac{\bar{\rho}\beta \ \Delta T d_c^3 g}{\eta^2} \quad (25)$$

give a measure of the importance of the temperature rise in a closed system. In these equations, $\bar{\rho}$ is the average fluid density, β is the thermal expansion coefficient, g the acceleration of gravity, and d_c the distance from the high temperature in the system to the closest lateral boundary (perpendicular to g). For vertical systems (those cooled from the sides) $Ra < 10^9$ implies laminar free convection. In general, Ra and Gr should be minimized, which usually means minimizing d_c. These numbers may be used to estimate a Nusselt number

$$Nu = \frac{hd_c}{k_t} \quad (26)$$

which gives the relative importance of convective heat transfer (h) to conductive heat transfer. Correlations for various geometries and flow regimes may be found in Perry and Chilton (21) and Parker et al.(22) It should be noted that the Rayleigh number is the product of the Grashof number and the Prandtl numbers, as the Prandtl number is frequently used in these correlations.

Heating in electrokinetic systems can result in mixing through natural convection. In this respect, heating is the single most important limitation to electrokinetic scale up. Heat must be removed from systems so as not to destroy heat sensitive components, but should not be removed in such a way that it leads to natural convection. One obvious way to accomplish this is to operate electrokinetic systems in high surface area / volume containers, such as capillaries (23,24) or between closely spaced flat plates. This is the opposite of scale up, however, and leads to enhanced electroosmotic transport. Flowing systems have been designed to overcome some of the problems of heating. In these, heat is removed from the system by convection, and the fluid may pass through a heat exchanger before being resubjected to the electric field. This concept has been used by Bier (25) to advantage with isoelectric focusing, and by Gobie et al (26,27) with recycling electrophoresis. Other ideas, such as rotating systems (28–31) and operation of electrokinetic units in space (32–34), have been tried with moderate success.

Electroosmosis occurs along fixed interfaces and results in bulk flow or a pressure gradient which contributes to mixing. In closed systems, where a pressure gradient is the result, a reverse flow occurs in the center of the unit, while electroosmotic flow occurs along the walls. The result is backmixing of any solute which spans the cross section of the unit. The effects of electroosmosis were demonstrated vividly by Micale et al. (35) in an electrophoresis unit operated in space. Electroosmosis is easier to overcome than heating, by the use of coatings on apparatus walls which do not adsorb ions preferentially (29,36).

SCALING of MASS TRANSFER in ELECTROPHORETIC SYSTEMS COMPARED TO CHROMATOGRAPHIC SYSTEMS. The thermodynamic parameter, ζ, is proportional to electrophoretic velocity of particles as shown in equations (19) and (20). In chromatography, the partitioning or adsorption of solutes on a resin is reflected by the capacity factor or distribution coefficient, K_d, which affects the mobility of a solute in the column as

$$v_c = \frac{v_s}{\alpha + (1-\alpha)K_d} \tag{27}$$

where v_c is the velocity of the solute, v_s is the superficial velocity of the mobile phase and α is the void fraction.

The approach to thermodynamic equilibrium is generally controlled by the effective mass transport from the surrounding bulk to the thermodynamic interface. The rate of mass transport to the surface of shear is rapid in electrophoresis, because all variations in the particle's electrostatic field take place within a few nanometers of the particle's surface. In comparison, chromatographic response to changes in the thermodynamic quantities occurs over the length scale of resin particles, which range from 2 to several hundred micrometers, and is likely to be sluggish.

Electrophoretic mobilities of proteins are typically on the order of 10^{-4} $cm^2/V \cdot s$, while interstitial velocities in chromatography columns may vary from 10^{-4} cm/s to nearly 1 cm/s. The diffusivity of a protein in solution is approximately 10^{-7} cm^2/s. Using these values, and the length scales cited above, we may calculate the ratio of the transport velocity to the rate of diffusive mass transport, or the Peclet number, for electrophoresis and chromatography. The electrophoretic Peclet number, $u_eE/D\kappa$ is approximately 0.0001 E, whereas the chromatographic Peclet number, vd_p/D is 10^3 to 10^7 times the particle size in centimeters, or more. The Peclet number compares the rate at which a solute migrates in a system to the rate at which it equilibrates with that system. The lower the Peclet number, the more closely equilibrated the system. The ratio of the electrophoretic to chromatographic Peclet numbers, $u_eE/vd_p\kappa$, indicates the relative extent to which each system is at equilibrium. The particle size in chromatography is rarely less than 1 micrometer, while the Debye length, $1/\kappa$, is on the order of 10^{-9} to 10^{-10} m. This means that the transport rate in electrophoresis may be three to four orders of magnitude greater than that in chromatography, while maintaining the same relative equilibrium. Some capillary electrophoresis applications have shown very rapid separations with very little dispersion (16,37), as this analysis suggests.

PROCESSING APPLICATIONS

Aqueous two phase extraction and cell tissue culture are two emerging technologies for the processing and production of large amounts of biological materials. Both systems lend themselves to applications of electric fields to improve processing or production selection. The current work in both of these fields is reviewed here.

DEMIXING of EMULSIONS and DISPERSIONS. Liquid liquid extraction is a popular purification method used heavily in the chemical industry today. The conventional technique has found limited popularity in bioprocessing owing to the damaging effects of organic solvents, which are used as an extracting phase, on most biomolecules and all biological cells. Two incompatible polymers dissolved in water will form aqueous two phase systems, which are biocompatible due to their high water content (38). Moreover, these systems are reported to have provided stability to biologically active substances, such as enzymes (39). Despite some 800 papers on this subject (40), large scale commercial applications of aqueous two phase systems are not widespread in the biotechnology industry. The high cost and low speed of the process apparently deter its widespread use.

The similarity of the physical properties of the two phases (small density difference, low interfacial tension) causes them to separate slowly in the absence of enhanced physical force, such as centrifugation. When internal circulation of the droplet is taken into account, the Stokes velocity at which a droplet of one phase falls (or rises) through another is given by

$$v_g = \frac{2}{9}\frac{a^2}{\eta_1} g \left[\frac{(\rho_2-\rho_1)(3\eta_2+3\eta_1)}{3\eta_2 + 2\eta_1}\right] \quad (28)$$

The electrophoretic velocity of the same droplet is (10)

$$v_e = \frac{a\,\sigma_e\,E}{3\eta_2+2\eta_1+\sigma_e^2/\lambda} \quad (29)$$

where σ_e is the charge density at the surface of shear (ρ_e @ v=0) and λ is a function of the conductivities of the droplet and continuous phases

$$\lambda = \frac{1-k_2/k_1}{2+k_2/k_1} \quad (30)$$

and the subscripts 1 and 2 refer to the continuous and droplet phases, respectively. For large droplets ($f(\kappa a) = 1$) the radius of the droplets for which the electrophoretic velocity will be greater than the gravitational velocity is

$$a < \frac{9}{2}\frac{\sigma_e}{(\rho_2-\rho_1)}\frac{3\eta_2+2\eta_1}{(3\eta_2+3\eta_1)(3\eta_2+2\eta_1+\sigma_e^2/\lambda)}\frac{E}{g} \quad (31)$$

Upon exploration in our laboratory, it was found that electrokinetic demixing methods could be appropriately applied to aqueous two phase systems (Rao, Stewart, and Todd, Sep. Sci. Technol., in press, 1990).

Since these are thermodynamically true two phase systems (41), it is appropriate that the thermodynamics discussed earlier should apply. Buffer ions, such as sodium and phosphate, are typically found in these systems as ions are required to solubilize most proteins and stabilize the pH. Sodium and phosphate partition according to their electrochemical potentials in each phase, and a surface charge results at the interface (42). The relative potential difference as an electrode is moved from one phase to the other has been measured for some systems (43), as has the electrophoretic mobility of the droplets (43). A discrepancy between the sign of the interface potential and the zeta potential has been observed, and explanations which assume the importance of an induced dipole potential on the droplet surface (10), have been offered (44,45). However, direct evidence of an induced dipole has not been found experimentally.

Since the droplets are fluid, with finite viscosities, their internal circulation will play a part in their charge distribution and their shapes. In some applications (46) the stress of the applied potential on the droplet surface causes necking, and, when the electrical stress exceeds the Rayleigh stability limits (electrical surface stress vs. surface tension), causes droplet disruption into smaller droplets. These types of phenomena are most often observed in highly resistive continuous phases, however, such as oil phases (47) and gas phases (see electrospray devices, below). In two phases which are aqueous the heating that results from applying a field high enough to induce Rayleigh instability would probably obscure the effect. However, droplet shape distortion undoubtedly occurs, and must be accounted for in predicting demixing. Electric fields have been used for breaking emulsions, however, predominantly in oil/water separations (48–50).

Finally, it should be noted that demixing occurs in these systems as they evolve from a one phase system, to a three phase system, and finally to a true two phase system. While each droplet is at near equilibrium at all times, compositions in the system are changing as droplets "ripen". One possible trajectory for this evolution is shown in Figure 5.

CELL SEPARATIONS. It is often desirable to preprocess a cell population before fermentation or extraction to enhance production of a target molecule. A common example of this is the use of antibiotics in a recombinant fermentation, to select for plasmid carrying cells. Some cell populations are not susceptible to genetic selection. Fortunately, cell populations are also found that are electrophoretically heterogeneous. These populations may be separated in electric fields to produce a subpopulation enriched in target molecule production capacity. The electrophoretic methods used for cell separations are generally methods that are also suitable for other biochemical (e.g. protein, polysaccharide) separations.

Several examples of the electrophoretic purification of specific mammalian cell types from populations of cells that occur naturally in suspension or in suspension cultures have been described. Examples include graft–vs.–host cells of bone marrow (51), mouse T–lymphocytes (52–54), human peripheral blood leucocytes (55), and ascites tumor cell subpopulations (56,57). Cells from solid tissues and monolayer culture have also been subjected to electrophoretic study and separation, including adult rat and rabbit kidney (58–60), virus infected cultured cells (61), canine pancreas (32), rat pituitary (62), cultured human embryonic kidney cells (63), and cells from human and animal tumors (64,65).

Electrophoretic isolation of cells depends on the same physical phenomena described earlier. Cells have surface potentials for physiological and/or thermodynamic reasons, and they may be considered a separate phase when suspended in aqueous media. As described previously, the partitioning of ions which generates interfacial electric potentials depends on the two phases, and the combination of ions present. Cells impose additional criteria for the use of electric fields for their separation.

1. The protocol used to disperse the cells should affect the potential of all cells equally, or not at all (66).
2. The electrophoresis buffer should affect the electrophoretic mobility of all cells equally, or not at all (67).
3. Other characteristics of cells, such as cell volume dependence on mitotic cell cycle position, must be considered or minimized (67–71).

There are four other operational criteria which must be considered before cell separation by electrophoresis is developed for a particular application

4. The starting population must be electrophoretically heterogeneous. A typical heterogeneous population is shown in histogram form in Figure 6, where the relative power of laser light scattering is proportional to the relative number of cells in each electrophoretic subpopulation.
5. The heterogeneity must be reproducible from cell batch to batch.
6. An enhanced function or product capacity must be shown reproducibly in a distribution of cells narrower than the total distribution (32).
7. The electrophoretic process must not alter or redistribute this function (32).

The two most frequently used electrophoretic methods to separate cells in free fluid are density gradient and free–flow (or continuous flow) electrophoresis. Both of these methods reduce the effects of heating on the separation, the first by imposing an artificial density barrier to natural convection, the second by allowing an external heat exchanger to cool circulating fluid in the absence of the electric field. The former is simple and inexpensive but applicable mainly to small samples and bench testing of separations, while the latter is a continuous process and hence more attractive for product–related applications. Owing to the absence of flow, density–gradient electrophoresis is the higher resolution method.

DENSITY GRADIENT COLUMN ELECTROPHORESIS. In a vertical density–gradient electrophoresis column (Figure 7) a density gradient that is tolerated by cells (72,73) is formed between a high–density "floor" solution below and a low–density "ceiling" solution above, connected by side–arm electrodes. The gradient is usually formed with a varying concentration of a simple carbohydrate, such as sucrose, or a carbohydrate polymer, such as Ficoll. A cell mixture is layered by serial flow at the top or bottom of this gradient.

Density gradient electrophoresis has been shown to separate cells on the basis of electrophoretic mobility as measured by microscopic electrophoresis (66,74). A review of several methods and applications of this technique has been published by Tulp (75). Downward electrophoresis in a commercially available device utilizing a Ficoll gradient has been used to separate immunological cell types (73,76,77). Similar gradient–buffer conditions are employed in the Boltz–Todd apparatus (78) in which electrophoresis is usually performed in the upward direction (79).

The migration rates of cells in density gradients are altered by the presence of the gradient forming molecule, and these alterations must be considered in column design. Since the density of the medium is changing as cells migrate from one density zone to another, their buoyancy forces change. The viscosity and conductivity of the medium also change, and the carbohydrate molecules often increase the cells' zeta potentials. The actual velocity of cells in a density gradient is (32)

$$v_p = u_e(c,T)E(c,T) - 2a^2g(\rho_p - \rho_g(c,T))/9\eta(c,T) \qquad (32)$$

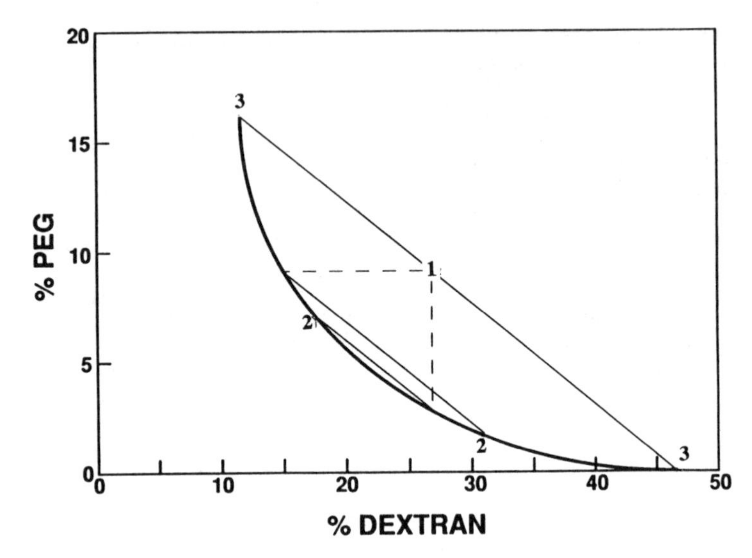

Figure 5. The ripening of polymer phases in aqueous two phase demixing. The initial composition of the polymer solution (1) decomposes to equilibrium compositions (3) through PEG and dextran droplets in near equilibrium with the bulk composition of the other polymer (2). The dashed lines show the equilibrium compositions of phases with the same concentration of one of the polymers in the bulk phase. The exact pathway between compositions (1) and (3) is poorly understood, and may vary from droplet to droplet.

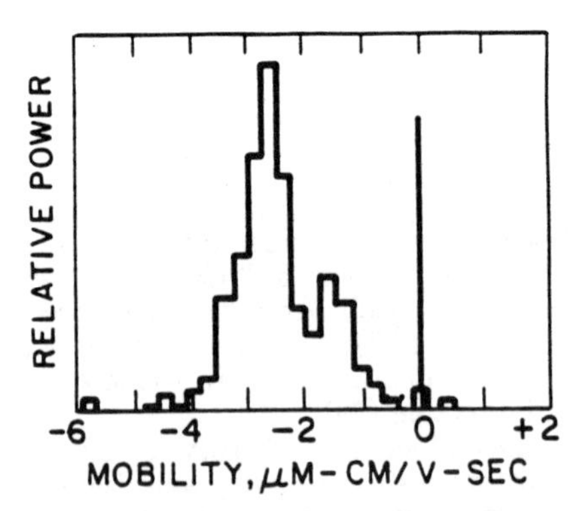

Figure 6. Electrophoretic mobility distribution of cultured human embryonic kidney cells in a low–conductivity buffer as determined by laser light scattering (Sarnoff, Kunze, Plank, and Todd, in preparation, 1990). These cells are heterogeneous with respect to surface charge density, and cells with different mobilities have different functions(67).

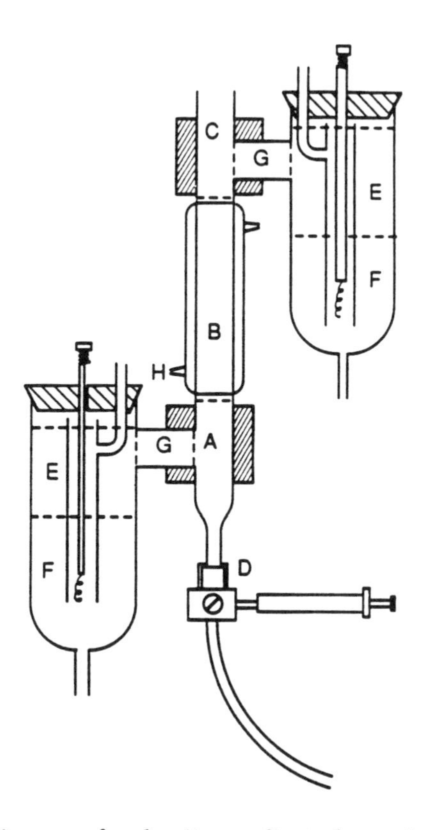

Figure 7.

Schematic diagram of a density gradient electrophoresis apparatus. A: dense "floor" solution that supports a sample zone and density gradient B, which is surrounded by a water jacket. C: low–density "ceiling" solution that provides electrical continuity to upper side–arm electrode vessel. D: inlet system for gradient, sample and floor. E: electrophoresis buffer (low conductivity) overlay to prevent ions from saturated salt solution F from entering column through gel plugs G. Bright platinum wire electrodes are submerged within a chimney that facilitates the evolution of electrolysis gases (adapted from Boltz and Todd, 78).

where the variables which are affected by the gradient concentration c, and medium temperature T, are noted, and subscript p refers to cell or particle properties. Most of these variables scale linearly with the concentration, but conductivity scales quadratically, and viscosity scales exponentially. The mobility and electric field scale according to the physical equations describing them, equation (19) for cells, and

$$E = I/k_1 \tag{33}$$

where k_1 is the conductivity of the medium. Heating effects don't appear to be important if column diameters are kept below 2.5 cm, and current densities below 5 mAmp/cm^2.

CONTINUOUS FLOW ELECTROPHORESIS. In continuous flow electrophoresis (Figure 8), carrier buffer is pumped upward or downward through a flat chamber 0.3 – 1.5 mm thick, 6.0 – 15 cm wide, and 25 – 110 cm long. Sample is injected parallel to the buffer flow in a narrow (0.1 – 0.3 mm) stream near the buffer inlet. Buffer and separands exit through 20 – 197 outlets, each leading to a fraction–collecting vessel (71,80).

Free flow electrophoresis has not found favor as an industrial–scale purification process. The process cannot be scaled up in dimensions because convection occurs as a consequence of inadequate heat removal from current–carrying electrolyte solutions. The process cannot be scaled up in mass throughput by increasing sample concentration, because sample zone sedimentation occurs when sample concentration is too high. Various approaches to this problem have been utilized. In one system the buffer curtain flows upward between a stationary and a rotating cylinder (28,29,31). Another method uses the microgravity environment of space flight (32,62). Of these two methods, the first sacrifices resolution, and the second is costly and seldom available. Conventional free flow electrophoretic separators have used downward flow and brine cooling. However, the Continuous Flow Electrophoresis System (CFES), designed for operation in low gravity and at 1 g, employs upward flow over a long distance with low field strength and the electrode buffer as a coolant (32,34,67,79). Figure 8 points out two major sources of zone dispersion: electroosmosis at the front and back walls, and residence time dispersion due to Poiseuille parabolic flow of the pumped buffer through the slit.

The problems of heating can be minimized by recycling separand streams through heat exchangers (25–27). Short exposure to the electric field, and subsequent removal of the heat before reexposure may be used to effect separations in continuous flow devices. However, resolution is automatically limited to the smallest volume of separand that is collectible at the electrophoresis chamber outlet. Miniaturization of recycle devices, rather than scale–up, seems to be the primary focus of this research area (M. Bier, personal communication).

ANALYTICAL APPLICATIONS of ELECTROKINETICS for PROCESS MONITORING

While the problem of scaling up electrokinetic methods for industrial downstream applications has received much attention, analytical applications of electric fields have been enjoying major advances. Analytical methods are important for monitoring and control of downstream processes, and for quality control and batch certification. In this section, capillary electrophoresis, a straightforward application of electrokinetics, electrospray devices, which depend on electrohydrodynamic instabilities in their operation, and pulsed field electrophoresis, a special application for elongated molecules, will be discussed.

CAPILLARY ELECTROPHORESIS. Capillary electrophoresis is developing into a method of choice in analytical applications. The process is driven by a very high electric field applied along the length of a capillary containing electrolyte solution. A typical system is shown in Figure 9. The capillary connects two buffer reservoirs which hold electrodes. The buffer reservoirs are typically open to atmospheric pressure, and at the same level hydrodynamically, so that no designed pressure drop exists over the capillary. After the capillary has been filled with electrolyte, sample may be applied, either gravimetrically or by exchanging a buffer reservoir for a sample reservoir, and drawing sample by brief application of the electric field. Slightly differing amounts of analyte will be drawn by this method, however, due to differences in analyte electrophoretic mobility (37,81). Analytes are detected by a variety of methods in the capillary, near the terminal buffer reservoir. The capillary is usually 10 to 100 cm long, and 20 to 200 μm in diameter (82) with volumes generally of a few microliters. The process was popularized by Jorgenson and Lukacs (23,37) in the early 1980's, and is characterized by high heat transfer, and rapid generation of hundreds of thousands of theoretical plates. Separation times are minutes or less, and robotics and other types of automation are easily adapted to simplify the process.

Analytes are carried in capillary electrophoresis either by electroosmotic flow, or by pressure driven flow, or by a combination. In this way, analytes of all charges are carried through the detector, even if they migrate electrophoretically towards the inlet electrode. An example of this is given by Gassman et al. (83) for separation of chiral amino acids. Electroosmosis would seem to be a preferable method of generating flow (84), since all the variations in flow profile occur within the double layer ($\kappa R >> 1$), as opposed to pressure driven flow, where a parabolic flow profile extends across the entire channel. Since the analyte migration rate is the sum of convective and electrophoretic velocities, the flow profile is important to resolution. Solute diffusivities in electrolytes are typically low ($\sim 10^{-7} cm^2/s$ for proteins), so solute distributed radially across the capillary cross section will elute from the capillary as a wide band if the flow profile varies substantially across the capillary cross section. This phenomenon is known as Taylor dispersion (85). The dispersion due to variation in flow profile in electroosmotically driven separations is currently an area of research interest (R.Datta, personal communication).

Adsorption of solutes on capillary walls is frequently a problem which limits the use of capillary electrophoresis. Since the rate at which double layers equilibrate is rapid, processes like the kinetics of adsorption/desorption on capillary walls can dominate the dispersion and resolution in capillary electrophoresis systems, especially in large diameter (> 50 μm) capillaries. Adsorption is usually found with high molecular weight solutes, such as proteins. The adsorption is generally weak, and can be overcome with solvents that are stronger in either ionic or hydrophobic character. The use of these solvents is limited by the electric field, however. Highly ionic solutions result in heating problems, highly hydrophobic solvents sometimes make dissolution of biological analytes more difficult. The problem has been partially addressed by using coatings on the capillary wall which do not adsorb analytes (82). Unfortunately, these coatings do not adsorb buffer ions, either, so pressure flow is used to drive the separation, leading to Taylor dispersion.

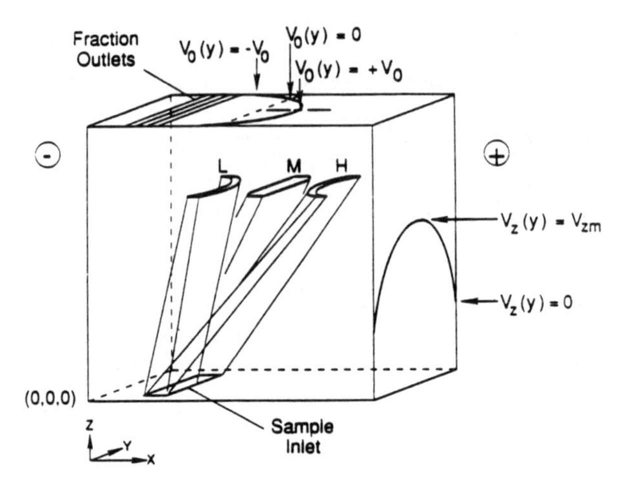

Figure 8. Schematic diagram (not to scale) of an upward–flowing free–flow electrophoresis apparatus. A coordinate system with (0,0,0) at the lower front left corner is shown. Sample stream is shown as a slit (circular sample streams are commonly used). Anodal migration of low, medium, and high (L,M,H) mobility separands results in crescent–shaped bands due to Poiseuille retardation of particles near the walls in the case of high–mobility particles and electroosmotic flow near the walls in the case of low mobility particles. These two parabolic distortions balance in the case of medium–mobility (M) separands.

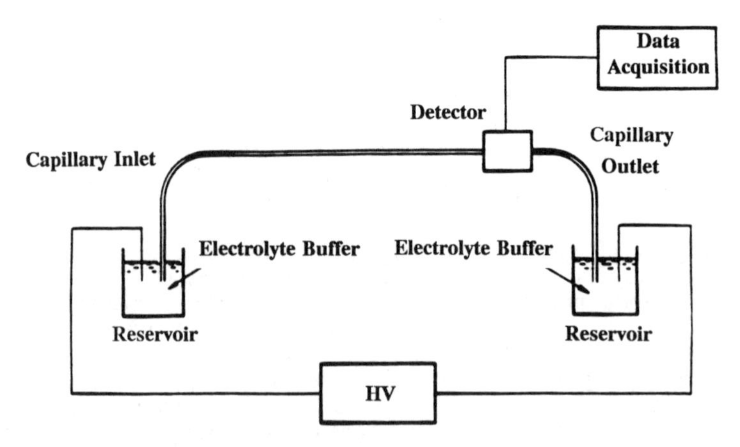

Figure 9. A capillary electrophoresis unit. The capillary, with on line detector, connects two hydrostatically equilibrated buffer reservoirs, which also contain high voltage electrodes. Sample may be injected by immersing the inlet end of the capillary and its electrode (left) into a sample reservoir and running the electric field. This may by done robotically. Some applications require a pump at the inlet (Reprinted with permission from ref 82, Copyright 1988 by the AAAS).

The popularization of capillary electrophoresis has led to the development of detectors sensitive enough to detect very small amounts of material (10^{-18} moles, 10^5 molecules) (86,87). This has led to the possibility that the constituents of individual cells may be analyzed. This and other new applications have recently been reviewed (24,88).

ELECTROSPRAY DEVICES. Recently, a great deal of interest has developed in using electric fields to generate, control and transport charged droplets (46). One of the applications of this technique is in electrospray devices for interfacing mass spectrometers with various analytical instruments, such as liquid chromatographs and capillary zone electrophoretic units (89). In this application, an electric potential maintained at the opening of a nozzle or needle will lead to the break off or emission of droplets of a very defined size, dependent on the field strength (46). A schematic of an electrospray chamber is shown in Figure 10 (89). As charged droplets are emitted from the needle, they are dried by the passage of dry gas through the chamber. The droplets, once part of a continuous, electroneutral solution, are now in a gas phase with which their electrochemical equilibrium very much favors the solvation of ions. Solvent associated with the droplet evaporates to the dry gas, however. As the solvent evaporates, the charge density in the droplet increases, thereby increasing the electrochemical potential in the droplet and the potential at the droplet surface, until either an ion evaporation (90,91) or a Rayleigh instability (92) event occurs, resulting in the eventual production of solvated ions in the gas phase. These ions are drawn through a capillary by a combination of electrostatic field and gas flow, through a series of skimmers or ion lenses, and into a quadrupole mass spectrometer (93). In this field application, charge is induced on droplets from an orifice with an electric potential. This charge density is concentrated by evaporation of solvent, until airborn ions result (from otherwise non volatile solutes) which can be analyzed by conventional techniques. The application of this technology to the study of proteins and other biopolymers shows great promise (89,93).

PULSED FIELD ELECTROPHORESIS. Electrophoretic separations are typically based on differences of zeta potential between particles. Although they are not strictly analogous, zeta potential and charge to mass ratio are frequently referred to interchangeably. The zeta potential is related to charge / mass ratio in that charge density determines electric potential, so that any charged mass has a charge density, and a small mass of equal charge and density has a higher charge density than a large mass. For example, two polymers of different length, but the same repeating charged unit will have similar zeta potentials, because the longer polymer will have a proportionately higher charge spread over a proportionately greater volume. This is the case for deoxyribonucleic acids (DNA), each unit having a charge of –1. However, separations by molecular weight are routinely observed for DNA in agarose gels, with low molecular weight fragments migrating more rapidly than high molecular weight fragments.

In the low molecular weight range this principle is exploited to separate DNA strands that differ by 1 nucleotide, thereby facilitating sequence determinations. The reason for this phenomenon appears to be related to long chain entanglements between the DNA and the gel matrix. Several theories have been proposed to describe this migration (94–96), most based on Flory–Huggins statistical mechanics, along with a Doi–Edwards (97) or DeGennes (98) model of polymer "reptation". Reptation refers to the mechanism by which a mobile polymer moves through a fixed polymer matrix, with its lead units constraining the path that all intervening units must follow.

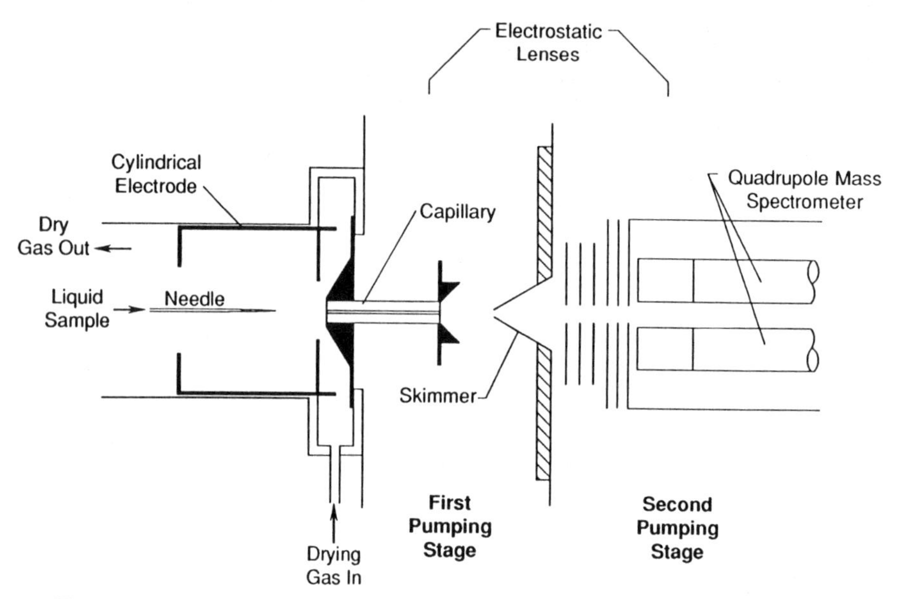

Figure 10.

A schematic of an electrospray mass spectrometer. Sample is sprayed into dry, atmospheric pressure (~ 1000 torr) gas with a high voltage (sub corona strength) electric field. The droplets are drawn (as they dry and disrupt) to the capillary by an electric field, and are transported through the capillary by gas flow, as the first chamber is maintained at ~ 10^{-3} torr or less by a high speed vacuum pump. The jet of droplets emerging from the capillary is skimmed by ion lenses, and a portion of it flows into the quadrupole mass spectrometer chamber, maintained at 10^{-6} torr (Reprinted with permission from ref 93, Copyright 1985 by the ACS).

In the case of DNA in agarose gel, a lead unit may occur anywhere along the molecule.

Pulsed field electrophoresis further exploits this mechanism by periodically reorienting the electric field. The pulsed field magnifies the entanglements for larger DNA molecules, and allows, after several reorientations, resolution of higher molecular weight strands. Several variations on this theme have appeared, including contour clamped homogeneous electric fields (99), periodically inverting electric fields (100), and heterogeneous pulsed fields (101), which have all been used to separate DNA segments the size of yeast chromosomes, which are two orders of magnitude larger than molecules that had been separable without reorientation of the field. This technique promises to become an increasingly important analytical technique as more genetically complicated organisms are screened for production of biochemicals, and as the sequencing of the human genome gets underway.

CONCLUSIONS

The thermodynamic basis for electric potentials at interfaces has been summarized. The concentration of ions near the interface, and resulting charge densities, may be calculated with the knowledge of bulk solution properties. This information is used to predict motion near the interface when an electric field is applied to a system.

The fact that charge separations exist near interfaces has been exploited in many types of separations. As examples we have discussed the demixing of liquid liquid emulsion and the complexities of cell separations. We have also discussed how electrokinetic forces have been used to separate infinitely small volumes of mixtures, generate charged aerosols for analysis, and sieve extended polymers according to their molecular weights.

We are optimistic about the prospects for electrokinetic methods in processing of biological materials. As a source of energy, electrons are cheap, and equilibrate quickly. As a force, electric fields are tunable to a very fine degree, and involve no moving parts. Small scale applications, such as capillary electrophoresis, are already making a major impact on analytical separation sciences. Pulsed field electrophoresis is becoming a routine application in sequencing labs, and electrospray mass spectrometry is showing the first possibilities of directly analyzing mass spectra of large biomolecules close to their native states. Large scale processes, while enjoying the same mass transfer advantages as their small scale counterparts, suffer from poor heat rejection. The prospects for aqueous biphasic demixing and cell culture electrophoresis depend primarily on the economic successes of their parent processes. Both applications out perform the respective parent process in the absence of a field, but it is not clear that the application of an electric field to either demixing polymer phases, or enriching cell cultures adds enough advantage to undertake aqueous extraction or cell culture. Only a few of the current applications of electric fields to downstream processing have been described here; however, the field for applications continues to grow at an increasing rate.

ACKNOWLEDGMENTS

The authors wish to acknowledge the support of the National Institute of Standards and Technology and its excellent library and computer facilities, and Ms. Robin Stewart and Mr. David Szlag for their useful comments during the preparation of this manuscript. One of the authors (SRR) wishes to acknowledge the helpful discussions with, and guidance of, Dr. Michael Ladisch prior to the production of this manuscript.

NOMENCLATURE

a	droplet radius	cm
C_p	heat capacity	$J/g \cdot {}^oK$
d	distance	cm
d_p	particle diameter	cm
D	diffusivity	cm^2/s
e	charge of an electron	$1.6 x 10^{-19}$ C
E	electric field strength	V/cm
F	Faraday's constant	96459 C/equiv.
g	acceleration of gravity	$9.8 m/s^2$
Gr	Grashof number	dim
h	convective heat transfer coefficient	$J/s \cdot cm^2 \cdot {}^oK$
I	current density	$amps/cm^2$
k	conductivity	mho/cm
k_t	thermal conductivity	$J/s \cdot cm \cdot {}^oK$
K	partition coefficient	dim
K_d	chromatographic distribution coefficient	dim
Nu	Nusselt number	dim
Q	volumetric gas flow rate	cm^3/s
r	radius	cm
R	gas constant	8.314 J/mol°K
Ra	Rayleigh number	dim
t	time	s
T	absolute temperature	^oK
ΔT	temperature difference over characteristic dimension	^oK
u	mobility	$cm^2/V \cdot s$
v	velocity	cm/s
v_c	chromatographic solute velocity	cm/s
v_g	sedimentation velocity	cm/s
v_s	superficial fluid velocity	cm/s
V	volume	cm^3
W	power density	W/cm^3
x	molar concentration	mol/cm^3
y	distance from the interface	cm
z	valence	equiv/mol

GREEK

α	void fraction	dim
β	thermal expansion coefficient	$cm^3/{}^oK$
ϵ	dielectric constant	dim
ϵ_0	permittivity of free space	$8.854 x 10^{-12}$ F/m
ϕ	electric potential	V
ϕ_0	potential at the interface compared to the bulk	V
$\Delta\phi$	$\phi^\beta - \phi^\alpha$	V
η	viscosity	$g/cm \cdot s$
κ	inverse Debye length	1/cm
λ	conductivity ratio	dim
μ	chemical or electrochemical potential	J/mol

$\bar{\mu}$	"uncharged" chemical potential	J/mol
$\Delta\mu$	$\mu^{\beta} - \mu^{\alpha}$	J/mol
$\Delta\bar{\mu}$	$\bar{\mu}^{\beta} - \bar{\mu}^{\alpha}$	J/mol
ρ	density	g/cm^3
$\bar{\rho}$	average density	g/cm^3
ρ_e	charge density	C/cm^3
σ_e	surface charge density	C/cm^2
ζ	zeta potential	V

SUPERSCRIPTS

o	standard potential
α	phase α
β	phase β

SUBSCRIPTS

c	characteristic
e	electrophoretic
g	gradient
i	ith solute
ib	ith solute, bulk concentration
p	particle
z	z direction
1	continuous phase property
2	droplet phase property

LITERATURE CITED

1. Prausnitz, J. M.; Lichtenhaler, R. N.; deAzevedo, E. G. Molecular Thermodynamics of Fluid–Phase Equilibria; Prentice Hall Inc.: Englewood Cliffs, NJ, 1986.
2. Newman, J. S. Electrochemical Systems; Prentice Hall, Inc.: Englewood Cliffs, NJ, 1973.
3. Robinson, R. A.; Stokes, R. H. Electrolyte Solutions; Butterworths Scientific Publications: London, 1959.
4. Patwardhan, V. S.; Kumar, A. AIChE. J. 1986, 32, 1419.
5. Patwardhan, V. S.; Kumar, A.; AIChE. J. 1986, 32, 1429.
6. Haghtalab, A.; Vera, J. H. AIChE. J. 1988, 34, 803.
7. Friedman, H. L.; Dale, W. D. T. In Modern Theoretical Chemistry; Berne, B. J. Ed.; Plenum Press: New York, 1977; 5, 85.
8. Atkinson, G.; Kumar, A.; Atkinson, B. L. AIChE. J. 1986, 32, 1561.
9. Pettitt, B. M.; Rossky, P. J. J. Chem. Phys. 1986, 84, 5836.
10. Hunter, R. J. Zeta Potential in Colloid Science, Principles and Applications; Ottewill, R.H.; Rowell, R.L. Eds.; Academic Press: New York, 1981.
11. Rice, C. L.; Whitehead, R. J. Phys. Chem. 1965, 69(11), 4017.
12. O'Brien, R. W.; White, L. R. J. Chem. Soc. Faraday, II 1978, 74, 1607.
13. Wiersema, P. H.; Loeb, A. L.; Overbeek, J.Th.G. J. Colloid Interface Sci. 1966, 22, 78.
14. Hunter, J. B. Sep. Sci. Technol. 1988, 23(8&9), 913.
15. Rudge, S. R.; Ladisch, M. R. Biotech. Prog. 1988, 4, 123.
16. Rudge, S. R.; Ladisch, M. R. In Separation and Purification in Biotechnology, Ascenjo, J.; Hong, J. Eds.; ACS Symposium Series No. 314; American Chemical Society: Washington, DC, 1986; 122.

17. Bier, M.; Palusinski, O. A.; Mosher, R. A.; Saville, D. A. Science, 1983, 219, 1281.
18. Saville, D. A.; Palusinski, O. A. AIChE. J. 1986, 32, 207.
19. Palusinski, O. A.; Graham, A.; Mosher, R. A.; Bier, M.; Saville, D. A. AIChE. J. 1986, 32, 215.
20. Ivory, C. F. Sep. Sci. Technol. 1988, 23(8&9), 875.
21. Chemical Engineer's Handbook; Perry, R. H.; Chilton, C. H. Eds; McGraw Hill Book Co.: New York, 1973.
22. Parker, J. D.; Boggs, J. H.; Blick, E. F. Introduction to Fluid Mechanics and Heat Transfer; Addison–Wesley Publishing: Reading, MA, 1970.
23. Jorgenson, J. W.; Lukacs, K. D. Science, 1983, 222, 266.
24. Kennedy, R. T.; Oates, M. D.; Cooper, B. R.; Nickerson, B.; Jorgenson, J. W. Science, 1989, 246, 57.
25. Bier, M. In Separation, Recovery, and Purification in Biotechnology, 1986, Asenjo, J. A.; Hong, J. Eds.; ACS Symposium Series No. 314, American Chemical Society: Washington, DC, 1986; 185.
26. Gobie, W. A.; Beckwith, J. B.; Ivory, C. F. Biotech. Prog. 1985, 1(1), 60.
27. Gobie, W. A.; Ivory, C. F. AIChE. J. 1988, 34, 474.
28. Philpot, J. St. L. Trans. Faraday Soc. 1940, 39, 38.
29. Hjertén, S. Free Zone Electrophoresis; Almqvist and Wiksells Boktr. AB, Uppsala, 1962.
30. Thomson, A. R. In Electrophoretic Techniques; Simpson C. F.; Whitaker, M. Eds.; Academic Press: New York, 1983; 253.
31. Mattock, P.; Aitchison, G. F.; Thomson, A. R. Sep. Purif. Meth. 1980, 9, 1.
32. Hymer, W. C.; Barlow, G. H.; Cleveland, C.; Farrington, M.; Grindeland, R.; Hatfield, J. M.; Kunze, M. E.; Lanham, J. W.; Lewis, M. L.; Morrison, D. R.; Olack, N.; Richman, D.; Rose, J.; Scharp, D.; Snyder, R. S.; Todd, P.; Wilfinger, W. Cell Biophysics 1987, 10, 61.
33. Hannig K.; Wirth, H.; Schoen, E. In Apollo–Soyuz Test Project Summary Science Report; NASA SP–412; Washington DC, 1977, 1, 335.
34. Snyder, R. S.; Rhodes, P. H. in Frontiers in Bioprocessing; Sikdar, S. K.; Bier, M.; Todd, P. Eds.; CRC Press: Boca Raton, FL, 1989; 245.
35. Micale, F. J.; Vanderhoff, J. W.; Snyder, R. S. Sep. Purif. Meth. 1976, 5(2), 361.
36. Vanderhoff, J.W.; Micale, F. J.; Krumrine, P. H. Sep. Purif. Meth. 1977, 6, 61.
37. Jorgenson, J. W.; Lukacs, K. D. Anal. Chem. 1981, 53, 1298.
38. Albertsson, P.–A. Partition of Cell Particles and Macromolecules; John Wiley and Sons: New York, 1986.
39. Shanbhag, V. P. Biochim. Biophys. Acta 1973, 320, 517.
40. Walter, H.; Brooks, D. E.; Fisher, D. Partitioning in Aqueous Two Phase Systems; Academic Press, Inc.: New York, 1985.
41. Cabezas, H. Jr.; Evans, J. D.; Szlag, D. C. In Downstream Processing and Bioseparation. Recovery and Purification of Biological Products; Hamel, J.–F. P.; Hunter, J. B.; Sikdar, S. K. Eds.; ACS Symposium Series 419; American Chemical Society: Washigton, DC, 1989; 38.
42. Bamberger, S.; Seaman, G. V. F.; Brown, J. A.; Brooks, D. E. J. Colloid Interface Sci. 1984, 99, 187.
43. Brooks, D. E.; Sharp, K. A.; Bamberger, S.; Tamblyn, C. H.; Seaman, G. V. F.; Walter, H. J. Colloid Interface Sci. 1984, 102, 1.

44. Levine, S. In Materials Processing in the Reduced Gravity Environment of Space; Rindone, G. E. Ed.; Elsevier Science Publishing Company, Inc.: Amsterdam, 1982; 241.
45. Pohl, H. A. J. Appl. Phys. 1915, 22, 869.
46. Scott, T. C. Sep. Purif. Meth. 1989, 18(1), 65.
47. Hendricks, C. D.; Sadek, S. IEEE Trans. Ind. Appl. 1977, 1A–13, 489.
48. Richards, K. J.; Clark, D. R. U.S. Patent 4 039 404, 1977.
49. Warren, K. W.; Prestridge, F. L. U.S. Patent 4 161 439, 1979.
50. Prestridge, F. L. U.S. Patent 3 772 180, 1973.
51. Nordling, S.; Anderson, L. C.; Häyry, P. Science, 1972, 178, 1001.
52. Platsoucas, C. D.; Griffith, A. L.; Catsimpoulas, N. J. Immunol. Meth. 1976, 13, 145.
53. Zeiller, K.; Holzberg, E.; Pascher, G.; Hannig, K. Hoppe–Zeylers Z. Physiol. Chem. 1972, 353, 101.
54. Mehrishi, J. N.; Zeiler, K. Brit. Meth. J. 1974, 1, 360.
55. Zeiller, K.; Hannig, K. Hoppe–Zeylers Z. Physiol. Chem. 1971, 352, 1162.
56. Pretlow, T. G. II; Pretlow, T. P. Int. Rev. Cytol. 1979, 61, 85.
57. Zarkower, D. A.; Hymer, W. C.; Plank, L. D.; Kunze, E.; Keith, A.; Todd, P. Cell Biophys. 1984, 6, 53.
58. Heidrich, G. G.; Dew, M. E. J. Cell Biol. 1983, 74, 780.
59. Kreisberg, J. I.; Sachs, G.; Pretlow T. G. II; McGuire, R. A. J. Cell Physiol. 1977, 93, 169.
60. Trump, B. F.; Sato, T.; Trifillis, A.; Hall–Craggs, M.; Kahng, M. W.; Smith, M. W. In Methods in Cell Biology; Harris, C. C.; Trump, B. F.; Stoner G. D. Eds.; Academic Press: New York, 1980; 309.
61. Thompson, C. J.; Docherty, J. J.; Boltz R. C. Jr.; Gaines, R. A.; Todd, P. J. Gen. Virol. 1978, 39, 449.
62. Hannig, K.; Wirth, H.; Schoen, E. In Apollo–Soyuz Test Project Summary Science Report, NASA SP–412: Washington, DC; 1977, 1, 335.
63. Morrison, D. R.; Lewis, M. L. In 33rd International Astronautical Federation Congress, Paper No. 82–152, 1983.
64. Mehrishi, J. N. Prog. Biophys. Molec. Biol. 1970, 25, 1.
65. Sherbet, G. V. The Biophysical Characterization of the Cell Surface; Academic Press: New York, 1978.
66. Boltz, R. C. Jr.; Todd, P.; Gaines, R. A.; Milito, R. P.; Docherty, J. J.; Thompson, C. J.; Notter, M. F. D.; Richardson, L. S.; Mortel, R. J. Histochem. Cytochem. 1976, 24, 16.
67. Todd, P.; Plank, L. D.; Kunze, M. E.; Lewis, M. L.; Morrison, D. R.; Barlow, G. H.; Lanham J. W.; Cleveland, C. J. Chromatog. 1986, 364, 11.
68. Strickler, A.; Sacks, T. Prep. Biochem. 1973, 3, 269.
69. Shank, B. B.; Burki, H. J. J. Cell. Physiol. 1971, 78, 243.
70. Mayhew E.; O'Grady, E. A. Nature 1965, 207, 86.
71. Hannig, K. In Modern Separation Methods of Macromolecules and Particles; Gerritsen, T., Ed.; Wiley Interscience: New York, 1969; 45.
72. Boltz, R. C. Jr.; Todd, P.; Streibel, M. J.; Louie, M. K. Prep. Biochem. 1973, 3, 383.
73. Griffith, A. L.; Catsimpoolas, N.; Wortis, H. H. Life Sci. 1975, 16, 1693.
74. Todd, P.; Szlag, D. C.; Plank, L. D.; Delcourt, S. D.; Kunze, M. E.; Kirkpatrick, F. H.; Pike, R. G. Adv. Space Res. 1989, 9(11), (11)97.
75. Tulp, A. Meth. Biochem. Anal. 1984, 30, 141.
76. Platsoucas, C. D.; Good, R. A.; Gupta, S. Proc. Natl. Acad. Sci. USA, 1979, 76, 1972.

77. Platsoucas, C. D.; Beck, J. D.; Kapoor, N.; Good, R. A.; Gupta, S. Cell. Immunol. 1981, 59, 345.
78. Boltz, R. C. Jr.; Todd, P. In Electrokinetic Separation Methods; Righetti, P. G.; Van Oss, C. J.; Vanderhoff, J. W. Eds.; Elsevier/North–Holland Press: Amsterdam, 1979, 229.
79. Plank, L. D.; Hymer, W. C.; Kunze, M. E.; Marks, G. M.; Lanham, J. W.; Todd, P. J. Biochem. Biophys. Meth. 1983, 8, 275.
80. Mel, H. C. J. Chem. Phys. 1959, 31, 559.
81. Tsuda, T.; Nomura, K.; Nakagawa, G. J. Chromatog. 1983, 264, 385.
82. Gordon, M. J.; Huang, X.; Pentoney, S. L. Jr.; Zare, R. N. Science 1988, 242, 224.
83. Gassman, E.; Kuo, J. E.; Zare, R. N. Science 1985, 230, 813.
84. Pretorius, V.; Hopkins, B. J.; Schieke, J. J. Chromatog. 1974, 99, 23.
85. Taylor, G. Proc. Royal Soc. 1953, A219, 186.
86. Cheng, Y.–F.; Dovichi, N. J. Science 1988, 242, 562.
87. Huang, X.; Pang, T.–K. J.; Gordon, M. J.; Zare, R. N. Anal. Chem. 1987, 59, 2747.
88. Eby, M. J. Bio/technology 1989, 7, 903.
89. Fenn, J. B.; Mann, M.; Meng, C. K.; Wong, S. F.; Whitehouse, C. M. Science 1989, 246, 64.
90. Iribarne, J. V.; Thomson, B. A. J. Chem. Phys. 1976, 64, 2287.
91. Thomson, B. A.; Iribarne, J. V. J. Chem. Phys. 1979, 71, 4451.
92. Röllgen, F. W.; Bramer–Weger, E.; Bütfering, L. J. Phys., Colloq. 1987, 48, C6–253.
93. Whitehouse, C. M.; Dreyer, R. N.; Yamashita, M.; Fenn, J. B. Anal. Chem. 1985, 57, 675.
94. Lumpkin, O. J.; Dejardin, P.; Zimm, B. H. Biopolymers 1985, 24, 1573.
95. Lerman, L. S.; Frisch, H. L. Biopolymers 1982, 21, 995.
96. Deutsch, J. M. Science 1988, 240, 922.
97. Doi. M.; Edwards, S. J. Chem. Soc., Farad. Trans. 2 1978, 74, 1789.
98. DeGennes, P. G. J. Chem. Phys. 1971, 55, 572.
99. Chu, G.; Volrath, D.; Davis, R. W. Science 1986, 234, 1502.
100. Carle, G. F.; Frank, M.; Olson, M. V. Science 1986, 232, 65.
101. Schwartz, D. C.; Cantor, C. R. Cell 1984, 37, 67.
102. Burrolla, V. P.; Pentoney, S. L.; Zare, R. Amer. Biotech. Lab. 1989, 7, 20.

RECEIVED February 13, 1990

Author Index

Affiliation Index

Subject Index

F

G

H

I

Production: Peggy D. Smith
Indexing: Deborah H. Steiner
Acquisition: Cheryl Shanks

Elements typeset by Hot Type Ltd., Washington, DC
Printed and bound by Maple Press, York, PA

Paper meets minimum requirements of American National Standard for Information Sciences—Permanence of Paper for Printed Library Materials, ANSI Z39.48–1984 ∞